BASIC METHODS OF LINEAR FUNCTIONAL ANALYSIS

Pure Mathematics

Editor
PROFESSOR J. C. SHEPHERDSON
M.A.
Professor of Pure Mathematics
in the University of Bristol

BASIC METHODS OF LINEAR FUNCTIONAL ANALYSIS

J. D. Pryce
Lecturer in Mathematics
in the University of Aberdeen

HUTCHINSON UNIVERSITY LIBRARY
LONDON

HUTCHINSON & CO (*Publishers*) LTD
3 Fitzroy Square, London W1

London Melbourne Sydney Auckland
Wellington Johannesburg Cape Town
and agencies throughout the world

First published 1973

This book has been set in cold type by E.W.C. Wilkins & Associates Ltd, London, printed in Great Britain by Anchor Press, and bound by Wm. Brendon, both of Tiptree, Essex

ISBN 0 09 113410 2 (cased)
0 09 113411 0 (paper)

CONTENTS

INTRODUCTION

This book is mainly for students in the final year of a British honours mathematics course or equivalent. Most of the recent books on functional analysis at this level are larger, or rather compressed, or devoted to obtaining one of the big theorems as fast as possible. It seemed useful to attempt a general-purpose, not over-detailed introduction to the themes that are most basic in current pure mathematical analysis. As I feel one should not be introduced to functional analysis till one has mastered enough preliminaries to allow a concise, uncluttered development (knowing good proof techniques is as important as knowing good theorems), I assume of the reader a fair knowledge of basic real analysis, metric space theory and linear algebra, and – more daringly perhaps – a minimum about measures and Lebesgue integration. These are surveyed in the first three sections.

A basic idea of the subject is that problems in analysis are often easier to solve if one considers (say) a function f, not on its own, but as a member of a whole *space of functions*, X. The space X usually carries both a natural metric and a natural linear space structure and it is profitable to treat X as a larger (usually infinite dimensional) version of Euclidean space: thus the metric brings in ideas of *topology*; linear operators on X bring in *linear algebra*; and we use *geometrical* language and methods involving convex sets, hyperplanes and the like. The intermingling of these approaches constitutes linear functional analysis.

The book is divided into sections; related sections are grouped, loosely, into chapters, of which the first three (§§1–10) contain, apart from preliminaries, the foundation results of the subject including the Hahn–Banach, Stone–Weierstrass, Uniform Boundedness and Open Mapping Theorems. Lecturers devising courses, and students with previous knowledge, may note that the remaining chapters can be read in any order *except*: in Chapter 5, §14 uses the Galois duality results from § 12; in Chapter 7, §§18, 19 and the very end of § 17 use the Hilbert space theory of §§ 11, 12.

Personal idiosyncrasy is bound to affect what one leaves out. Nonlinear analysis is too big a subject to be done justice here. Also I felt that though, say, generalised functions or Banach algebras or differential equations in Banach spaces are important to analysts, even more so are weak topologies, infinite products and evaluation maps (§ 16). Topology comes into its own here, and for completeness I have included a proof of Tychonov's Theorem on the way to the main w^*-compactness theorem.

No course in functional analysis should ignore the famous classical theorems in which the subject has its roots. Proofs of some of these are given as applications of the basic theory – e.g. a simple version of Runge's Theorem, and Lerch's Uniqueness Theorem. The sign † marks these sections, and others which may be omitted or skimped on first reading by students fearing mental indigestion. The power of the Spectral Theorem for compact self-adjoint operators is brought out by a quite detailed discussion of the eigenfunctions of a regular Sturm–Liouville system.

Notation is, I hope, standard. The numbering system used in the text should be self-explanatory. As for names, Theorem connotes an important result with wide implications; Propositions are easier results and usually less important, though interesting in their own right; Lemmas are technical results of little interest except as a preliminary to a Theorem.

My thanks are due to Ian Craw for reading the manuscript and making many helpful suggestions; to Miss Ann Collie for her excellent typing; to F.F. Bonsall, G. Cook, J. Duncan and J.H. Williamson and others for help and advice; and to my wife for her salutary impatience over the past two years and her work on the index.

I

PRELIMINARIES

1. Metric spaces and topological spaces

This section covers concisely those portions of metric space theory needed in elementary functional analysis, as well as some very elementary theory of topological spaces. The latter are not really needed outside §16, where they are essential; in §6, where they also occur, none of the interest is lost if the reader prefers to substitute 'metric space' for 'topological space' wherever the latter occurs.

1.1 A **metric** on a set X is a non-negative real-valued function d on $X \times X$ obeying the rules:

$$d(x, y) = 0 \text{ iff } x = y,$$
$$d(y, x) = d(x, y),$$
$$d(x, z) \leqslant d(x, y) + d(y, z) \text{ (the triangle inequality).}$$

The number $d(x, y)$ is called the **distance** from x to y. A pair (X, d), where d is a metric on X, is called a **metric space.** Often one suppresses mention of d and speaks of 'the metric space X'. We sometimes use the same symbol d for the metric on different spaces.

A set of the form $\{x \in X : d(a, x) \leqslant r\}$, where $a \in X$ and $r > 0$, is a **closed ball** (of radius r, round a); replacing $\leqslant$ by $<$ gives the corresponding **open ball.** We sometimes denote these by $B(a, r)$ and $U(a, r)$ respectively. (Some authors use the word 'sphere' which will be used in this book to mean a set $\{x \in X : d(a, x) = r\}$.)

The real line R and the complex plane C are metric spaces under the **usual metric** $d(x, y) = |x - y|$. More generally R^n (and C^n, which can be thought of for this purpose as R^{2n}) are metric spaces under the **Euclidean metric**

$$d(x, y) = \left(\sum_{j=1}^{n} |x_j - y_j|^2\right)^{\frac{1}{2}},$$

where $x = (x_1, \ldots, x_n)$, $y = (y_1, \ldots, y_n)$; the reader probably knows this but it follows from results proved later.

1.2 We now define the basic topological concepts in a metric space (X, d). A subset N of X is a **neighbourhood** of a point $a \in X$ (and a is an **interior point** of N) if N contains some ball round a. A sequence $\{x_n\}$ in X **converges** to a point $a \in X$ – one writes $x_n \to a$ as $n \to \infty$ – if given $\epsilon > 0$ there exists n_0 such that $d(x_n, a) < \epsilon$ whenever $n \geqslant n_0$, that is, if $d(x_n, a) \to 0$ as $n \to \infty$; equivalently, if given any neighbourhood N of a there exists n_0 such that $x_n \in N$ for $n \geqslant n_0$ (we say that x_n is **eventually** in N). A sequence can converge to at most one point a, called the **limit** of $\{x_n\}$, $a = \lim_n x_n$.

A point a is a **closure point** of a subset A if each neighbourhood of a meets A. A set is **open** if each of its points is an interior point; **closed** if it contains all its closure points. The **interior** int(A) of A is the set of interior points of A, the **closure** A^- of A is the set of closure points of A. The **boundary** of A is $A^- \sim \text{int}(A)$ and consists of those points each of whose neighbourhoods meets both A and $X \sim A$. A subset A is **dense** in X if $A^- = X$.

Let f be a map from a metric space X to a metric space Y, and let $a \in X$; then f is **continuous at** a if for each neighbourhood N of $f(a)$ there exists a neighbourhood M of a such that $f(M) \subset N$. If f is continuous at each point of X it is called **continuous on** X or just continuous. If f is one-to-one and onto, and both f and f^{-1} are continuous then f is a **homeomorphism**; one says X and Y are **homeomorphic** under f.

We shall assume the reader is thoroughly familiar with the elementary relations between these concepts, for instance: that open balls are open and closed balls are closed; that int(A) is the largest open set inside A and A^- the smallest closed set containing A, the equivalence of the neighbourhood, sequential, ϵ–δ and open/closed set definitions of continuity; and the fact that a homeomorphism $f: X \to Y$ sets up a one-to-one correspondenc between the classes of open sets, closed sets and convergent sequences in X, and the corresponding classes in Y. Note that according to some authors a neighbourhood is required to be open; otherwise all the above terms are standard.

Two metrics d, e on the same set X are **equivalent** if they give the same open sets; that is, if the identity map $x \mapsto x$ of (X, d) to (X, e) is a homeomorphism. It is easy to see that this happens iff each ball in the d-metric ('d-ball') contains a concentric e-ball and vice versa.

1.3 Any subset A of a metric space (X, d) becomes a metric space (a **subspace** of (X, d)) under the **relative metric** d_A obtained by restricting d to $A \times A$. Balls in (A, d_A) are intersections of A with balls in (X, d): denoting the former by B_A, U_A one has $B_A(a, r) = A \cap B(a, r)$, $U_A(a, r) = A \cap U(a, r)$ – here of course a must be in A. From this it follows easily that the open (closed) sets in (A, d_A) are just the sets $A \cap S$ where S is open (closed) in (X, d) and that if f is a continuous map from (X, d) to (Y, e) then the **restriction** $f|_A$ is continuous from (A, d_A) to (Y, e). A subspace can have properties very different from those of the whole space.

Let (X, d) and (Y, e) be metric spaces. The two **standard metrics** for the **product** $X \times Y$ are

$$\rho((x, y), (x', y')) = d(x, x') + e(y, y'),$$

$$\sigma((x, y), (x', y')) = \max\{d(x, x'), e(y, y')\}.$$

Each σ-ball contains the concentric ρ-ball of the same radius; this in turn contains the concentric σ-ball of half the radius, so the metrics ρ, σ are equivalent.

1.4 Proposition A sequence $\{(x_n, y_n)\}$ converges to (a, b) in $(X \times Y, \rho)$ (or $(X \times Y, \sigma)$) iff $x_n \to a$ in (X, d) and $y_n \to b$ in (Y, e).

Proof It is clear from the definition of ρ that $\rho((x_n, y_n), (a, b)) \to 0$ iff $d(x_n, a) \to 0$ and $e(y_n, b) \to 0$.

The similar result holds for the analogous metrics ρ, σ on an arbitrary finite product $X_1 \times \ldots \times X_n$ of metric spaces. Briefly 'convergence in the product is coordinatewise convergence'.

1.5 Let (X, d) and (Y, e) be metric spaces. The map $f: X \to Y$ is **uniformly continuous** if given $\epsilon > 0$ there exists $\delta > 0$ such that $e(f(x), f(y)) < \epsilon$ whenever $x, y \in X$ and $d(x, y) < \delta$. For an example of a continuous but not uniformly continuous function consider $x \mapsto x^2$ on $\mathbf{R}$ with the usual metric.

Metric properties of subsets

1.6 Let A, B be nonempty subsets of (X, d). The **distance** from A to B is defined by $d(A, B) = \inf\{d(a, b): a \in A, b \in B\}$. When A is a single point $\{x\}$ we obtain the quantity $d(x, B) = \inf\{d(x, b): b \in B\}$.

1.7 Lemma The real-valued function $x \mapsto d(x, B)$ is uniformly continuous on X.

Proof Since $d(x, b) \leqslant d(x, y) + d(y, b)$ one easily obtains $d(x, B) \leqslant d(x, y) + d(y, B)$; swapping x and y and combining the two inequalities yields $|d(x, B) - d(y, B)| \leqslant d(x, y)$ from which the result is immediate.

Taking B to be a single point shows that the map $x \mapsto d(x, b)$ is uniformly continuous.

A subset A is **bounded** if its **diameter** $\operatorname{diam}(A) = \sup\{d(x, y): x, y \in A\}$ is finite, equivalently if A lies inside some ball.

Completeness

A sequence $\{x_n\}$ in a metric space (X, d) is **Cauchy** if given $\epsilon > 0$ there exists n_0 such that $d(x_m, x_n) < \epsilon$ whenever $m, n \geqslant n_0$ We leave as an exercise the following useful fact:

1.8 Lemma If a Cauchy sequence $\{x_n\}$ has a subsequence $\{x_{n_k}\}$ convergent to a point x, then also $\{x_n\}$ converges to x.

Every convergent sequence is Cauchy; (X, d) is **complete** if every Cauchy sequence is convergent. It is of course a basic theorem of analysis that $\boldsymbol{R}$ and $\mathbf{C}$ are complete. Note that $\{x_n\}$ is Cauchy $\Leftrightarrow \operatorname{diam}\{x_k : k \geqslant n\} \to 0$ as $n \to \infty$.

1.9 Cantor's Theorem Let $\{A_n\}$ be a decreasing sequence of nonempty closed sets in a complete metric space (X, d) such that $\operatorname{diam}(A_n) \to 0$. Then $\bigcap_n A_n$ consists of exactly one point.

Proof Choose an arbitrary point x_n in A_n for each n. Then $\{x_k : k \geqslant n\} \subset A_n$, so $\{x_n\}$ is easily seen to be Cauchy. Let a be its limit. For each n, a is the limit of the subsequence $\{x_k : k \geqslant n\} \subset A_n$, so $a \in A_n^- = A_n$ for all n. If any other point b were in all the A_n then $\operatorname{diam}(A_n) \geqslant d(a, b) > 0$ for all n, contrary to hypothesis. Hence $\bigcap_n A_n = \{a\}$.

1.10 Proposition Let A be a complete subset of a metric space (X, d) (that is, suppose (A, d_A) is complete). Then A is closed. Conversely every closed subset of a complete metric space is complete.

Proof Let (A, d_A) be complete. Any point $a \in A^-$ is the limit (in (X, d)) of a sequence $\{x_n\}$ in A. This sequence is Cauchy in (A, d_A) so it has a limit $b \in A$. But sequences that converge in (A, d_A) also converge to the same point in (X, d), so $a = b$, $a \in A^-$, $A^- \subset A$, showing A is closed in X. The second assertion is left as an exercise.

This proposition is often the most convenient way to show that a given metric space is *not* complete: namely, embed it as a subset of a larger metric space (with the metrics matching up) and show that it fails to be closed in this space.

1.11 Proposition The product of complete metric spaces (X, d), (Y, e) is complete under either of the two standard metrics ρ, σ of 1.3.

Proof The definitions of ρ, σ imply that if $\{(x_n, y_n)\}$ is an arbitrary Cauchy sequence in $X \times Y$ with either metric, then $\{x_n\}$ and $\{y_n\}$ are Cauchy in X, Y respectively, and so have limits a, b. By 1.4, $\{(x_n, y_n)\}$ has a limit, namely (a, b).

Compactness

1.12 We take as our definition of compactness the one that is most suited for generalizing to topological spaces. A subset C of a metric space X is **compact** if every family of open sets which covers C (that is, whose union contains C) has a finite subfamily which also covers C. Since open sets in C are precisely the sets $U \cap C$ where U is open in X, it is easy to see that C is compact in X iff C is compact in itself, i.e. in the metric space (C, d_C), so that compactness is a property of C itself and not of the way C sits inside X.

A useful consequence of the definition is the following analogue of Cantor's Theorem (a case of the 'finite intersection property').

1.13 Lemma If $\{A_n\}$ is a decreasing sequence of nonempty closed sets in a compact metric space then $\bigcap_n A_n$ is nonempty.

Proof If $\bigcap_n A_n$ were empty then $\{X \sim A_n\}$ would be a sequence of open sets covering X. But clearly no finite subfamily of it can cover X, contrary to hypothesis.

The next theorem, which will be used frequently, derives the two most important conditions equivalent to compactness in a metric space; the reader may have encountered condition (ii) below as the definition. A finite set of points $\{x_1, \ldots, x_n\}$ is an **ϵ-net** for a set A in a metric space (X, d) if each point of A is distant at most ϵ from some x_j; equivalently, if the (closed) balls of radius ϵ round the x_j cover A. If A possesses an ϵ-net for each $\epsilon > 0$ it is **totally bounded** (or **precompact**); since ϵ-balls have diameter at most 2ϵ an equivalent condition is that for each $\epsilon > 0$, A can be covered by finitely many sets $A_1, \ldots, A_n$ of diameter $\leqslant \epsilon$; one may take these closed (replacing A_k by A_k^-) or disjoint (replacing A_k by $A_k \sim (A_1 \cup \ldots \cup A_{k-1})$) – though not generally both at once. An ϵ-net for A is clearly one for A^-; thus is A is totally bounded so is A^-.

1.14 Theorem The following are equivalent for a metric space (X, d):

(i) X is compact;

(ii) Every sequence in X has a convergent subsequence (the **Bolzano–Weierstrass Property**);

(iii) X is complete and totally bounded.

Proof (i) ⇒ (ii) Assume X is compact and $\{x_n\}$ a sequence in X. Define $S_n = \{x_k : k \geqslant n\}$ for $n = 1, 2, \ldots$; the sets S_n^- form a decreasing sequence of nonempty closed sets, so there is a point a such that $a \in S_n^-$ for all n by 1.13. That is, for all n and all $\epsilon > 0$ there exists $k \geqslant n$ such that $d(a, x_k) < \epsilon$. An easy inductive construction yields an increasing sequence of numbers k_j with $d(a, x_{k_j}) < 1/j$ for all j, hence a subsequence of $\{x_n\}$ convergent to a.

(ii) ⇒ (iii). Assume every sequence in X has a convergent subsequence. That X is complete follows from 1.8. Suppose that X were not totally bounded. Then there exists $\epsilon > 0$ such that no **ϵ-net** exists for X. Choose any $x_1 \in X$; $\{x_1\}$ is not an ϵ-net so there must be $x_2 \in X$ such that $d(x_2, x_1) > \epsilon$. Now $\{x_1, x_2\}$ is not an ϵ-net so there is $x_3 \in X$ with $d(x_3, x_1) > \epsilon$ and $d(x_3, x_2) > \epsilon$. Continuing gives a sequence $\{x_n\}$ with $d(x_m, x_n) > \epsilon$ whenever $m \neq n$. Clearly no subsequence of $\{x_n\}$ can converge, a contradiction; so X is totally bounded.

(iii) ⇒ (i). Assume (iii), and suppose there exists a family $\mathfrak{U}$ of open sets which covers X, but no finite subfamily of which covers X. For the purpose of the proof, say that a subset A of X is *intractable* if there is no finite subfamily of $\mathfrak{U}$ which covers A. We can express X as the union of finitely many closed subsets of diameter < 1; at least one of these, call it A_1, must be intractable. A_1 is the union of finitely many closed subsets of diameter $< \frac{1}{2}$; one of these, say A_2, is intractable. Continuing, one obtains a decreasing sequence of intractable closed subsets A_n with $\operatorname{diam}(A_n) \to 0$. By Cantor's Theorem there is a (unique) point a in all the A_n. Now a lies in one of the members U of $\mathfrak{U}$. Since U is open it contains a ball B round a, of radius r, say. But we can choose n so that $\operatorname{diam}(A_n) < r$, and then $A_n \subset B \subset U$, contradicting the intractability of A_n. Thus some finite subfamily of $\mathfrak{U}$ must cover X.

From the remark just before the theorem, together with 1.10, one has at once:

Corollary A subset of a complete metric space has compact closure iff it is totally bounded.

Since also it is clear that every bounded subset of $\mathbf{R}$ (or for that matter of $\mathbf{R}^n$) has an ϵ-net for each $\epsilon > 0$, this implies at once the Heine–Borel Theorem that a subset of $\mathbf{R}$ (or $\mathbf{R}^n$) is compact iff it is closed and bounded. Note that in *any* metric space, a compact subset must be closed (because complete) and bounded (totally bounded sets are obviously bounded).

1.15 Theorem Let (X, d) and (Y, e) be compact metric spaces. Then $X \times Y$ is compact (under either of the standard metrics).

Proof 1 Let $\{(x_n, y_n)\}$ be a sequence in $X \times Y$. By 1.14 (i $\Rightarrow$ ii) we can extract a sequence of indices $\{n_j\}$ such that $\{x_{n_j}\}$ converges in X. From $\{n_j\}$ we extract a further sequence, $\{n_j'\}$ say, such that $\{y_{n_j'}\}$ converges in Y. By 1.4, $\{(x_{n_j'}, y_{n_j'})\}$ converges in $X \times Y$, so the result follows by 1.14 (ii $\Rightarrow$ i).

Proof 2 Note that $X \times Y$ is complete by 1.11 and 1.14 (i $\Rightarrow$ iii) and that if F and G are ϵ-nets in X and Y then $F \times G$ is an ϵ-net in $X \times Y$, with respect to the metric σ.

The generalization of this result to more general spaces and infinite products is taken up in §16.

Topological spaces

1.16 Rather arbitrarily we choose this point to make the transition from metric to topological spaces. The following easily verified properties of the class of open subsets in a metric space serve as a starting point:

(1) X and $\emptyset$ are open (the latter being open 'by convention' so as to make property 3 work).
(2) The union of any family of open sets is open.
(3) The intersection of two (hence, by induction, finitely many) open sets is open.
(4) Given distinct points x, y of X there exist disjoint open sets U, V with $x \in U$, $y \in V$.

It is both useful and enlightening to replace 'metric' by 'open set' as the basis of a theory of limits and continuity. A **topology** for a nonempty set X is a family $\mathbf{T}$ of subsets of X, designated **open** sets (or $\mathbf{T}$-open if one has to distinguish) such that properties (1), (2), (3) hold. A pair $(X, \mathbf{T})$, where $\mathbf{T}$ is a topology for X, is a **topological space.** As with metric spaces, one often omits to mention $\mathbf{T}$ and speaks of 'a topological space X'. If also (4) holds the space and the topology are called **Hausdorff.** Thus every metric space X is a Hausdorff topological space when one endows X with the **metric** (or **usual**) **topology**, that is the class of sets that are open in the sense defined in 1.2.

In a topological space $(X, \mathbf{T})$ one defines a set N to be a **neighbourhood** of a point a (and a to be **interior to** N) if there exists $U \in \mathbf{T}$ (that is, an open set U), with $a \in U \subset N$. It then follows easily from property (2) above that the sets that are **open** in the sense of 1.2 are exactly the open sets one started with. Exactly as in 1.2 one defines the following concepts: **closure point, closure, interior, closed, boundary, dense, continuous at a point** for maps from one topological space to another, **continuous, homeomorphism**; and the usual elementary relations between these notions are unaltered. In particular one has the key property of continuous maps: $f: X \to Y$ is continuous $\Leftrightarrow f^{-1}(U)$ is open in X whenever U is open in $Y \Leftrightarrow f^{-1}(F)$ is closed in X whenever F is closed in Y.

Convergence of a sequence is also defined as in 1.2. In general a sequence may converge to more than one point; it is easy to see that the Hausdorff property (4) is just what is needed to ensure that limits are unique, for if U, V are disjoint open neighbourhoods of points x, y, and if $x_n \to x$, then x_n is eventually in U, so cannot eventually be in V, so $x_n \not\to y$. It is no longer true, as it is in a metric space, that A^- is the set of limits of sequences in A: one often expresses this by saying that 'sequences fail to describe the topology', and this is one reason why sequence arguments, which we use frequently in metric spaces, play a less important role in general topological spaces.

Bases and sub-bases

1.17 A family $\mathcal{B}$ of open sets is a *base* for a topological space $(X, \mathbf{T})$ (or for the topology $\mathbf{T}$) if every open set can be expressed as a union of members of $\mathcal{B}$, equivalently if given $U \in \mathbf{T}$ and $x \in U$ there exists $B \in \mathcal{B}$ with $x \in B \subset U$. A family $\mathcal{S}$ of open sets is a *sub-base* if the family of all sets obtained by taking finite intersections of members of $\mathcal{S}$ is a base, equivalently if given $U \in \mathbf{T}$ and $x \in U$ there exist $S_1, \ldots, S_n \in \mathcal{S}$ with $x \in S_1 \cap \ldots \cap S_n \subset U$. For instance the open balls form a base in any metric space; in particular the open intervals form a base in $\mathbf{R}$. The intervals of the form $(-\infty, b)$ and (a, ∞) form a sub-base in $\mathbf{R}$ since any open interval is the intersection of two such intervals. It is useful to find a base or sub-base whose members have an especially simple form in view of the next result, whose proof is left as an exercise.

1.18 Proposition Let X, Y be topological spaces and let $f: X \to Y$. Let $\mathcal{B}$ and $\mathcal{S}$ be respectively a base and a sub-base for Y. Then f is continuous $\Leftrightarrow$ $f^{-1}(B)$ is open in X for each $B \in \mathcal{B}$ $\Leftrightarrow$ $f^{-1}(S)$ is open in X for each $S \in \mathcal{S}$.

For instance, a real-valued function on X is continuous iff the sets $\{x \in X : f(x) > a\}$, $\{x \in X : f(x) < a\}$ are open for each $a \in \mathbf{R}$, since these are respectively $f^{-1}(a, \infty)$, $f^{-1}(-\infty, a)$.

Let $\mathbf{T}_1$, $\mathbf{T}_2$ be topologies on the same set X. One says $\mathbf{T}_1$ is **larger (stronger, finer)** than $\mathbf{T}_2$ if $\mathbf{T}_1 \supset \mathbf{T}_2$, that is if every $\mathbf{T}_2$-open set is $\mathbf{T}_1$-open; equivalently if the identity map of $(X, \mathbf{T}_1)$ onto $(X, \mathbf{T}_2)$ is continuous.

1.19 Proposition Let $\mathcal{S}$ be an arbitrary family of subsets of a nonempty set X. Form the family $\mathcal{B}$ consisting of X and all finite intersections of members of $\mathcal{S}$; then form the family $\mathbf{T}$ consisting of $\emptyset$ and all possible unions of members of $\mathcal{B}$. Then, $\mathbf{T}$ is topology for which $\mathcal{B}$ is a base and $\mathcal{S}$ a sub-base.

Proof Trivial, except for the fact that $U, V \in \mathbf{T} \Rightarrow U \cap V \in \mathbf{T}$. To see this, suppose $U = \bigcup_i B_i$, $V = \bigcup_j C_j$ where the B_i and C_j are in $\mathcal{B}$, and apply the identity $\bigcup_i B_i \cap \bigcup_j C_j = \bigcup_i \bigcup_j (B_i \cap C_j)$.

It is clear that $\mathbf{T}$ is the smallest topology containing $\mathcal{S}$ in the sense that any other topology containing $\mathcal{S}$ is larger than $\mathbf{T}$.

Subspace, product of topological spaces

1.20 A nonempty subset A of a topological space (X, T) is turned into a **subspace** by giving it the **relative topology** T_A consisting of all sets $A \cap U$ where U is T-open. The product $X \times Y$, where (X, T_1) and (Y, T_2) are topological spaces, is given the **product topology** consisting of arbitrary unions of members of the family $\mathfrak{B}$ of all sets of the form $U \times V$ where $U \in T_1$ and $V \in T_2$; these are called **open rectangles.** The reader should verify that the relative topology and product topology are indeed topologies; that $\mathfrak{B}$ is a base for the latter; and that in the case of metric spaces these definitions match up with those of the *relative metric* and *product metrics.* The definition and properties extend in the obvious way to any *finite* product $Z = X_1 \times \ldots \times X_n$ of topological spaces. The **coordinate projections** $p_i : Z \to X_i$ defined by $p_i(x_1, \ldots, x_n) = x_i$ are clearly continuous because the inverse image of an open set U in X_i is a member of the defining base for the product topology. Infinite products are treated in §16.

Compactness in topological spaces

1.21 The concept of a **compact** subset of a topological space X is defined in terms of covers by open sets exactly as in 1.12; we note that 1.13 continues to hold in this more general situation. However 1.14, which is a key result in metric space theory, is false in topological spaces: indeed, condition (iii) has no meaning for topological spaces, and, in general, neither of conditions (i) and (ii) implies the other.

We now derive some of the fundamental properties of compact spaces. One or two of them are definitely metric space results, and for the rest the reader can think 'metric' for 'topological' throughout if he desires.

The first result should be compared with 1.10.

1.22 Proposition A compact subset of a Hausdorff topological space is closed. Conversely a closed subset of a compact topological space is compact.

Proof Let C be compact in X and choose $x_0 \in X \sim C$. For each $c \in C$, there exist disjoint open neighbourhoods $U(c)$ round x_0 and $V(c)$ round c. The sets $V(c)$ cover C so some finite subset $V(c_1), \ldots, V(c_n)$ covers C. The intersection of the corresponding $U(c_1), \ldots, U(c_n)$ is a neighbourhood of x_0 not meeting C, so $x_0 \notin C^-$. Hence all the closure points of C lie in C.

The second assertion is proved by noting that a family of open sets covering a closed set C in X can be turned into a family covering X, by adding the extra open set $X \sim C$.

1.23 Proposition The continuous image of a compact set is compact.

Proof Let X, Y be topological spaces and let $f: X \to Y$ be continuous and C a compact subset of X. If the family $\{U_i\}$ of open sets in Y covers $f(C)$ then the sets $f^{-1}(U_i)$ are open sets covering C. Choose a finite subfamily of these covering C, then the corresponding U_i are a finite subfamily covering $f(C)$, so $f(C)$ is compact.

1.24 Theorem Let X, Y be topological spaces and f a continuous map from X to Y. Suppose X is compact, Y is Hausdorff (metric, for instance) and f is one-to-one and onto. Then f is a homeomorphism.

Proof By assumption f has an inverse $g: Y \to X$. For any closed set F in X, F is compact by 1.22, so $f(F)$ is compact by 1.23, so $g^{-1}(F) = f(F)$ is closed by 1.22. Thus $f^{-1} = g$ is continuous.

This theorem is very often useful as a work-saver: for instance consider the map f of $\boldsymbol{R}^2$ into itself defined by

$$f(\mathbf{x}) = \frac{\max\{|x|, |y|\}}{(x^2 + y^2)^{1/2}}\, \mathbf{x} \quad \text{where } \mathbf{x} = (x, y),$$

(agree that $f(0, 0) = (0, 0)$). This has been constructed so as to map the square $Q: -1 \leqslant x \leqslant 1, -1 \leqslant y \leqslant 1$ one-to-one onto the unit disc $D: x^2 + y^2 \leqslant 1$, as the reader can check. It is simple to verify that f is continuous; we conclude, without computing f^{-1}, that f is a homeomorphism of Q with D.

One reason why compact Hausdorff spaces are among the pleasantest topological spaces is the next result, which follows at once from 1.24 if one takes f to be the identity map from $(X, \boldsymbol{T}_1)$ to $(X, \boldsymbol{T}_2)$.

1.25 Corollary If X is Hausdorff under a topology $\boldsymbol{T}_2$ and compact under a larger topology $\boldsymbol{T}_1$ then $\boldsymbol{T}_1 = \boldsymbol{T}_2$.
Put another way: if $(X, \boldsymbol{T})$ is a compact Hausdorff space then one cannot make $\boldsymbol{T}$ larger without losing compactness; nor smaller without losing Hausdorff-ness.

1.26 Theorem A continuous real-valued function f on a compact topological space X is bounded and attains its bounds in the sense that there exist a, $b \in X$ such that $f(a) \leqslant f(x) \leqslant f(b)$ for all $x \in X$.

Proof By 1.23 and the Heine–Borel theorem, $f(X)$ is a closed bounded subset of $\boldsymbol{R}$ and so contains its supremum and infimum.

1.27 In particular, if $f(x) > 0 (x \in X)$ then there exists $\delta > 0$ such that $f(x) \geqslant \delta$ $(x \in X)$. In particular again, if A, B are disjoint nonempty compact subsets of a metric space then $d(A, B) > 0$ and there exist $a \in A$, $b \in B$ such that $d(a, b) = d(A, B)$. For $(x, y) \mapsto d(x, y)$ is easily seen to be continuous and positive on the compact (by 1.15) metric space $A \times B$.

Separability

A topological space T is **separable** if there exists a countable dense subset $C = \{x_1, x_2, \dots\}$ of T. Clearly $\boldsymbol{R}$ is separable (consider the rationals), and $\boldsymbol{C}$, $\boldsymbol{R}^n$ and $\boldsymbol{C}^n$ are easily seen to be separable. Any compact metric space X is separable, for if F_n is a $1/n$-net in X $(n = 1, 2, \dots)$ (see 1.13) then $C = \bigcup_{n=1}^{\infty} F_n$ is countable and dense. The next result is subtler than one might think (and is false in a general topological space, though to construct examples is a bit tricky).

1.28 Proposition Each subspace of a separable metric space is separable.

Proof Let $C = \{x_1, x_2, \dots\}$ be countable and dense in (X, d) and let $Y \subset X$. Now $C^- = X$, so for each $y \in Y$ and each m there exists $x_n \in C$ such that $d(y, x_n) \leqslant 1/m$, that is such that $y \in B(x_n, 1/m)$. Let

$$\Lambda = \{(n, m) : Y \cap B(x_n, 1/m) \neq \phi\}$$

which is countable (being a subset of $\boldsymbol{N} \times \boldsymbol{N}$), and nonempty as we have just seen. For each $(n, m) \in \Lambda$ choose a point $y_{nm} \in Y \cap B(x_n, 1/m)$, and let B be the set of all these. Then B is a countable subset of Y. Given $y \in Y$ and $\epsilon > 0$ choose m so that $1/m \leqslant \epsilon/2$, and then n as above so that $y \in B(x_n, 1/m)$. Then $(n, m) \in \Lambda$, and $d(y, y_{nm}) \leqslant d(y, x_n) + d(x_n, y_{nm}) \leqslant 1/m + 1/m \leqslant \epsilon$, proving that B is dense in Y.

Scalar functions

We use the term 'scalar function' to mean a real or complex-valued function on a set S. Scalar functions f, g on S are combined by the **pointwise operations** to form $f+g$, fg, λf for any scalar λ, where $(f+g)(x)=f(x)+g(x)$ and so on. A sequence of scalar functions $\{f_n\}$ **converges pointwise** to f if given $\epsilon > 0$ and $x \in S$ there exists n_0 such that $|f_n(x)-f(x)| < \epsilon$ for $n \geqslant n_0$; and **converges uniformly** if the same n_0 will do for all $x \in S$, that is given $\epsilon > 0$ there exists n_0 such that $|f_n(x)-f(x)| < \epsilon$ for $n \geqslant n_0$, $x \in S$. Pointwise convergence is equivalent to $\lim_n |f_n(x) - f(x)| = 0 (x \in X)$; uniform convergence is equivalent to $\lim_n \sup_{x \in X} |f_n(x) - f(x)| = 0$.

1.29 Theorem If f, g are continuous scalar functions on a topological space X, and λ a scalar, then $f+g$, fg, λf are continuous. If the sequence $\{f_n\}$ of continuous scalar functions converges uniformly to a function f, then f is also continuous.

Proof The same, possibly with slight modifications, as proofs with which the reader will be familiar from elementary analysis.

1.30 Theorem Let f be a continuous scalar function on a compact metric space (X, d). Then f is uniformly continuous.

Proof Given $\epsilon > 0$ the sets $E_n = \{(x, y): |f(x) - f(y)| \geqslant \epsilon, d(x, y) \leqslant 1/n\}$ are easily seen to form a decreasing sequence of closed sets in the compact metric space $X \times X$, having empty intersection. By 1.13, $E_k = \emptyset$ for some k; thus $|f(x) - f(y)| < \epsilon$ whenever $d(x, y) \leqslant 1/k$.

We shall have occasional use for one other result on scalar functions. The **usual** (or **pointwise**) **ordering** for real functions on a set S is defined by saying that $f \leqslant g$ means $f(s) \leqslant g(s)$ for all $s \in S$. A sequence of functions $\{f_n\}$ is **increasing** if $f_1 \leqslant f_2 \leqslant f_3 \dots$; **decreasing** if $f_1 \geqslant f_2 \geqslant f_3 \dots$.

1.31 Dini's Theorem Let $\{f_n\}$ be an increasing sequence of continuous real functions on a compact topological space X converging pointwise to a continuous real function f. Then $\{f_n\}$ converges uniformly to f. (Of course a similar result holds for decreasing sequences.)

Proof The functions $f - f_n$ are continuous and form a decreasing sequence converging pointwise to 0. Thus given $\epsilon > 0$ the sets $E_n = \{x \in X: (f - f_n)(x) \geqslant \epsilon\}$ form a decreasing sequence of closed sets with empty intersection; by 1.13 some E_{n_0} is empty and so are the E_n for all $n \geqslant n_0$. Thus $0 \leqslant f(x) - f_n(x) < \epsilon$ $(n \geqslant n_0, x \in X)$ and the result follows.

Connectedness
A topological space X is **connected** if it cannot be expressed as the union of two disjoint nonempty open subsets (which, being complements of each other, are also closed subsets).

1.32 Theorem Each interval I (closed, open or half-open) in $\mathbf{R}$ is connected.

Proof Assume that I is all of $\mathbf{R}$; the general case is similar. If $\mathbf{R}$ is the union of two disjoint nonempty open-and-closed sets then some closed bounded interval meets both of them, say in sets A, B which, being closed and bounded, are compact; by 1.27 there exist $a \in A$, $b \in B$ such that $d(a, b) \leqslant d(x, y)$ $(x \in A, y \in B)$. A contradiction now arises on assigning the point midway between a and b to either A or B: this completes the proof.

We shall not discuss the subtleties of this important notion, since the only use made of it in the book is the fact that $\mathbf{R}^2 \sim \{0\}$ is connected: this we leave as an exercise.

2. Integration

This section is a quick revision course in the definitions, main results and *raison d'être* of Lebesgue integration on an abstract measure space.

The feature which most sharply distinguishes the Lebesgue definition of the integral from the elementary Riemann definition is the scope and versatility of its *convergence theorems*. The defects of the Riemann integral are: that not enough functions are integrable, so that the pointwise limit of a sequence of integrable functions f_n may fail to be integrable; and that even if $f_n \to f$ and f is integrable, there is no simple condition which guarantees that $\int f_n$ converges to $\int f$ apart from the too stringent one that the interval of integration is finite and the f_n converge uniformly. The Lebesgue integral largely overcomes these restrictions, and the resulting flexibility eliminates many technicalities from analysis. Many of the interesting spaces of analysis are spaces of integrable functions of one sort or another (the most useful Hilbert spaces for example), and the Lebesgue integral imparts to them a pleasing completeness, both in the metric-space and the aesthetic sense.

2.1 The raw material for integration over an abstract set S is a large class of functions on S called **measurable**, a class that is closed under all the usual pointwise operations of analysis: addition, multiplication; pointwise limits, sups and lim sups of sequences. In the case of the ordinary integral on $\boldsymbol{R}$ this class contains all continuous functions, indeed all Riemann-integrable functions. The **integrable** functions are a subclass characterized roughly by 'not being too big too often': in particular, on a bounded interval in $\boldsymbol{R}$, any bounded measurable function is integrable; more generally if g is integrable so is any measurable f such that $|f| \leqslant |g|$.

The basic convergence theorems for a sequence of integrable functions $\{f_n\}$ are:

Monotone Convergence Theorem If $f_1 \leqslant f_2 \leqslant \ldots$ and $\sup \int f_n < \infty$ then $f = \sup f_n$ (possibly after adjustment on a null set – see below and 2.10) is integrable and $\int f = \sup \int f_n$.

Fatou's Lemma If $f_n \geqslant 0$ for all n and $\liminf \int f_n < \infty$ then $f = \liminf f_n$ (possibly after adjustment as above) is integrable and $\int f \leqslant \liminf \int f_n$.

Dominated Convergence Theorem If there is an integrable function g such that $|f_n| \leqslant g$ for all n, and if the f_n converge pointwise to f, then f is integrable and $\int f = \lim \int f_n$.

There is a class of sets in S called *null*, such that any function vanishing outside a null set has zero integral. A property of points of S that only fails on a null set is said to hold *almost everywhere* (a.e.). The three theorems above remain valid if a.e. is inserted at relevant points – for instance $|f_n| \leqslant g$ a.e. and $f_n \to f$ a.e. are allowed in the last theorem.

These few facts, together with the results on approximation by simple functions (2.19), will arm the reader for most of the applications of integration theory in the book, though there are times when more technical details are needed.

Measurable functions may be either real or complex; it seems to be in the nature of things that the complex case is most conveniently treated as an appendage to the real case, which we do in 2.20.

Review of measure and integration

We follow Bartle's treatment and notation in (**1**).

2.2 A *σ-algebra* over a set S is a family $\mathbf{S}$ of subsets of S such that

(i) $\emptyset, S \in \mathbf{S}$;

(ii) If $A \in \mathbf{S}$ then the complement $A^c = S \sim A$ is in $\mathbf{S}$;

(iii) If $\{A_n\}$ is a sequence of sets in $\mathbf{S}$ then $\bigcup_1^\infty A_n$ is in $\mathbf{S}$.

We say $\mathbf{S}$ **admits** (or is closed under) **countable unions.** It is immediate that $\mathbf{S}$ admits finite unions, since $A_1, \ldots, A_k$ can be padded out to $A_1, \ldots, A_k, \emptyset, \emptyset, \ldots$ By complements, it admits finite and countable intersections as well; and if $A, B \in \mathbf{S}$ the relative complement $A \sim B = A \cap B^c$ is in $\mathbf{S}$. A pair $(S, \mathbf{S})$ where $\mathbf{S}$ is a σ-algebra over S is called a **measurable space,** and the sets in $\mathbf{S}$ are called **S-measurable sets,** or simply **measurable sets.**

2.3 If $\mathbf{S}_1$ and $\mathbf{S}_2$ are σ-algebras of subsets of S it is easy to see that $\mathbf{S}_1 \cap \mathbf{S}_2$, that is the class of all subsets A of S which are in both $\mathbf{S}_1$ and $\mathbf{S}_2$, is a σ-algebra. More generally it is clear that the intersection of *any* family of σ-algebras $\{\mathbf{S}_i\}$ over S is a σ-algebra.

If $\mathbf{A}$ is any class of subsets of S there is a smallest σ-algebra containing $\mathbf{A}$. For the class of *all* subsets of S is a σ-algebra containing $\mathbf{A}$, and the intersection of all σ-algebras containing $\mathbf{A}$ is a σ-algebra containing $\mathbf{A}$, which is clearly the required σ-algebra. It is called the σ-algebra **generated** by $\mathbf{A}$.

In $\mathbf{R}$, and more generally in any metric space, a particularly important σ-algebra is the class $\mathbf{B}$ of **Borel sets** which is the σ-algebra generated by all open sets.

Observe that in $\mathbf{R}$, $\mathbf{B}$ is generated by the (much smaller) class of all intervals of the form (a, ∞). For $(a, b) = \bigcup_{n=1}^\infty \{(a, \infty) \sim (b - \frac{1}{n}, \infty)\}$ and any open set is a countable union of open intervals (a, b).

2.4 A real-valued function f on S is called an **S-measurable** (or simply, **measurable**) function if for each real α the set $\{s \in S : f(s) > \alpha\}$ – in other words the set $f^{-1}(\alpha, \infty)$ – is measurable. The definition looks unsymmetrical but isn't, since one easily proves: f is measurable $\Leftrightarrow f^{-1}(U)$ is a measurable set for each open $U \subset \mathbf{R} \Leftrightarrow f^{-1}(F)$ is measurable for each closed $F \subset \mathbf{R} \Leftrightarrow f^{-1}(B)$ is measurable for each Borel set $B \subset \mathbf{R}$.

Any constant function is measurable, as is the characteristic function χ_E (defined to be 1 on E and 0 elsewhere) of a measurable set E. If S is $\mathbf{R}$ (or any metric space) and $\mathbf{S}$ is the class of Borel sets in S then any continuous function $f: S \to \mathbf{R}$ is measurable.

If $f, g: S \to \boldsymbol{R}$ are measurable and if $c \in \boldsymbol{R}$ then the functions

$$cf,\ f^2,\ f+g,\ fg,\ |f|,\ f \vee g,\ f \wedge g,\ f^{\frac{1}{2}} \text{ if } f \geqslant 0$$

are measurable; so also is g/f if f is never zero. Here $\vee$, $\wedge$ are the pointwise max and min operations, $(f \vee g)(s) = \max\{f(s), g(s)\}$, $(f \wedge g)(s) = \min\{f(s), g(s)\}$. Of these the only one that presents any difficulty is the measurability of $f+g$, which relies on the fact that $\boldsymbol{R}$ has a countable dense set.

In particular the functions $f_+ = f \vee 0$, $f_- = (-f) \vee 0$ are measurable. They are called the **positive** and **negative parts** of the measurable function f and it is clear that $f = f_+ - f_-$ and $|f| = f_+ + f_-$.

2.5 It is convenient when dealing with limit operations on sequences of functions to introduce the **extended real number system** $\bar{\boldsymbol{R}}$, which consists of $\boldsymbol{R}$ together with two symbols $+\infty$, $-\infty$ (which are *not* real numbers). We assume the reader is familiar with the basic algebraic and topological properties of $\bar{\boldsymbol{R}}$, reminding him only that one agrees that 0 times $\pm\infty$ is taken as 0.

As with real-valued functions one calls an extended-real-valued function f on S **measurable** if for each real α the set $\{s \in S : f(s) > \alpha\}$ is measurable. Note that this set is now $f^{-1}(a, \infty]$ rather than $f^{-1}(a, \infty)$. To distinguish $\boldsymbol{R}$-valued functions from $\bar{\boldsymbol{R}}$-valued ones one often calls the former **finite-valued,** or just **finite.** It is easily seen that an $\bar{\boldsymbol{R}}$-valued function f is measurable iff (i) the finite function f_1 obtained by 'topping and tailing' f, that is

$$f_1(x) = \begin{cases} f(x) & \text{if this is finite} \\ 0 & \text{otherwise} \end{cases}$$

is measurable and (ii) the sets $f^{-1}\{+\infty\}$, $f^{-1}\{-\infty\}$ are measurable.

The class of $\mathcal{S}$-measurable extended-real-valued functions on S is denoted $M(S, \mathcal{S})$, and the class $\{f \in M(S, \mathcal{S}) : f \geqslant 0\}$ is denoted $M^+(S, \mathcal{S})$. If $\{f_n\}$ is a sequence of functions in $M(S, \mathcal{S})$ then the supremum function $f = \sup f_n$ defined by $f(s) = \sup f_n(s)$ $(s \in S)$ is measurable in view of the equation $\{s : f(s) \geqslant \alpha\} = \bigcap_n \{s : f_n(s) \geqslant \alpha\}$, and similar arguments establish the measurability of $\inf f_n$, $\limsup f_n$ and $\liminf f_n$.

In particular if $\{f_n\}$ converges pointwise to f then f is measurable since $f = \limsup f_n = \liminf f_n$.

2.6 Integration of measurable functions depends on introducing some notion of the 'size' (= length, area, mass, etc.) of a measurable set.

A **measure** on a measurable space $(S, \mathbf{S})$ is a function μ assigning to each A in $\mathbf{S}$ an extended real number $\mu(A)$ (often called the **measure of** A) such that (i) $\mu(\emptyset) = 0$, (ii) $\mu(A) \geqslant 0$ for all A, (iii) μ is **countably additive** in the sense that if $\{A_1, A_2, \ldots\}$ is a pairwise disjoint sequence of sets in $\mathbf{S}$ then $\mu(\bigcup_n A_n) = \sum_n \mu(A_n)$.

This includes the case of the union of finitely many disjoint sets since we can pad out with empty sets. Note that the right hand side can take the value $+\infty$ in two ways: either $\mu(A_n) = +\infty$ for some m, or the series consists of finite terms but diverges.

Simple consequences of the definition:

(i) If $A, B \in \mathbf{S}$ and $A \subset B$ then $\mu(B \sim A) = \mu(B) - \mu(A)$.

(ii) If $A_1, A_2, \ldots \in \mathbf{S}$ and $A_1 \subset A_2 \subset A_3 \ldots$ then $\mu(\bigcup_n A_n) = \lim_n \mu(A_n)$.

(iii) If $A_1, A_2, \ldots \in \mathbf{S}$, $\mu(A_n) < \infty$ for all n, and $A_1 \supset A_2 \supset A_3 \ldots$ then $\mu(\bigcup_n A_n) = \lim_n \mu(A_n)$.

For (i) note that B is the disjoint union of $(B \sim A)$ and A, so $\mu(B) = \mu(B \sim A) + \mu(A)$. Applying this to (ii) and noting that $A = \bigcup_n A_n$ is the disjoint union of $A_1 \cup (A_2 \sim A_1) \cup (A_3 \sim A_2) \cup \ldots$ gives $\mu(A) = \mu(A_1) + \sum_1^\infty (\mu(A_{n+1}) - \mu(A_n)) = \lim_n \mu(A_n)$, and (iii) is proved similarly.

If μ never takes the value $+\infty$, equivalently if $\mu(S) < \infty$, it is called **finite**; if there is a sequence of sets A_n such that $\mu(A_n) < \infty$ for all n and $\bigcup_n A_n = S$ then μ is called **σ-finite.** Clearly a finite measure is σ-finite.

2.7 Examples (1) Let $(S, \mathbf{S})$ be a measurable space and let s be a fixed point of S. Define $\mu(A)$ to be 0 if $s \notin A$ and 1 if $s \in A$. This is called the **point measure concentrated at** s, and will be denoted ϵ_s. It is a finite measure.

(2) Let $\mathbf{S}$ be the class of all subsets of a set S and define $\mu(A)$ to be the number of points in A if this is finite, and $+\infty$ otherwise. This is **counting measure.** It is finite iff S is finite, and σ-finite iff S is countable.

(3) Let μ, ν be measures on the same measurable space $(S, \mathbf{S})$ and let c be a non-negative real number. Then $c\mu$ and $\mu + \nu$ defined (in the obvious way) by

$$(c\mu)(A) = c\mu(A)$$

$$(\mu + \nu)(A) = \mu(A) + \nu(A)$$

are measures on $(S, \mathbf{S})$. (These definitions do not make the set of measures on $(S, \mathbf{S})$ into a linear space since negative c are not allowed; for this the idea of a 'charge', which takes negative as well as positive values, needs to be introduced, which we do not do here.)

(4) Let μ be a measure on $(S, \mathbf{S})$ and let $E \in \mathbf{S}$. Define $\mathbf{E} = \{A \in \mathbf{S} : A \subset E\}$. Then $\mathbf{E}$ is a σ-algebra over E and the restriction of μ to the subclass $\mathbf{E}$ of $\mathbf{S}$ is a measure on $(E, \mathbf{E})$, usually termed the **restriction of μ to E**.

Examples (3), (4) are frequently-used constructions.

(5) A vastly more interesting measure is **Lebesgue measure** λ on $\mathbf{R}$, that is the unique measure on a certain σ-algebra, the **Lebesgue-measurable subsets**, whose value for open, closed or half-open intervals coincides with the ordinary idea of length of an interval. This measure of course is the most important one in analysis, and half the labour in integration theory is in proving that Lebesgue measure exists: a long and quite subtle argument, compared with which the definition and properties of the integral are straightforward.

2.8 A **simple** function is a measurable function taking only finitely many values, all of them finite: i.e. one that can be represented as a finite linear combination $\sum_1^n c_j \chi_{E_j}$ of characteristic functions of measurable sets E_j, with the c_j in $\mathbf{R}$. This representation is not unique. If ϕ is a non-negative simple function represented as above, we define the **integral of ϕ with respect to μ** to be

$$\int \phi \, d\mu = \sum_1^n c_j \mu(E_j)$$

The restriction $\phi \geqslant 0$ ensures that a meaningless $((+\infty) - (+\infty))$ does not occur in the sum on the right. An elementary combinatorial argument shows $\int \phi \, d\mu$ to be independent of the representation of ϕ, and so well-defined. Its value is a non-negative real number or $+\infty$.

2.9 It is clear from the definition that for simple functions one has the restricted **linearity properties**:

$$\int(\phi+\psi)\,d\mu = \int\phi\,d\mu + \int\psi\,d\mu \text{ and } \int c\phi\,d\mu = c\int\phi\,d\mu \ (c \geqslant 0)$$

(note the convention $0.\,(+\infty) = 0$ here).

2.10 If now f is any non-negative measurable function we define the **integral of** f **with respect to** μ to be

$$\int f\,d\mu = \sup\left\{\int \phi\,d\mu : \phi \text{ simple, } 0 \leqslant \phi \leqslant f\right\},$$

again a non-negative real number or $+\infty$. The term **integrable** is reserved for those *finite*-valued f such that $\int f\,d\mu$ is finite.

We define the integral of f **over a measurable set** E with respect to μ to be

$$\int_E f\,d\mu = \int f\chi_E\,d\mu.$$

It is easy to see that it coincides with $\int f'\,d\mu'$ where f', μ' denote the restrictions of f, μ to the subset E (2.7 Example 4).

The key result to all the convergence theorems, established by a simple but most ingenious argument, (see Bartle **1**, p. 31), is

2.11 B. Levi's Monotone Convergence Theorem If $\{f_n\}$ is a monotone increasing sequence of functions in $M^+(S, \mathbf{S})$ which converges to f, then

$$\int f\,d\mu = \lim \int f_n\,d\mu.$$

Using this one easily shows that the linearity properties 2.9 of simple functions are valid also for functions in $M^+(S, \mathbf{S})$.

The next result, whose proof we sketch, is

2.12 Fatou's Lemma If $\{f_n\}$ is a sequence in $M^+(S, \mathbf{S})$ then

$$\int(\liminf f_n)\,d\mu \leqslant \liminf \int f_n\,d\mu.$$

Proof Let $g_m = \inf\{f_m, f_{m+1}, \ldots\}$, so that $g_m \leqslant f_n$ whenever $m \leqslant n$. Then $\int g_m\,d\mu \leqslant \int f_n\,d\mu\,(m \leqslant n)$, so $\int g_m\,d\mu \leqslant \liminf \int f_n\,d\mu$. Since $\{g_m\}$ increases and $\lim g_n = \liminf f_n$ the result follows from the last theorem.

(Note that 2.11 and 2.12 are stated in a form slightly different from that in 2.1, since here our functions can take infinite, but not negative, values whereas in 2.1 the reverse is the case: see 2.13.)

2.13 An arbitrary real-valued measurable function f is called **integrable** (or **summable**) if its positive and negative parts have finite integrals, and the **integral of** f **with respect to** μ is defined as

$$\int f\,d\mu = \int f_+\,d\mu - \int f_-\,d\mu.$$

The set of all integrable functions is denoted $\mathcal{L} = \mathcal{L}(S, \mathbf{S}, \mu)$.

The notations $\int f\,d\mu$, $\int f$, $\int f(s)\,d\mu(s)$ are synonymous; the second is used when it is clear what measure is under consideration and the third when one wants to display f by a formula. When μ is *Lebesgue measure* one writes $\int f\,dt$, $\int f(t)\,dt$ in the usual way.

From the fact that we explicitly exclude functions taking the values $\pm\infty$, and from the restricted linearity properties of the integral for non-negative functions, one easily sees that the integral is linear in the following sense:

2.14 Theorem A constant multiple cf and sum $f + g$ of functions in $\mathcal{L}$ is in $\mathcal{L}$, and

$$\int cf\,d\mu = c\int f\,d\mu, \quad \int (f+g)\,d\mu = \int f\,d\mu + \int g\,d\mu.$$

In other words $\mathcal{L}$ is a linear space and $f \mapsto \int f\,d\mu$ is a linear functional.

Since $f \geqslant 0$ implies $\int f\,d\mu \geqslant 0$ we also see that if $f, g \in \mathcal{L}$ and $f \leqslant g$ then $\int f\,d\mu \leqslant \int g\,d\mu$.

2.15 Proposition If $f \in M(S, \mathbf{S})$ and $|f| \leqslant g$ for some $g \in \mathcal{L}$ then $f \in \mathcal{L}$. In particular $f \in \mathcal{L} \Leftrightarrow |f| \in \mathcal{L}$.

Proof One has $\int f_+ \leqslant \int g < \infty$, $\int f_- \leqslant \int g < \infty$. The second assertion follows from the fact that $|f| = f_+ + f_-$.

From this and Fatou's Lemma easily follows:

2.16 Dominated Convergence Theorem Let $\{f_n\}$ be a sequence of integrable functions converging pointwise to a function f. If there is an integrable function g such that $|f_n| \leqslant g$ for all n then f is integrable and $\int f\,d\mu = \lim \int f_n\,d\mu$.

Proof f is measurable by 2.5, and $|f| \leqslant g$, so f is integrable. Now Fatou's Lemma applied to the non-negative functions $g + f_n$ gives

$$\int g\,d\mu + \int f\,d\mu \leqslant \liminf \int (g+f_n)\,d\mu = \int g\,d\mu + \liminf \int f_n\,d\mu$$

and applied to the non-negative functions $g - f_n$ gives by a similar argument

$$\int g\,d\mu - \int f\,d\mu \leqslant \int g\,d\mu - \limsup \int f_n\,d\mu.$$

Combining these gives $\limsup \int f_n\,d\mu \leqslant \int f\,d\mu \leqslant \liminf \int f_n\,d\mu$, and the result follows.

An example of this result failing when no such dominating function g exists is the sequence of functions on $[0, 1]$ defined by $f_n = n\chi_{(0,1/n]}$ for which $f_n \to 0$ pointwise, $\int_0^1 f_n\,dt = 1$ for all n.

Null sets, equivalent functions and a.e. *convergence*

2.17 The class $\mathbf{N}$ of sets $N \in \mathbf{S}$ such that $\mu(N) = 0$ (easily seen to be a σ-algebra) is important in that its members are usually ignored. They are called the **μ-null**, or just **null**, sets. If $P(s)$ denotes a property of points s of S, and the set $\{s : P(s)$ does not hold$\}$ is null, one says $P(s)$ holds **μ-almost everywhere**, shortened to **μ-a.e.** or just **a.e.** For instance if $f \in M^+(S, \mathbf{S})$ and $\int f\,d\mu < \infty$ then $\mu\{s : f(s) = +\infty\}$ must be zero, that is 'f is finite almost everywhere'. Two functions are called **μ-equivalent**, or just **equivalent**, if $f = g$ μ-a.e; one writes $f \sim_\mu g$. It is clear that if $f, g \in M^+(S, \mathbf{S})$ and $f \sim_\mu g$ then $\int f\,d\mu = \int g\,d\mu$; similarly for functions in $\mathfrak{L}(S, \mathbf{S}, \mu)$.

If $\{f_n\}$ is a sequence in $M(S, \mathbf{S})$ and $f_n \to f$ μ-a.e. then by redefining the values of f_n and f to be (say) 0 on a suitable null set one obtains functions $f_n' \sim_\mu f_n$ and $f' \sim_\mu f$ such that $f_n' \to f'$ everywhere. Similarly a sequence such that $f_n \leqslant f_{n+1}$ μ-a.e. for each n can, by suitable redefinition, be turned into a sequence $\{f_n'\}$ for which $f_n' \leqslant f_{n+1}'$ everywhere. In this way the more general forms of the basic convergence theorems, mentioned in 2.1, can be derived. They are in fact the forms most often used.

When μ is Lebesgue measure on $\mathbf{R}$ the null sets include all countable subsets of $\mathbf{R}$ as well as more complicated examples such as Cantor's 'middle third' set. Some authors take an even more cavalier attitude to null sets, allowing integrable functions to be infinite, or even undefined, on a null set: thus one may speak of the function $f(t) = |t-1|^{-\frac{1}{2}}$ as integrable on $[0, 2]$ although $f(1)$ is not defined. We shall not go as far as this, however.

Approximation by simple functions

2.18 We shall denote by $\mathfrak{L}_0(S, \mathbf{S}, \mu)$, or just $\mathfrak{L}_0$, the set of **μ-integrable simple functions.** It is clear from the definition 2.8 that these are precisely the simple functions which vanish outside some set E such that $\mu(E) < \infty$, equivalently the functions of the form $\sum_1^n c_j \chi_{E_j}$ where $\mu(E_j) < \infty$ for each j. That is, $\mathfrak{L}_0$ is the linear subspace of $\mathfrak{L}$ spanned by the functions χ_E where

$\mu(E) < \infty$. The importance of $\mathfrak{L}_0$ is that, at least when μ is σ-finite, it is (in a sense) dense in $M(S, \mathbf{S})$ as the next result shows.

2.19 Proposition

(i) Let μ be σ-finite and let $f \in M(S, \mathbf{S})$. Then there is a sequence $\{f_n\}$ in $\mathfrak{L}_0(S, \mathbf{S}, \mu)$ such that $|f_n| \leqslant |f_{n+1}|$ for all n and $f_n \to f$ pointwise. In particular $|f_n| \leqslant |f|$ for all n.

(ii) If μ is finite and f bounded then the sequence $\{f_n\}$ can be chosen to converge uniformly.

Proof Let $\{A_n\}$ be an increasing sequence of measurable sets of finite measure whose union is S. Suppose first that $f \geqslant 0$. Define

$$f_n(s) = \begin{cases} k/2^n & \text{if } k/2^n \leqslant f(s) \leqslant (k+1)/2^n \text{ for some } k \leqslant n\,2^n - 1; \\ n & \text{if } f(s) \geqslant n. \end{cases}$$

(More succinctly, $f_n(s) = \min\{n, [2^n f(s)]/2^n\}$ where $[\ \ ]$ denotes 'integer part of'.) It is left as an exercise to verify that $\{f_n \chi_{A_n}\}$ has the property required in (i) and $\{f_n\}$ the property required in (ii).

When f takes positive and negative values consider $\{g_n - h_n\}$ where $\{g_n\}$ and $\{h_n\}$ are sequences chosen in the above way for f_+ and f_- respectively.

Complex functions

2.20 A complex-valued function f on a measurable space $(S, \mathbf{S})$ is called measurable if its real and imaginary parts are measurable real functions. It is then easily verified that if f, g are complex measurable functions and c a complex scalar then also cf, $f + g$, fg, $\bar{f}$, $|f|$ are measurable (the last because $|f| = (f\bar{f})^{\frac{1}{2}}$). So also is g/f if f is never zero. The function $\overline{\text{sgn}}\, f$ defined by

$$\overline{\text{sgn}}\, f(s) = \begin{cases} \overline{f(s)}/|f(s)| & (f(s) \neq 0) \\ 0 & (f(s) = 0) \end{cases}$$

is also easily seen to be measurable: it is important because of the property $|f| = f\, \overline{\text{sgn}}\, f$.

The integral is extended in the obvious way by defining

$$\int (g + ih)\, d\mu = \int g\, d\mu + i \int h\, d\mu$$

whenever g and h are integrable. We leave it to the reader to verify that the obvious analogues of 2.14, 2.15, the Dominated Convergence Theorem 2.16, the generalizations to a.e. convergence in 2.17 and the approximation result 2.19 are valid. There is however one fact that is almost trivial for real functions but is not so obvious for complex functions.

2.21 Proposition Let f be integrable. Then $|\int f\,d\mu| \leqslant \int |f|\,d\mu$.

Proof Let α be a (complex) number of modulus 1 such that $\alpha \int f\,d\mu = |\int f\,d\mu|$, and write $\alpha f = g + ih$ where g, h are real functions. Since $\int \alpha f\,d\mu$ is then real, $\int h\,d\mu$ must be zero and so, since $g \leqslant |g| \leqslant |\alpha f| = |f|$, $|\int f\,d\mu| = \int \alpha f\,d\mu = \int g\,d\mu \leqslant \int |f|\,d\mu$.

Special properties of Lebesgue measure on $\boldsymbol{R}$

2.22 In an abstract measure space there is no topology and so no notion of continuous function. In $\boldsymbol{R}$ however, Lebesgue measure λ is set up in a way that ensures the measurability of intervals, and hence of all open sets. Since, for a continuous function f on $\boldsymbol{R}$, $\{t \in \boldsymbol{R} : f(t) > \alpha\}$ is an open set it follows that continuous functions are measurable. Furthermore, λ has the very important **Regularity Property:** Given a Lebesgue measurable set $A \subset \boldsymbol{R}$ there is for each $\epsilon > 0$ an open set U such that $U \supset A$ and $\lambda(U \sim A) < \epsilon$. In other words the measurable sets are 'nearly open'. The significance of this, which may not be obvious at first sight, will become clear in §9.

We note that the construction of Lebesgue measure on $\boldsymbol{R}$ can be derived (by generalizing and then specializing! – the problems suggest how) from the Riesz Representation Theorem of §15.

Counting measure on $\boldsymbol{N}$

2.23 Functions on $\boldsymbol{N}$ can be thought of as sequences $f = (f(1), f(2), \ldots)$. When μ is counting measure (2.7 Example 2) all functions are measurable; it is easily seen that simple functions are those for which there exists n_0 such that $f(n) = f(n_0)(n \geqslant n_0)$, and that the integral of a non-negative function is just

$$\sum_{n=1}^{\infty} f(n)$$

so that $\mathfrak{L}$ consists of those functions for which $\sum_1^\infty |f(n)| < \infty$, with

$$\int f\,d\mu = \sum_1^\infty f(n).$$

In this way, with appropriate changes of notation, one can interpret the convergence theorems as results about double sequences and series. This idea will be useful when we deal with the ℓ_p spaces in §5.

3. Linear spaces

We assume the reader is fairly well acquainted with the elementary properties of linear (or vector) spaces as taught in first courses of linear algebra, so in this section the basic notions are reviewed very tersely, mostly without proofs.

3.1 A linear space consists of a set X of objects called **vectors** (denoted in this section by small letters $x, y, \ldots$) and a field $\boldsymbol{F}$ whose elements are called **scalars** (denoted usually by small Greek letters $\lambda, \mu, \ldots$). Though in general $\boldsymbol{F}$ can be any field, we shall only be interested in the two basic number fields of analysis, the real numbers $\boldsymbol{R}$ or the complex numbers $\mathbf{C}$.

There is in X a **zero vector** denoted 0, and two operations are defined whereby one can **add** two vectors x, y to produce the vector $x + y$ and can **multiply** a vector x by a scalar λ to produce the vector λx, in such a way that the following properties hold:

Addition laws

$$\begin{aligned} x + y &= y + x; \\ (x + y) + z &= x + (y + z); \\ x + 0 &= x; \end{aligned}$$

For each x there is a vector $-x$ such that $x + (-x) = 0$.

Scalar multiplication laws

$$\begin{aligned} \lambda(x + y) &= \lambda x + \lambda y \\ \lambda(\mu x) &= (\lambda\mu) x \\ (\lambda + \mu) x &= \lambda x + \mu x \\ 1x &= x \end{aligned}$$

X is called a **real linear space** or a **complex linear space** according as the scalar field $\boldsymbol{F}$ is $\boldsymbol{R}$ or $\mathbf{C}$.

We shall take for granted many facts which follow at once from the definitions: for instance $\lambda 0 = 0$ (0 here denoting the zero vector); $0x = 0$ (0 being the zero scalar on the left side and the zero vector on the right); $(-1)x = -x$; any finite sum like

$u + v + w + x$ is unambiguously defined because however we bracket it the answer is always the same. The expression $x - y$ is short for $x + (-y)$. One has $\lambda(x - y) = \lambda x - \lambda y$ and similarly for more complicated expressions like $\lambda(x - y + z)$.

If a nonvoid subset M of X, under the original addition and scalar multiplication operations, forms a linear space in its own right, it is called a **linear subspace** of X. This happens iff $\lambda x + \mu y$ is in M for any x, y in M and any scalars λ, μ. The largest subspace is X itself; the smallest is the singleton subset $\{0\}$.

Given vectors $x_1, \ldots, x_n$, any vector y of the form $y = \lambda_1 x_1 + \ldots + \lambda_n x_n$ for suitable scalars $\lambda_1, \ldots, \lambda_n$ is a **linear combination** of the x_j. It is easy to prove the:

3.2 Proposition Let A be any nonempty (not necessarily finite) subset of X. Then the set of all possible linear combinations $\lambda_1 x_1 + \ldots + \lambda_n x_n$, where n is arbitrary, of vectors $x_1, \ldots, x_n$ chosen from A is a linear subspace. It is the smallest linear subspace containing A.

3.3 The subspace referred to is called the subspace **generated** or **spanned** by A, or the **linear span** of A, and denoted $\text{lin}(A)$ or $\text{lin}\, A$. The linear span of a set A together with a finite number of extra points crops up often enough to deserve a convenient notation, so for instance we write $\text{lin}(x, y, A)$ as short for $\text{lin}(\{x, y\} \cup A)$. The smallest subspace containing the empty set is $\{0\}$, so one defines $\text{lin}(\emptyset)$ to be $\{0\}$.

If there is a *finite* set of vectors $e_1, \ldots, e_n$ whose linear span is the whole of X then X is called **finite-dimensional**; otherwise X is **infinite-dimensional.** Possibly the reader has so far only done serious linear algebra in finite dimensional spaces, and since most of the spaces (usually spaces of functions) encountered in functional analysis are infinite-dimensional, we point out that even when A is an infinite set only *finite* sums are featured in the definition of $\text{lin}(A)$. A linear combination with infinitely many terms is (at this stage) totally meaningless.

Example The set $\mathcal{C}[0, 1]$ of all continuous (real or complex) functions on the closed interval $[0, 1]$ in $\boldsymbol{R}$ becomes a linear space if addition and scalar multiplication are defined in the usual (pointwise) sense. Among the vectors (= functions) in this space are the functions f_n where $f_0(t) = 1$, $f_1(t) = t, \ldots, f_n(t) = t^n, \ldots (0 \leqslant t \leqslant 1)$. Their linear span $\text{lin}\{f_1, f_2, \ldots\}$ consists precisely of the *polynomial functions* of the form

$$p(t) = c_0 + c_1 t + \dots + c_n t^n$$

and does not contain any functions that can be expressed as infinite series such as $e^t = \sum_0^\infty t^n/n!$, since this involves notions of convergence that have not yet entered the scene.

3.4 A nonempty subset A of X is **linearly independent** if it is not possible to express the vector 0 as a linear combination of vectors in A, except trivially: that is if the situation

$$\lambda_1 x_1 + \dots + \lambda_n x_n = 0$$

with $x_1, \dots, x_n$ in A, implies

$$\lambda_1 = \dots = \lambda_n = 0.$$

Otherwise A is called a **linearly dependent** subset. If $\{x_1, x_2, \dots\}$ is a linearly dependent *sequence* of vectors it is easy to see that there is a vector x_n which is a linear combination of the ones before it, which we offer as a hint to the:

Exercise Show that X is infinite-dimensional iff it has an infinite linearly independent subset.

Purely as a matter of notation it is convenient, given a (possibly infinite) set $A = \{x_j\}_{j \in J}$ of vectors, to write linear combinations of vectors in A in the form of sums $\Sigma \lambda_j x_j$ extended over all the j's in J, it being understood that only a finite number of the λ_j are nonzero. This notation allows a nice concise statement of the:

3.5 Principle of equating coefficients If $\{x_j\}$ is a linearly independent set of vectors then each $y \in \text{lin}\{x_j\}$ has a *unique* representation as $\Sigma \lambda_j x_j$. Equivalently, if $\Sigma \lambda_j x_j = \Sigma \mu_j x_j$ then $\lambda_j = \mu_j$ for each j.

Proof $\Sigma \lambda_j x_j = \Sigma \mu_j x_j \Rightarrow \Sigma (\lambda_j - \mu_j) x_j = 0 \Rightarrow \lambda_j - \mu_j = 0$ for each j, by linear independence.

Determining whether a set of vectors is linearly independent in specific cases is often a matter of elementary calculus or algebra. For instance in the last example the question 'are the functions f_n linearly independent?' reduces to 'if a polynomial $p(t) = \lambda_0 + \lambda_1 t + \dots + \lambda_n t^n$ vanishes identically on $[0, 1]$ (i.e. equals the zero element of $\mathcal{C}[0, 1]$) do all its coefficients vanish?' to which elementary calculus gives the answer, 'Yes' (how?).

3.6 A subset A of X such that $\text{lin}\, A = X$ is called a **spanning set**; thus a finite dimensional space is one with a finite spanning set. A set which is both linearly independent and a spanning set is called a **basis** for X. It is therefore a set B such that every vector is *uniquely* expressible as a linear combination of vectors in B.

We assume the reader is familiar with the fundamental facts about finite-dimensional linear spaces contained in the following:

3.7 Theorem Every finite-dimensional space X has a basis; indeed every linearly independent subset can be enlarged to form a basis, and any spanning set contains a basis.

All bases for X have the same number n of elements; n is the **dimension** of X, written $\dim X$.

What the reader may not know is that this theorem also holds for infinite-dimensional spaces, if 'number of elements' is interpreted as 'cardinal number'. This is *Hamel's basis theorem.* Surprisingly, in view of the important role played by bases in finite-dimensional spaces, they are of negligible use in infinite-dimensional spaces, because there is hardly ever any tie-up between a basis and the topological properties of the space. Thus we do not stop to prove Hamel's theorem.

Product and quotient

3.8 The **Cartesian product** $X \times Y$ of two linear spaces (over the same scalars) is made into a linear space in the obvious way by defining

$$(x_1, y_1) + (x_2, y_2) = (x_1 + x_2, y_1 + y_2)$$

and

$$\lambda(x, y) = (\lambda x, \lambda y).$$

We leave it to the reader to check that the vector space properties are satisfied. Similarly one can turn the product of any finite (or indeed infinite) family of linear spaces into a linear space.

Next, suppose N is a linear subspace of a linear space X. We write $x + N$ for the set $\{x + u : u \in N\}$ and call it the **coset** of N containing x, or the **translate** of N by x (see 3.22 for the geometrical significance of this idea).

The **quotient space** of X by N, written X/N and often called the space obtained by **factoring out** N, is defined to be the class of all the cosets $x + N$ (note that of course distinct x's may give rise to the same coset) with linear operations defined by

$$(x + N) + (y + N) = (x + y) + N$$
$$\lambda(x + N) = (\lambda x) + N$$

3.9 Lemma X/N with the above operations is a linear space.

Proof The main thing to check is that these operations are well-defined. Define $x \sim x'$ to mean $x - x' \in N$. It is easy to check that $\sim$ is an equivalence relation and $x + N$ is exactly the $\sim$-equivalence class containing x, so that

$$x + N = x' + N \Leftrightarrow x \sim x' \Leftrightarrow x - x' \in N$$

If $x + N = x' + N$, $y + N = y' + N$ then $x - x' \in N$, $y - y' \in N$, so $(x + y) - (x' + y') = (x - x') + (y - y') \in N$, showing $(x + y) + N = (x' + y') + N$ and therefore addition is well defined. Scalar multiplication is similar, and we leave the reader to check that the operations obey the linear space axioms.

It is useful to think of vectors in N as in some sense 'negligible', and X/N is then obtained from X by considering two vectors as the same if their difference is negligible.

3.10 Example If $\mathcal{L}$ denotes the space of integrable functions on some measure space we can, and do, regard two functions f, g in $\mathcal{L}$ as the same if $f(s) = g(s)$ a.e. The result is the space L, obtained from $\mathcal{L}$ by factoring out the subspace N of functions which vanish almost everywhere.

Linear maps

3.11 Let X and Z be linear spaces with the same scalars. A mapping $T : X \to Z$ is **linear** if for all x, y in X and for all scalars λ one has

$$T(x + y) = T(x) + T(y), \quad T(\lambda x) = \lambda T(x)$$

or equivalently, if $T(\lambda x + \mu y) = \lambda T(x) + \mu T(y)$ for all x, y in X and all scalars λ, μ. The **zero map** 0, defined by $0(x) = 0$ for every x, is a (rather trivial) example.

Isomorphisms

3.12 A (linear) **isomorphism** of a linear space X with a linear space Y, with the same scalars, is a linear map T which is one-to-one and onto. Two linear spaces X, Y are **isomorphic** if there exists an isomorphism $T : X \to Y$. It is easy to prove the

3.13 Lemma The inverse of an isomorphism, and the composition of two isomorphisms, is an isomorphism. So also is the identity map $I: x \mapsto x$ on any linear space. Consequently isomorphism is an equivalence relation.

All properties of a space that can be expressed purely in linear space terms are preserved by an isomorphism – for instance it maps a linearly independent set in X to one in Y, and a spanning set in X to one in Y. In a certain sense then, isomorphic spaces can be regarded as identical (but see p. 56 for a word of caution about taking this too seriously.)

3.14 Proposition Two linear spaces of finite dimension (with the same scalars) are isomorphic $\Leftrightarrow$ they have the same dimension.

Proof Suppose Y is isomorphic to X under the map T. It follows from the remarks above that if $\{e_1, \ldots, e_n\}$ is a basis in X then $\{T(e_1), \ldots, T(e_n)\}$ is a basis in Y so $\dim X = n = \dim Y$. Conversely if X and Y have the same dimension n, choose a basis $e_1, \ldots, e_n$ in X and a basis $f_1, \ldots, f_n$ in Y. Every $x \in X$ can be uniquely written $x = \lambda_1 e_1 + \ldots + \lambda_n e_n$. The map associating to x the element $y = \lambda_1 f_1 + \ldots + \lambda_n f_n$ in Y is easily seen to be linear, one-to-one and onto, in other words an isomorphism.

Image and inverse image of a linear subspace

3.15 Proposition Let $T: X \to Y$ be a linear mapping. Then the image of any linear subspace of X is a linear subspace of Y, and the inverse image of any linear subspace of Y is a linear subspace of X.

Proof Left as an exercise.

In particular the **image** or **range** $T(X)$ of T, and its **kernel** or **nullspace** $\{x : Tx = 0\} = T^{-1}\{0\}$, are linear subspace of Y and X respectively.

The nullspace is denoted by $\ker T$; less often, the image is denoted by $\operatorname{im} T$. We have the important

3.16 Proposition T is one-to-one $\Leftrightarrow \ker T$ is the zero subspace of X.

Proof Let $\ker T = \{0\}$, then $T(x) = T(y) \Rightarrow 0 = T(x) - T(y) = T(x - y)$ so $x - y = 0$, $x = y$. The converse is obvious.

A linear map $T: X \to Y$ which is one-to-one is often called an *isomorphism into* (rather than onto) Y.

Quotient map

3.17 An important example of a linear map is the map Q of a space X onto the quotient X/N defined by $x \mapsto x + N$. It is obvious that Q is linear. Note that

$$N = \ker Q$$

because $Qx = 0$ (the zero in X/N) $\Leftrightarrow x + N = 0 + N \Leftrightarrow x \in N$. The next little fact is a prototype of several more powerful results we shall meet later.

3.18 Lemma If S and T are linear maps from a linear space X into a linear space Y, and if S and T agree on a spanning set A in X then they are equal. (The stated condition means that $S(x) = T(x)$ for all $x \in A$.)

Proof If $S(x) = T(x)$ and $S(y) = T(y)$ then it is clear by linearity that $S(\alpha x + \beta y) = T(\alpha x + \beta y)$ for any scalars α, β. Thus the set $\{x \in X: S(x) = T(x)\}$ is a linear subspace. It contains the spanning set A and so must be the whole of X.

Subsets of linear spaces

3.19 One of the characteristics of a modern analyst is that he likes to look at large numbers of functions or transformations simultaneously. Most sets of functions lie in a linear space of some sort, and it is useful to have a notation for manipulating subsets of a linear space in quite a 'geometrical' way, often using language borrowed from everyday Euclidean geometry. In this section we introduce the ideas of a **convex** set, and of **addition** and **scalar multiplication of sets** in a linear space.

Let X be a (real or complex) linear space and A a subset of X (see Diagram 1). Then A is **convex** if for any $x, y \in A$ and any scalar α with $0 \leqslant \alpha \leqslant 1$ we have $\alpha x + (1 - \alpha)y \in A$. The set $\{\alpha x + (1 - \alpha)y : 0 \leqslant \alpha \leqslant 1\}$ is often called the **line segment** joining x to y – when X is ordinary 3-space it is precisely that – and one can restate the definition by saying that a convex set contains the line segment between any two of its points.

We agree to call the empty set convex, and subject to this convention, the intersection of any family of convex sets $A_i \subset X$ is convex. For suppose $x, y \in \bigcap_i A_i$; then the segment joining x to y must be inside A_i for each i (since $x, y \in A_i$ for each i) and so lies inside $\bigcap_i A_i$, proving that $\bigcap_i A_i$ is convex. The first part of the next result follows at once from this.

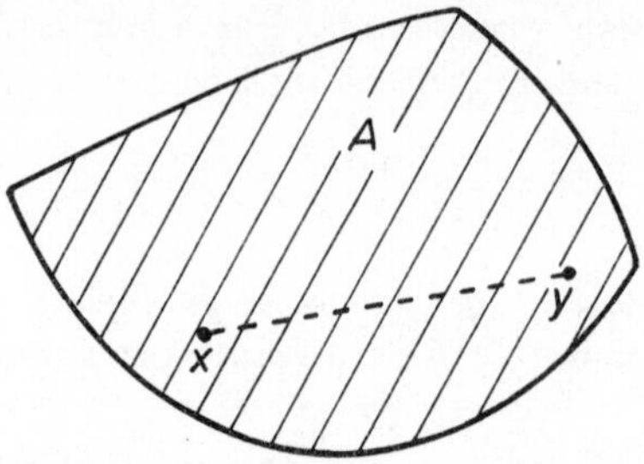

Diagram 1.

3.20 Proposition Let A be any subset of X. Then

(i) There is a unique smallest convex set containing A, called the **convex hull** or **convex cover** of A and written $\mathrm{co}(A)$ or $\mathrm{co}\, A$.
(ii) $\mathrm{co}(A)$ consists precisely of all finite linear combinations $\sum_1^n \alpha_i x_i$ of elements of A, where the α_i are nonnegative and add up to 1. Such linear combinations are called **convex combinations** of elements of A.

Proof (i) There is at least one convex set containing A, namely X itself, and the intersection of all such sets is clearly the smallest convex set containing A. (Note $\mathrm{co}(\emptyset) = \emptyset$ in contrast to the operator 'linear span' for which $\mathrm{lin}(\emptyset) = \{0\}$.)
(ii) Let B denote the set of all convex combinations of vectors in A.

B is convex: For if $u = \alpha_1 x_1 + \dots + \alpha_n x_n$ and $v = \beta_1 y_1 + \dots + \beta_m y_m$ are convex combinations of vectors in A, and $0 \leqslant \gamma \leqslant 1$ then all the coefficients in the representation

$$\gamma u + (1-\gamma)v = \gamma\alpha_1 x_1 + \dots + \gamma\alpha_n x_n + (1-\gamma)\beta_1 y_1 + \dots + (1-\gamma)\beta_m y_m$$

are clearly nonnegative and their sum is

$$\sum_1^m \gamma\alpha_i + \sum_1^n (1-\gamma)\beta_j = \gamma \sum \alpha_i + (1-\gamma)\sum \beta_i = \gamma + 1 - \gamma = 1,$$

showing $\gamma u + (1-\gamma)v$ is a convex combination of vectors in A. We now prove by induction that any convex set C that contains A contains B. Suppose C contains all convex combinations of at most n vectors of A (which is obviously true for $n = 1$), and let $u = \alpha_1 x_1 + \ldots + \alpha_n x_n + \alpha_{n+1} x_{n+1}$ be a convex combination of $n + 1$ vectors of A. Clearly by relabelling we can suppose $\alpha_{n+1} \neq 1$. Denoting $1 - \alpha_{n+1}$ by β we have $u = \beta v + (1-\beta)x_{n+1}$ where

$$v = \frac{\alpha_1}{\beta} x_1 + \ldots + \frac{\alpha_n}{\beta} x_n.$$

Since $\dfrac{\alpha_1}{\beta} + \ldots + \dfrac{\alpha_n}{\beta} = \dfrac{1-\alpha_{n+1}}{1-\alpha_{n+1}} = 1$, v is a convex combination of n vectors in A, and so is in C by the inductive hypothesis. Since C is convex and $v, x_{n+1} \in C$ it follows that $u \in C$, completing the induction step. The result of this is to show that B is a convex set which is contained in any convex set that contains A, and hence $B = \operatorname{co}(A)$ by the definition of $\operatorname{co}(A)$.

Convex sets are important because they are fairly tractable objects and because, fortunately, many sets that crop up in analysis turn out to be convex. Here are some:

3.21 Examples

(1) Any linear subspace, and any coset of a linear subspace, is convex.

(2) In the (real or complex) space $\mathcal{C}[0, 1]$, the following are examples of convex sets:

$C_1 = \{f : 0 \leqslant f(t) \leqslant 100 \ (0 \leqslant t \leqslant 1)\}$;
$C_2 = \{f : f$ is differentiable and satisfies the differential equation $y' + 2y = \sin t\}$;
$C_3 = \{f : f(0) = 0,\ f(1) = 1$ and f is strictly increasing on $[0, 1]\}$;
$C_4 = \{f : |\int_0^t f(s)\,ds| \leqslant t$ for $0 \leqslant t \leqslant 1\}$;
$C_5 = \{f : f$ satisfies the Lipschitz condition $|f(s) - f(t)|$ $|s - t|$ for $s, t \in [0, 1]\}$.

We have already met, in the special case of cosets, the *translate* of a set A by a vector x, defined by $x + A = \{x + a : a \in A\}$. To add to our notations for moving sets around in linear spaces, expanding, contracting and sometimes distorting them we make the

3.22 Definitions Let A, B be subsets of a linear space X, and λ be a scalar. Then

(i) λA means the set $\{\lambda a : a \in A\}$;
(ii) $A + B$, the **vector sum** of A and B, is the set $\{a + b : a \in A, b \in B\}$;
(iii) $-A$ is the set $\{-a : a \in A\}$, in other words $(-1)A$.

A set A is **symmetric** if $A = -A$, in other words if $-a \in A$ whenever $a \in A$. Geometrically speaking, $x + A$ has the same size, shape and orientation as A but has been shifted; λA is obtained by 'blowing up' A in the ratio $\lambda : 1$ and it is a similar figure to A, 'similarly situated' in the sense that corresponding lines on A and λA are parallel (see Diagram 2).

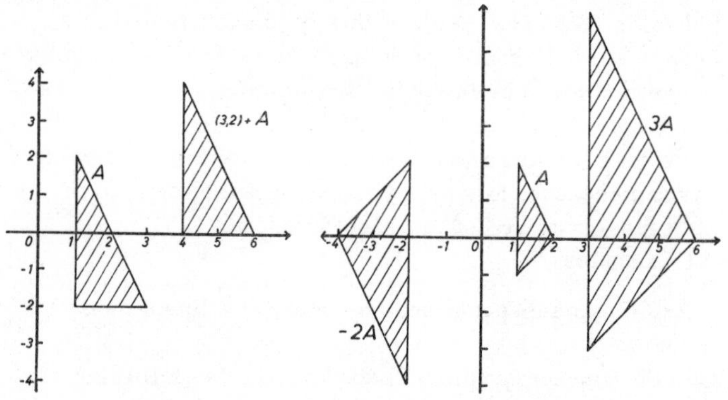

Diagram 2.

To get a visual idea of $A + B$, note that it consists of all points (= vectors) which are in $a + B$ for some a in A; equally, it consists of all vectors which are in $A + b$ for some b in B. In symbols,

$$A + B = \bigcup_{a \in A} a + B = \bigcup_{b \in B} A + b$$

This makes it easy in simple cases to sketch the sum of two sets in the plane. In particular if A is an open disc of radius ϵ round the origin in $\mathbf{R}^2$, $A + B$ is roughly speaking 'B with an open border of width ϵ' in other words the open set $\{x \in \mathbf{R}^2 : d(x, B) < \epsilon\}$ where d is the usual metric. We mention this little fact because the analogous result is true in any normed space, and we hope the reader will find that this sort of geometrical notation gives quite a vivid picture of the topological properties of sets in normed spaces (see Diagram 3).

Of course not all problems in functional analysis are suited to a geometrical approach, but it is a worthwhile attitude to cultivate.

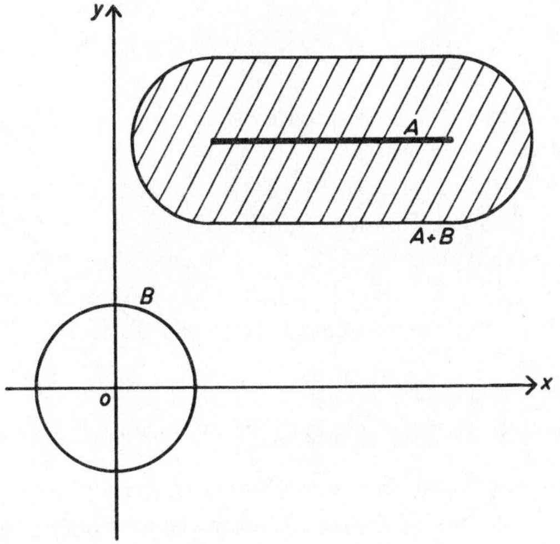

Diagram 3.

The algebra of subsets

3.23 The addition and scalar multiplication of subsets of X mirror the linear operations in X *up to a point*: for instance one has

$$A + B = B + A$$

$$A + (B + C) = (A + B) + C$$

so that one can unambiguously write such an expression as $A + B - C - D$ ($-C$ in such a case being short for $+(-C)$). The reader should supply the (trivial) proofs, as well as proofs or counter-examples as required for the properties below which are intended as exercises and will be used frequently in the sequel without explicit mention.

(1) $A + 2B + 3C$ consists of all vectors $a + 2b + 3c$ with $a \in A$, $b \in B$, $c \in C$; similarly for any similar expression.
(2) $\lambda(A \pm B) = \lambda A \pm \lambda B$ and $B - A = -(A - B)$.
(3) $A + (B \cup C) = (A + B) \cup (A + C)$; but the analogous equation with $\cap$ is generally untrue.
$\lambda(A \cup B) = \lambda A \cup \lambda B$; and this works for $\cap$ also, provided $\lambda \neq 0$.
(4) M is a subspace iff: $M + M \subset M$, and $\lambda M \subset M$ for every scalar λ.
In fact when M is a subspace, $x + M = M = \lambda M$ for any $x \in M$ and any nonzero λ.
(5) $T(A + B) = T(A) + T(B)$, $T(\lambda A) = \lambda T(A)$ if T is a linear map on X to some linear space Y.
(6) Similarly, given $C, D \subset Y$, it is true that $T^{-1}(C + D) = T^{-1}(C) + T^{-1}(D)$, $T^{-1}(\lambda C) = \lambda T^{-1}(C)$ provided $C, D \subset \operatorname{im} T$, but not in general (this one is not as simple as it looks).
(7) $A \cap B \neq \emptyset \Leftrightarrow 0 \in A - B$.
(8) C meets $A + B \Leftrightarrow A$ meets $C - B \Leftrightarrow B$ meets $C - A$.

Here are some things that don't work: In general $A + A \neq 2A$, and clearly $A - A \neq \{0\}$ when A has more than one point. The *cancellation law* $A + C = B + C \Rightarrow A = B$ is generally false, as is the equation $(\lambda + \mu)A = \lambda A + \mu A$ (for a counter-example with positive λ, μ, A must be nonconvex because of the next but one Proposition).

An amusing fact is that the cancellation law does hold if one restricts oneself to *compact convex* subsets of the plane. Using this, Shephard has shown that the class of all such subsets, with the $A + B$ and λA operations, can be isomorphically embedded in a linear space which becomes, under a natural norm, a Banach space whose structure has yet to be explored and looks most interesting.

The first two of the three assertions below are standard facts of group and ring theory.

3.24 Proposition

(i) If M, N are linear subspaces so is $M + N$; indeed $M + N = \operatorname{lin}\{M \cup N\}$.

(ii) If H, K are cosets and λ is a scalar then $H + K$ and λH are cosets.

(iii) If C, D are convex sets and λ is a scalar then $C + D$ and λC are convex.

Proof We leave (i) to the reader. To prove (ii), by definition we can write $H = x + M$, $K = y + N$ for suitable vectors x, y and subspaces M, N. Then $H + K = (x + M) + (y + N) = (x + y) + (M + N)$ which represents $H + K$ as a coset. When $\lambda = 0$, λH is the trivial coset $\{0\}$, and otherwise $\lambda H = \lambda(x + M) = \lambda x + \lambda M = \lambda x + M$ by 3.23 (2) and (4).

To prove (iii), let $x = c + d$, $y = c' + d'$ be in $C + D$. An arbitrary point on the line joining them is represented as an element of $C + D$ by the formula $\alpha x + (1 - \alpha)y = (\alpha c + (1 - \alpha)c') + (\alpha d + (1 - \alpha)d')$, showing $C + D$ is convex. The proof for λC is similar.

The next result is an illuminating property of convex sets. The reader should try to get the feel of it by drawing sketches in the plane, taking C to be a square or disc. In the case of the disc, note the connexion with the triangle inequality in the Euclidean metric. This is not coincidence, for with C as the unit ball in a normed space (to be defined shortly) the result can be thought of as the geometric restatement of the triangle inequality for such spaces.

3.25 Proposition If C is convex and $\lambda, \mu \geqslant 0$ then $\lambda C + \mu C = (\lambda + \mu)C$.

Proof The force of this is the assertion that $\lambda C + \mu C \subset (\lambda + \mu)C$, since the reverse inclusion is trivially true for any set C. To avoid trivialities we may assume $\lambda, \mu > 0$. Let $x \in \lambda C + \mu C$, then for some $y, z \in C$,

$$\begin{aligned} x &= \lambda y + \mu z \\ &= (\lambda + \mu)w \end{aligned}$$

where $w = \dfrac{\lambda}{\lambda + \mu} y + \dfrac{\mu}{\lambda + \mu} z$ is in C because C is convex. This shows $x \in (\lambda + \mu)C$ and hence $\lambda C + \mu C \subset (\lambda + \mu)C$, proving the proposition.

2

NORMED SPACES—BASIC PROPERTIES AND EXAMPLES

4. Normed linear spaces

4.1 **A normed linear space**, or more briefly a **normed space**, is a linear space X over the real or complex field on which is defined a real-valued function called the **norm**, whose value at x is usually written $\|x\|$, with the following properties:

(1) $\|x\| \geqslant 0$;

(2) $\|x\| = 0$ iff $x = 0$;

(3) $\|\alpha x\| = |\alpha|\ \|x\|$;

(4) $\|x + y\| \leqslant \|x\| + \|y\|$ (the **triangle inequality**).

One thinks of the number $\|x\|$ as being the *length* of the vector x. It is easy to see, using property (4), that $d(x, y) = \|x - y\|$ defines a metric on X, and we shall always assume that a normed space carries this metric and the associated topology, which we call the **norm topology**. Thus

$$x_n \to x \text{ in the norm topology if and only if } \|x_n - x\| \to 0.$$

A **Banach space** is a normed space that, regarded as a metric space, is complete.

The variety of normed spaces that crop up in analysis is vast, and it is presumptuous to label one class of space as more important than another, because one's preferences depend on the field of analysis one is interested in. Here however are some of the normed spaces which this book is most concerned. Others are described in the problems.

4.2 **Example 1** $\boldsymbol{R}$ and $\boldsymbol{C}$, with norm equal to the ordinary absolute value, are Banach spaces. (That they are complete is of course a fundamentally important fact of analysis.)

Example 2 R^n and $\mathbf{C}^n$, under the usual **Euclidean norm**,

$$\|(x_1, \dots, x_n)\| = (|x_1|^2 + \dots + |x_n|^2)^{\frac{1}{2}}$$

are Banach spaces. The associated metric is the usual Euclidean metric $d(x, y) = (\Sigma |x_j - y_j|^2)^{\frac{1}{2}}$, where $x = (x_1, \dots, x_n)$, $y = (y_1, \dots, y_n)$, and we refer to these spaces as **real** or **complex Euclidean n-space.**

Many normed spaces consist of n-tuples of scalars, or sequences of scalars, or scalar-valued functions on some set; in the next examples the scalars may be either real or complex, and it is customary not to give different symbols to the two versions of each space but to distinguish them verbally, and say, for instance 'the complex space ℓ_p'.

Example 3 The space ℓ_p^n is defined for each positive integer n and each real p with $1 \leqslant p < \infty$, to be the linear space R^n or $\mathbf{C}^n$ with the norm $\|x\|_p = (|x_1|^p + \dots + |x_n|^p)^{1/p}$.

We emphasize that when $p \neq q$, the spaces ℓ_p^n and ℓ_q^n are different normed spaces, in spite of the fact that they consist of the same vectors with the same linear operations. Observe that Euclidean n-space is the special case $p = 2$.

Example 4 The space ℓ_∞^n is defined to be R^n or $\mathbf{C}^n$ with the norm $\|x\|_\infty = \max_{1 \leqslant j \leqslant n} |x_j|$. The reason for this notation is the easily proved fact that $\|x\|_\infty = \lim_{p \to \infty} \|x\|_p$.

Example 5 The space ℓ_∞ is defined to consist of all bounded sequences of scalars $x = \{x_1, x_2, x_3, \dots\}$ with the obvious coordinatewise definitions of addition and scalar multiplication and the norm $\|x\|_\infty = \sup_j |x_j|$. Inside ℓ_∞ are two important linear subspaces:

$$\mathcal{c} = \{\text{all convergent sequences}\}.$$
$$\mathcal{c}_0 = \{\text{all sequences convergent to zero}\},$$

both of which are regarded as carrying the norm $\|x\|_\infty$, restricted to the space in question.

Example 6 $\mathcal{C}[0, 1]$ with the norm $\|f\| = \sup\{|f(t)| : 0 \leqslant t \leqslant 1\}$. More generally, if T is any compact topological space, the real or complex space $\mathcal{C}(T)$ of continuous scalar functions on T, with the norm

$$\|f\| = \sup\{|f(t)| : t \in T\},$$

is a normed space.

(More generally still, if T is *any* topological space, the set $\mathcal{BC}(T)$ of all *bounded* continuous scalar functions forms a normed space with the norm defined by the same formula.) This norm is called the **supremum norm**, or **sup-norm**, and by analogy with the last example it is often denoted by $\|f\|_\infty$.

Example 7 $\mathcal{C}[0, 1]$ with the norm $\|f\| = \int_0^1 |f(t)|\, dt$. The only slightly non-trivial point here is that $\int_0^1 |f(t)|\, dt = 0$ does indeed imply $f = 0$.

Example 8 The L_p spaces to be introduced in §5, of which the spaces ℓ_p^n and ℓ_∞ defined in Examples 3–5 are special cases.

Example 9 The space $\mathcal{C}^n[a, b]$ of functions with continuous first n derivatives $f', f'', \dots, f^{(n)}$ on $[a, b]$ with the norm

$$\|f\| = \max\{\|f\|_\infty, \|f'\|_\infty, \dots, \|f^{(n)}\|_\infty\}.$$

All of these spaces are complete except for Example 7, as will be shown either in the text or the problems.

In the rest of this chapter we study first of all the basic properties of convergence, the topology of subsets of a normed space, and the idea of an infinite series of vectors in a normed space; then we look at some concrete examples of normed spaces including those above. If the reader would rather see the concrete examples first he can skip to §5 provided he is willing to refer back in the fairly few cases where the general theory is used. The space $\mathcal{C}(T)$, because it is so important, has the whole of §6 to itself.

Basic topology and geometry in normed spaces

Unless stated to the contrary, X denotes a real or complex normed space.

4.3 Proposition For any $x, y \in X$, $|\,\|x\| - \|y\|\,| \leqslant \|x - y\|$.

Proof By the triangle inequality,

$$\|x\| = \|(x - y) + y\| \leqslant \|x - y\| + \|y\|$$

so that $\|x\| - \|y\| \leqslant \|x - y\|$.
In the same way

$$\|y\| - \|x\| \leqslant \|y - x\| = \|(-1)(x - y)\| = \|x - y\|.$$

Combining these gives

$$-\|x - y\| \leqslant \|x\| - \|y\| \leqslant \|x - y\|$$

and the result follows.

4.4 Proposition The norm is a continuous function on X, and addition and scalar multiplication are jointly continuous functions on X.

Proof These statements mean, by definition, that

$$x_n \to x \Rightarrow \|x_n\| \to \|x\|,$$
$$x_n \to x, y_n \to y \Rightarrow x_n + y_n \to x + y,$$
$$\lambda_n \to \lambda, x_n \to x \Rightarrow \lambda_n x_n \to \lambda x;$$

which follow from the inequalities

$$|\,\|x_n\| - \|x\|\,| \leqslant \|x_n - x\|,$$
$$\|(x_n + y_n) - (x + y)\| = \|(x_n - x) + (y_n - y)\| \leqslant \|x_n - x\| + \|y_n - y\|,$$
$$\|\lambda_n x_n - \lambda x\| = \|\lambda_n(x_n - x) + (\lambda_n - \lambda)x\| \leqslant |\lambda_n|\,\|x_n - x\| + |\lambda_n - \lambda|\,\|x\|.$$

4.5 Many important properties of a normed space are connected with the shape of its **unit ball**, that is the ball of radius one round the origin or equivalently the set of elements of norm at most one. We shall use, fairly consistently, the notation

$B = \{x \in X : \|x\| \leqslant 1\}$, the **closed unit ball** in X,
$U = \{x \in X : \|x\| < 1\}$, the **open unit ball** in X.

(B is for ball, U is intended to suggest an open set.) It is clear that U is an open set, and B is a closed set.

An important fact about normed spaces is the:

4.6 Proposition B and U are convex.

Proof If $\|x\|, \|y\| \leqslant 1$ and $0 \leqslant \alpha \leqslant 1$ then

$$\|\alpha x + (1 - \alpha)y\| \leqslant \alpha\|x\| + (1 - \alpha)\|y\| \leqslant \alpha + (1 - \alpha) = 1$$

so that B is convex. Similarly with U.

As an illustration it is illuminating to look at the shape of the unit ball in one of the simplest types of space, namely the two-dimensional spaces ℓ_p^2 where $1 \leqslant p \leqslant \infty$, (see Diagram 4). Here the underlying space is the set R^2 of pairs (x, y) of real numbers, and B is the set of points on and inside the curve $|x|^p + |y|^p = 1$. When $p = 1$ the 'curve' $|x| + |y| = 1$ is the diamond joining the points $(1, 0)$, $(0, 1)$, $(-1, 0)$, $(0, -1)$ and as p increases the curve gradually flattens out, passing through an exact circular shape when $p = 2$, until for large p it approaches the outer square in the

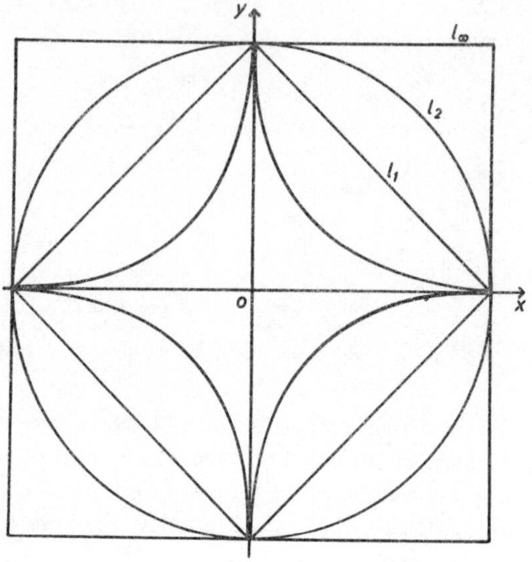

Diagram 4.

figure which represents the limiting case of the surface of the ℓ_∞ sphere, namely $\max\{|x|, |y|\} = 1$. When $p < 1$ the curve becomes star-shaped and no longer encloses a convex region; this is the case in higher dimensions too, and is the reason why one always assumes $p \geqslant 1$.

In recent years, shapes having the slightly more general form $|x/a|^p + |y/b|^p = 1$ have found favour with architects as an aesthetic compromise between a rectangle ($p = \infty$) and an ellipse ($p = 2$). The most popular value of p seems to be $5/2$.

4.7 We shall be making continual use of the fact that the closed ball of radius r ($r > 0$) round x_0 is $x_0 + rB$. (Geometrically, this means that all closed balls in a normed space have the same shape.) The proof is a simple, but not quite trivial, application of the basic properties of the norm, as follows.

Suppose x is such that $\|x - x_0\| \leqslant r$, and let $y = r^{-1}(x - x_0)$.

Then $\|y\| \leqslant 1$ and $x = x_0 + ry$, which proves $\{x: \|x - x_0\| \leqslant r\} \subset x_0 + rB$. The reverse inclusion follows from the fact that if $y \in B$ and $x = x_0 + ry$ then $\|x - x_0\| = r\|y\| \leqslant r$.

Similarly, every open ball has the form $x_0 + rU$.

These facts lead to a useful topological property of normed spaces:

4.8 Proposition The closure of any subset A of X is equal to each of the two sets

$$\bigcap_{n=1}^{\infty}\left(A + \frac{1}{n}B\right), \bigcap_{n=1}^{\infty}\left(A + \frac{1}{n}U\right).$$

Proof The equality of these two sets follows from the fact that $\frac{1}{n+1}B \subset \frac{1}{n}U \subset \frac{1}{n}B$, and that they equal A^- follows from the computation

$$\begin{aligned} x \in A^- &\Leftrightarrow \text{for each } n \text{ there is } a_n \in A \text{ with } \|x - a_n\| \leqslant \frac{1}{n} \\ &\Leftrightarrow \text{for each } n \text{ there is } a_n \in A \text{ with } x \in a_n + \frac{1}{n}B \\ &\Leftrightarrow x \in A + \frac{1}{n}B \text{ for each } n. \end{aligned}$$

The numbers $1/n$ could be replaced by any sequence of positive numbers tending to 0.

It is clear from the last paragraph or two that the unit ball B determines the norm topology completely; indeed, a set V in X is open iff for each $x \in V$ there is $r > 0$ such that $x + rB \subset V$. In fact B also determines the norm itself, as Problem 4J shows.

4.9 Proposition The closure of a linear subspace, or of a coset of a linear subspace, or of a convex set, is again a set of the same sort.

Proof We shall give the proof for the case of a convex set C, and leave the others as an exercise since they are very similar. Let $x, y \in C^-$ and let $0 \leqslant \alpha \leqslant 1$. Then there are sequences $\{x_n\}$, $\{y_n\}$ in C converging to x and y. Since C is convex the sequence $\{\alpha x_n + (1 - \alpha)y_n\}$ lies in C and by the continuity of the linear operations in X (Proposition 4.4) this sequence converges to $\alpha x + (1 - \alpha)y$. The latter vector is therefore in C^-, showing that C^- is convex.

4.10 Let A be a subset of X. We recall that the proof that there exists a smallest convex set co A containing A depended simply on the fact that X itself is a convex set, and the intersection of any family of convex sets is again a convex set. By exactly the same reasoning, there is a unique smallest closed convex set containing A, and there is a unique smallest closed linear subspace containing A. These are called the **closed convex hull** and the **closed linear span** of A respectively, and are denoted by $\overline{\text{co}}\, A$ and $\overline{\text{lin}}\, A$.

The following facts aren't quite so utterly obvious as they look:

4.11 Lemma $\overline{\text{co}}\, A = (\text{co}\, A)^-$ and $\overline{\text{lin}}\, A = (\text{lin}\, A)^-$.

Proof Since $\overline{\text{co}}\, A$ is convex and contains A it contains $\text{co}\, A$; since it is closed it therefore contains $(\text{co}\, A)^-$. On the other hand $(\text{co}\, A)^-$ is convex by the last Proposition, so it is a closed convex set containing A and hence, by definition, contains $\overline{\text{co}}\, A$.

The proof for $\overline{\text{lin}}\, A$ is the same.

On completeness

It is possible to do a certain amount of functional analysis without bothering about the completeness or otherwise of the normed spaces one deals with. It will become clear as we go on, however, that much of the theory, particularly that of linear operators, only works smoothly in a Banach space. This is why we devote several theorems in this and following chapters, as well as a number of the problems, to proving that certain spaces are complete. Because of the many possible types of convergence on an infinite-dimensional space it is vital for the student to see quite clearly what is involved in such a proof.

4.12 Consider the commonest situation: X is a normed space whose elements are real or complex functions on some set S, with the linear operations defined pointwise, and some norm $\|f\|$. A completeness proof will typically start by considering an arbitrary sequence $\{f_n\}$ which is Cauchy in the norm of X and follow these steps:

(i) Find a candidate for the limit function f. Usually this is done by showing that the *pointwise* limit $f(s) = \lim f_n(s)$ exists for all (or, in a measure space, almost all) $s \in S$.

(ii) Show that f is a member of the space X, not merely of some larger space.

(iii) Show that $f_n \to f$ in the norm of X, not merely in some weaker sense.

The notation will differ if X is a space of sequences or of linear operators, and we sometimes start with a series instead of a

Cauchy sequence (see Theorem 4.21) but the basic steps are always the same. We illustrate this by two examples.

4.13 Example 1 Let T be a compact topological space. Then $\mathcal{C}(T)$, with the sup-norm $\|f\| = \sup\{|f(t)| : t \in T\}$, is complete, and is therefore a Banach space. Since the proof is merely a restatement of ideas that should be familiar to the student from a first course of real analysis, we sketch it very briefly. To say that $\{f_n\}$ is Cauchy in the sup-norm means that given $\epsilon > 0$ there is n_0 such that

$$|f_m(t) - f_n(t)| \leqslant \epsilon \quad (m, n \geqslant n_0\,;\ t \in T). \tag{*}$$

From (*) and the completeness of $\mathbf{R}$ or $\mathbf{C}$ it follows at once that $f(t) = \lim f_n(t)$ exists for all $t \in T$, which achieves step (i). Fixing n and letting $m \to \infty$ in (1) shows that

$$|f(t) - f_n(t)| \leqslant \epsilon \quad (n \geqslant n_0\,;\ t \in T)$$

so that $f_n \to f$ uniformly on T. Steps (ii) and (iii) are simply restatements of the facts that the uniform limit of continuous functions is continuous, and that $\|f - f_n\| \to 0 \Leftrightarrow f_n \to f$ uniformly.

Example 2 $X = \mathcal{C}[0,1]$, with the norm $\|f\| = \int_0^1 |f|$ (this is short for $\int_0^1 |f(t)|\,dt$) is not complete.

We shall show, for a suitably chosen Cauchy sequence, that no possible candidate for a limit function in X can be the right one.

Let $f_n (n = 2, 3, \ldots)$ – a 'diagonal step function' – be defined by:

$$f_n(t) = \begin{cases} 0, & 0 \leqslant t \leqslant \frac{1}{2} \\ n\left(t - \frac{1}{2}\right), & \frac{1}{2} < t < \frac{1}{2} + \frac{1}{n} \\ 1, & \frac{1}{2} + \frac{1}{n} \leqslant t \leqslant 1. \end{cases}$$

A simple calculation shows that for $m, n \geqslant n_0$,

$$\|f_m - f_n\| = \int_0^1 |f_m - f_n| = \int_{\frac{1}{2}}^{\frac{1}{2}+\frac{1}{n_0}} |f_m - f_n| \leqslant \frac{1}{n_0}$$

so that $\{f_n\}$ is Cauchy. Suppose now there were some continuous f for which $\|f - f_n\| \to 0$. For any n_0 define $J = [0, 1] \sim (\frac{1}{2}, \frac{1}{2} + \frac{1}{n_0})$. Then for $n \geqslant n_0$.

$$\int_0^{\frac{1}{2}} |f| + \int_{\frac{1}{2}+\frac{1}{n_0}}^1 |f - 1| = \int_J |f - f_n| \leqslant \int_0^1 |f - f_n| = \|f - f_n\|.$$

Letting $n \to \infty$ shows that

$$\int_0^{\frac{1}{2}} |f| = 0 = \int_{\frac{1}{2}+\frac{1}{n_0}}^{1} |f-1|,$$

so by the assumed continuity of f we have (how?) that $f = 0$ on $[0, \frac{1}{2}]$ and 1 on $[\frac{1}{2} + \frac{1}{n_0}, 1]$. Since n_0 is arbitrary we conclude that f must be the discontinuous function $\chi_{(\frac{1}{2}, 1]}$, a contradiction. Hence $\{f_n\}$ has no limit in X and X is not complete.

Exercise Show that, with respect to the sup-norm, $\{f_n\}$ is *not* Cauchy, by evaluating $\|f_m - f_n\|_\infty$ explicitly.

Compactness

4.14 Compact sets in normed spaces, as in more general metric and topological spaces, have especially tractable behaviour and can be regarded as a first step away from finite sets. We shall encounter them later in connexion with *Ascoli's theorem* 6.11 which gives a complete characterization of compact subsets of the space $\mathcal{C}(T)$, and with the theory of *compact operators* which is a basic tool in the study of differential and integral equations.

To avoid unnecessary generality we shall assume X is a Banach space, and we restrict ourselves to one trivial observation, a few simple but useful facts, and one important and non-trivial theorem. The trivial observation is that a finite subset $F = \{x_1, \ldots, x_n\}$ is an ϵ-net for a subset A of X iff

$$A \subset F + \epsilon B \quad (B \text{ being the unit ball}). \tag{1}$$

For $A \subset F + \epsilon B \Leftrightarrow A \subset \bigcup_{i=1}^{n} (x_i + \epsilon B) \Leftrightarrow A$ is covered by the ϵ-balls round the points x_i, and this is the definition of an ϵ-net. Since in a complete metric space, A has compact closure $\Leftrightarrow A$ is totally bounded, one can deduce immediately that for a subset A of a Banach space X,

$$\bar{A} \text{ is compact} \Leftrightarrow \text{for each } \epsilon > 0 \text{ there is a finite set } F \text{ with } A \subset F + \epsilon B. \tag{2}$$

It is obvious from the continuity of the algebraic operations that a translate $x + A$ and scalar multiple λA of a compact set A are compact. In the same way, if C, D are both compact so is $C + D$ because it is the image of the compact set $C \times D$ under the continuous map $(x, y) \mapsto x + y$. An alternative proof, which the reader may find more illuminating, is to note that if F is an $\epsilon/2$-net for C and G is an $\epsilon/2$-net for D then the (finite!) set $F + G$ is an ϵ-net for $C + D$, because

$$C + D \subset (F + \frac{\epsilon}{2}B) + (G + \frac{\epsilon}{2}B) = (F + G) + \epsilon B, \text{ by 3.25.}$$

4.15 Theorem Let A be a compact subset of a Banach space X. Then $\overline{\text{co}}\, A$ is compact.

Proof Given $\epsilon > 0$ we can choose an $\epsilon/2$-net $F = \{x_1, \dots, x_k\}$ for A. The line segments $S_j = \{\alpha x_j : 0 \leqslant \alpha \leqslant 1\}$ joining 0 to x_j are trivially compact and convex, and therefore by the remarks above, 3.24, and induction, so is the set $S = S_1 + \dots + S_k$. S contains F, because for instance x_2 can be represented as $0x_1 + 1x_2 + 0x_3 + \dots + 0x_k$ and similarly with each of the x_j

Since S is compact we can choose a finite set G with $S \subset G + \frac{1}{2}\epsilon B$. Then

$$A \subset F + \tfrac{1}{2}\epsilon B \subset S + \tfrac{1}{2}\epsilon B \subset G + \tfrac{1}{2}\epsilon B + \tfrac{1}{2}\epsilon B = G + \epsilon B.$$

But $S + \frac{1}{2}\epsilon B$ is convex and contains A, so that it contains co$A$. We deduce that co $A \subset G + \epsilon B$. Thus co$A$ has an ϵ-net for any $\epsilon > 0$, so that $\overline{\text{co}}\, A = (\text{co}A)^-$ is compact by 4.14 (2).

To help the reader catch the flavour of this theorem we point out that it includes the generalized Heine–Borel theorem, that closed bounded subsets of $\mathbf{R}^n$ are compact, as a special case. For every such set A is a closed subset of some cube

$$C = \{y = (y_1, \dots, y_n) \in \mathbf{R}^n : -c \leqslant y_j \leqslant c \text{ for each } j\}$$

and it is easily verified that such a cube is the closed convex hull of the finite, and compact, set consisting of the 2^n points of the form $(\pm c, \pm c, \dots, \pm c)$. Hence C is compact, which shows A is compact.

Congruences

4.16 A frequently used concept, which is to *normed* linear spaces what a linear isomorphism is to plain linear spaces, is that of an **isometric linear isomorphism**, or **congruence** (we shall use the latter term). If X and Y are normed spaces with the same scalars, a congruence from X to Y is a linear map $T: X \to Y$ which is onto and is such that $\|T(x)\| = \|x\|$ for all x. (We shall generally use the same symbol $\|\ \|$ for the norm on different spaces.

This is customary and does not lead to confusion.) A congruence is necessarily one-to-one, for $T(x) = T(y) \Rightarrow \|x - y\| = \|T(x - y)\| = \|0\| = 0 \Rightarrow x = y$. It is clear that T preserves the metric, i.e. that $d(T(x), T(y)) = d(x, y)$ and hence that it is a homemorphism between the two spaces. Two normed spaces are **congruent** if there exists a congruence from one to the other. Thus, from the point of view of the norm and linear-space structure alone, isometric spaces are indistinguishable, and one may regard them as identical. Clearly (cf. 3.13) congruence is an equivalence relation.

4.17 Example 1 If the compact space T is a finite set $\{t_1, \ldots, t_n\}$, with the discrete topology, then $\mathcal{C}(T)$ may be regarded as identical with ℓ_∞^n. For the map which associates to each $f \in \mathcal{C}(T)$ the vector $(f(t_1), \ldots, f(t_n))$ is clearly a congruence.

Example 2 In the Euclidean plane with the usual ℓ_2 norm, any rotation round the origin, and any reflection in a line through the origin, is a congruence of the space with itself. These are in fact precisely the congruences, using the word as it is used in elementary geometry, which keep the origin fixed.

These simple examples may have persuaded the reader that congruences are fairly trivial operators, but this is by no means the case. For one thing the congruences of a space *with itself* (which tell one, in a sense, how symmetrical the unit ball of the space is) often have a rich and complex structure. Secondly, the establishment of a congruence between two spaces may involve some deep analysis and issue in fruitful applications; the same is true of the slightly weaker notion of **topological isomorphism** between normed spaces defined on p. 101.

Example 3 Any normed space whatsoever is congruent to a linear subspace of $\mathcal{C}(T)$ for a suitable compact Hausdorff space T. This is Theorem 16.21. Even more mind-bendingly, any *separable* normed space is congruent to a subspace of $\mathcal{C}[0, 1]$ – which illustrates how difficult it would be to find a complete description of the linear structure of $\mathcal{C}[0, 1]$ in terms of its subspaces. We shall not prove this result but a proof is sketched in the problems on p. 257.

Infinite sums in Banach spaces, elementary grade

4.18 If we have – as we frequently do – a sequence of vectors x_n in a normed space X we can investigate the convergence of the sequence of **partial sums**

$$s_n = x_1 + \ldots + x_n \qquad (n = 1,2,\ldots).$$

If $\lim s_n = x$ exists in the metric of X we say the **series** $\sum_1^\infty x_n$ **converges** to x. This is the same as saying that

$$\|x - \sum_1^n x_r\| \to 0 \quad (n \to \infty)$$

4.19 Example 1 If X is any of the spaces ℓ_p^k $(1 \leqslant p \leqslant \infty)$ where the underlying linear space is $\boldsymbol{R}^k$ or $\mathbf{C}^k$, then $\sum_1^\infty x_n$ converges to x iff for each $j = 1, \ldots, k$ the jth coordinates of the x_n form a series (of real or complex numbers) converging to the jth coordinate of x.

Example 2 If $X = \mathcal{C}(T)$, T a compact space, with the usual sup-norm, then a series $\sum_1^\infty f_n$ in X converges to $f \Leftrightarrow \sum_1^\infty f_n(t)$ converges to $f(t)$, uniformly for t in T. This is a situation of frequent occurrence in analysis, for instance in the study of power series, T being a disc in the complex plane.

We shall only prove two basic results in this section, but it is worth saying that much of the elementary theory of series in $\boldsymbol{R}$ or $\mathbf{C}$ can be taken over verbation to series of vectors, with absolute value replaced by norm, provided that we assume, as we shall do, that X is complete. Generalizing from the elementary case we say that the **series** $\sum_1^\infty x_n$ **is absolutely convergent** provided $\sum_1^\infty \|x_n\| < \infty$. As in the case of real or complex series one must take this phrase as a whole, as describing a certain property of a given *sequence* of vectors $\{x_n\}$ – the symbols $\sum_1^\infty x_n$, taken by themselves, are entirely meaningless.

4.20 Theorem Let $\sum_1^\infty x_n$ be an absolutely convergent series in a Banach space X. Then $\sum_1^\infty x_n$ converges.

Proof By assumption $\sum_1^\infty \|x_n\|$ is a convergent series of non-negative numbers, so given $\epsilon > 0$ we can choose n_0 such that $\sum_{m+1}^n \|x_i\| < \epsilon$ whenever $n > m \geqslant n_0$. Consider the sequence of partial sums $\{s_n\}$. If $n > m$ we have $s_n - s_m = (x_1 + \ldots + x_n) - (x_1 + \ldots x_m) = x_{m+1} + \ldots + x_n$ so that by the triangle inequality,

$$\|s_n - s_m\| \leqslant \|x_{m+1}\| + \ldots + \|x_n\| < \epsilon \quad (n > m \geqslant n_0).$$

This shows $\{s_n\}$ is a Cauchy sequence, and hence $x = \lim s_n = \sum_{n=1}^\infty x_n$ exists since X was assumed complete.

The next result can be thought of as a sort of converse to the last one. It contains the essence of certain arguments used from time to time in proving that a normed space is complete, and we shall use it in 4.24 and 5.9.

4.21 Theorem Let X be a normed space in which every absolutely convergent series converges. Then X is a Banach space.

Proof Let $\{x_n\}$ be a Cauchy sequence in X. For each number k there is a number n_k such that $\|x_m - x_n\| < 2^{-k}$ $(m, n \geqslant n_k)$, and without loss we may and do assume that $n_1 < n_2 < n_3 \ldots$ We shall construct a series Σy_k having the x_{n_k} as its partial sums. Let s_k denote x_{n_k}, let y_1 denote s_1 and let y_k denote $s_k - s_{k-1}$ For $k \geqslant 2$. Then it is clear that the s_k are the partial sums of $\sum_1^\infty y_k$ and that

$$\|y_k\| = \|s_k - s_{k-1}\| = \|x_{n_k} - x_{n_{k-1}}\| < 2^{-(k-1)}$$

by the choice of the numbers n_k. Hence $\sum_1^\infty \|y_k\| < \sum_1^\infty 2^{-(k-1)} < \infty$, and hence by the hypothesis of the theorem $\sum_1^\infty y_k$ converges. This is the same as saying that $x = \lim s_k = \lim x_{n_k}$ exists. Lemma 1.8 now implies that $x = \lim x_n$. Hence every Cauchy sequence has a limit, so X is complete.

Subspace, product, quotient

The title of the section describes three of the mathematician's standard ways of producing new spaces from old. It is clear, to take the simplest case first, that if M is a linear subspace of a normed space X then M becomes a normed space if we think of the norm as restricted to M, and that the norm on M generates the relative metric and the relative norm topology. Also, since a subspace of a complete metric space is complete iff it is closed we have

4.22 Lemma A linear subspace of a Banach space is a Banach space iff it is closed.

4.23 Turning now to products let us take two normed spaces X and Y over the same scalars. Since there are several natural norms on $X \times Y$ this is a good place to introduce the following important notion: two norms on the same linear space are called **equivalent** if they give the same norm topology. We shall prove in 7.7 that two norms, say $\| \ \|$ and $\| \ \|'$, are equivalent iff there exist positive constants k, K such that

$$k\|x\| \leqslant \|x\|' \leqslant K\|x\| \text{ for all } x.$$

The most commonly used norms on $X \times Y$ are

$$\|(x, y)\|_\infty = \max\{\|x\|, \|y\|\}$$

and

$$\|(x, y)\|_1 = \|x\| + \|y\|,$$

the notation being intended to suggest the similarity with the definitions of the ℓ_∞ and ℓ_1 norms on $\boldsymbol{R}^n$. We leave it to the reader to verify that these are indeed norms. Since the associated metrics are just the most commonly used metrics for the product of two metric spaces (1.3) both norms give the product topology on $X \times Y$ and are therefore equivalent – a fact that follows also from the criterion for equivalence mentioned above and the simple inequality

$$\|(x, y)\|_\infty \leqslant \|(x, y)\|_1 \leqslant 2\|(x, y)\|_\infty$$

which the reader should verify. The norm $\| \quad \|_\infty$ is especially convenient, and easy to visualize, in that its closed or open unit ball is just the Cartesian product of the corresponding unit balls in X and Y. This follows from

$$\|(x, y)\|_\infty \leqslant 1 \Leftrightarrow \max\{\|x\|, \|y\|\} \leqslant 1 \Leftrightarrow \|x\| \leqslant 1 \text{ and } \|y\| \leqslant 1$$

for the closed ball, and similarly for the open ball. In symbols,

$$B_{X \times Y} = B_X \times B_Y \text{ and } U_{X \times Y} = U_X \times U_Y.$$

No deep ideas were involved in the last few paragraphs. With quotient spaces on the other hand we have a meatier proof and a result with much deeper consequences.

4.24 Theorem Let M be a closed linear subspace of a normed space X. Then X/M becomes a normed space if we define the **quotient norm**

$$\|x + M\| = \inf_{m \in M} \|x + m\|.$$

Moreover if X is a Banach space so also is X/M.

Proof Recall that X/M is made up of the cosets $x + M$ as x runs over X, with linear operations defined by $(x + M) + (y + M) = (x + y) + M$ and $\alpha(x + M) = (\alpha x) + M$. For any $x, y \in X$ and $m, n \in M$ one has $m + n \in M$ and so by the definition

$$\|(x + y) + M\| \leqslant \|x + y + m + n\| \leqslant \|x + m\| + \|y + n\|.$$

Taking the infimum over all m, n in M gives

$$\|(x + y) + M\| \leqslant \|x + M\| + \|y + M\|.$$

That $\|(\alpha x) + M\| = |\alpha| \, \|x + M\|$ is obvious. To complete the proof that the quotient norm is indeed a norm, suppose x is such that $\|x + M\| = 0$. That is, there is a sequence $\{m_n\}$ in M with $\|x + m_n\| \to 0$. Since M was assumed closed this implies $x = \lim(-m_n) \in M^- = M$, so that $x + M = M = 0 + M$, the zero element of X/M. This shows we do have a norm.

We now change the notation and denote members of X/M by y. Each $y \in X/M$ is one of the cosets $x + N$, and the definition of the quotient norm can be rephrased as

$$\|y\| = \inf_{x \in y} \|x\|.$$

To show that X/M is complete it is sufficient by Theorem 4.21 to show that if $\{y_n\}$ is a sequence in X/M and $\Sigma\|y_n\| < \infty$ then there is a member y of X/M such that Σy_n converges to y. We can choose a representative vector $x_n \in y_n$ such that $\|x_n\| < \|y_n\| + 2^{-n}$ for each n. Then Σx_n is an absolutely convergent series in X, for $\Sigma\|x_n\| < \Sigma\|y_n\| + \Sigma 2^{-n} < \infty$. If X is a Banach space it follows from Theorem 4.20 that there is a vector x such that Σx_n converges to x. Let y be the coset $x + M$. Then $(\sum_1^n y_r) - y = \sum_1^n (x_r + M) - (x + M) = (\sum_1^n x_r - x) + M$ and so by the definition of the quotient norm

$$\|\sum_1^n y_r - y\| \leqslant \|\sum_1^n x_r - x\| \to 0 \quad (n \to \infty),$$

proving Σy_n converges to y, as required.

The reason for changing the notation in the above proof was to emphasize the need for care in choosing the x_n. If a sequence of cosets is represented $\{x_n + M\}$ with the x_n chosen at random, there is no need for $\Sigma\|x_n + M\| < \infty$ to imply $\Sigma\|x_n\| < \infty$, as the reader can easily verify.

There is a simple geometrical description of the quotient norm: $\|x + M\|$ is just the distance from the origin to $x + M$, or equivalently the distance from x to M (see Diagram 5).

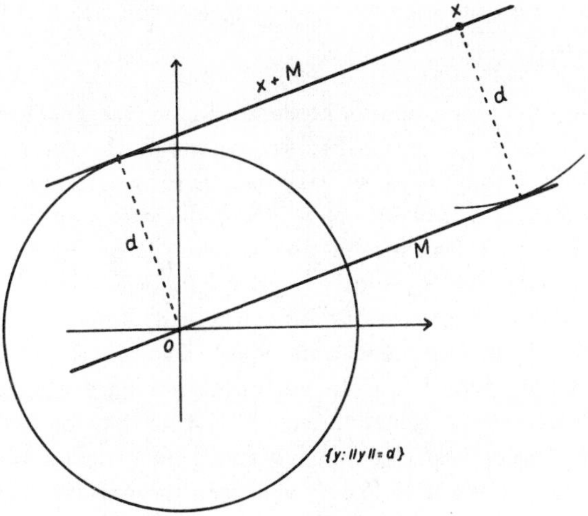

Diagram 5.

Application

A number of common operations in analysis can be interpreted as being constructions of a quotient space or norm. We shall show, as a simple example, how the operation 'lim sup' can be interpreted as a quotient of the operation 'sup'. Let X be the space ℓ_∞ of bounded sequences $x = \{x_n\}$ with the norm $\|x\| = \sup_n |x_n|$, and let M be the subspace c_0 consisting of sequences with $\lim x_n = 0$. By Problem 40, c_0 is closed.

4.25 Proposition The norm on X/M is given by $\|x + M\| = \limsup |x_n|$.

Proof The coset $x + M$ consists of all $y = \{y_n\}$ in ℓ_∞ with $\lim (x_n - y_n) = 0$. It is clear from the definition of limit superior that for all such y, $\|y\| \geqslant \limsup |y_n| = \limsup |x_n|$ and hence

$$\|x + M\| \geqslant \limsup |x_n|.$$

Conversely, given any $\epsilon \geqslant 0$ there are only finitely many indices n for which $|x_n| > \limsup |x_n| + \epsilon$. If we define y_n to be 0 for these indices and x_n otherwise then clearly $y = \{y_n\} \in x + M$, and $\|x + M\| \leqslant \|y\| \leqslant \limsup |x_n| + \epsilon$. Since ϵ was arbitrary, combine this with the reverse inequality to obtain $\|x + M\| = \limsup |x_n|$.

The space X/M can loosely be described as the space of bounded sequences, counting two sequences as the same if their difference tends to 0 or, as we may say, if they are *asymptotically equal.* Since ℓ_∞ is a Banach space, we deduce as a corollary that the norm $\limsup |x_n|$ makes the above space into a Banach space. This fact, though not especially significant, illustrates an important idea. Suppose there is a problem where it is natural to regard asymptotically equal sequences as equal. The functional analyst immediately asks 'can I turn the set of (equivalence classes of) such sequences into a Banach space?' Noting that the set in question is the quotient of a Banach space by a closed subspace he sees that the answer is 'Yes', and goes through the computation above. We shall see in §10 that in a sense, to be made precise, any two 'natural' norms making a space into a Banach space must be equivalent, so that the norm just described can be called *the natural norm* on the space, unique up to equivalence.

Problems

Norm and metric properties

4A Show that the interior of the closed ball $B(x, r)$ of radius r round a point x in a normed space is the corresponding open ball $U(x, r)$; conversely the closure of $U(x, r)$ is $B(x, r)$. Give examples to show that this is false in a general metric space.

4B Sketch the unit balls of ℓ_1^3 and ℓ_∞^3 (see 4.2).

4C Verify that $(x, y) \mapsto (|x|^p + |y|^p)^{1/p}$ fails to be a norm on $\boldsymbol{R}^n$ when $p < 1$ (see 4.6).

4D Show that the maps $x \mapsto x + a$ and $x \mapsto \alpha x$ are homeomorphisms of a normed space X onto itself for each $a \in X$ and each nonzero scalar α.

4E Show that every normed space is homeomorphic to its open unit ball.

4F Let X be a linear space and d a metric on X with $d(x + a, y + a) = d(x, y)$ and $d(\alpha x, 0) = |\alpha|\, d(x, 0)$ for all x, y, a in X and all scalars α. Show that $x \mapsto d(x, 0)$ is a norm on X and d the metric defined by this norm.

4G Show that the norm defined in 4.2 Example 7 is indeed a norm on $\mathcal{C}[0,1]$.

Topology and geometry

4H Let A, B be nonempty sets in a normed space X, x a point of X and α a nonzero scalar. Show that $(x + A)^- = x + A^-$, $(\alpha A)^- = \alpha A^-$ and $(A + B)^- \supset A^- + B^-$; and that if A is open so is $A + B$.

4I Use 4.8 to give an alternative proof that the closure of a convex set is convex.

4J Show two norms on a linear space are equal iff they give the same closed unit ball.

4K Prove that a nonempty convex set in a normed space is connected.

4L Show that if A is a bounded set in a normed space then A and $\overline{\text{co}}\, A$ have the same diameter. [Let $\text{diam}(A) = r$; then $A \subset x + rB$ for all x in A, where B is the unit ball. Deduce that $\{x\colon \overline{\text{co}}\, A \subset x + rB\}$ is closed, convex and contains A. An alternative proof uses 3.20(ii).].

4M Show that if C is a convex set in a normed space and 0 is an interior point of C then $C^- = \bigcap \{\alpha C\colon \alpha > 1\}$. [Use 4.8 and 3.25].

Completeness

4N Prove that the space $B(S)$ of all bounded scalar functions on a nonvoid set S is a Banach space under the usual pointwise operations and the sup-norm $\|f\|_\infty$. (Note that the space ℓ_∞ of 4.2 Example 5 is the case where $S = \mathbf{N}$.)

4O Prove the completeness of

(i) the spaces c and c_0 of 4.2 Example 5;

(ii) the space ℓ_1 of all sequences $x = (x_1, x_2, \ldots)$ of scalars such that $\sum_n |x_n| < \infty$, with the natural linear operations and the norm $\|x\| = \sum_{n=1}^{\infty} |x_n|$.

4P Show that a normed space is complete iff each decreasing sequence of (not necessarily concentric) closed balls has nonvoid intersection.

Compactness

4Q Verify that the cube C mentioned after Theorem 4.15 is indeed the convex hull of the 2^n points $(\pm c, \pm c, \ldots, \pm c)$.

4R Show that the vector sum of a compact set and a closed set in a normed space is closed.

4S Let $\{x_n\}$ be a sequence in a Banach space, converging to zero. Show that $\overline{\text{co}}\,\{x_1, x_2, \ldots\}$ is compact and consists precisely of all points of the form $\sum_{n=1}^{\infty} \alpha_n x_n$ where $\alpha_n \geqslant 0$ and $\sum_{n=1}^{\infty} \alpha_n = 1$.

Congruences

4T Show that $\mathcal{C}[0,1]$ is congruent to the linear subspace of $\mathcal{C}[0,1]$ consisting of those functions f for which $f(t) = f(1-t)$, $0 \leqslant t \leqslant 1$.

4U Show that $\mathcal{C}(S)$ and $\mathcal{C}(T)$ are congruent whenever S and T are homeomorphic compact topological spaces.

Series

4V Show that if the series $\sum_1^\infty x_n$ and $\sum_1^\infty y_n$ in a normed space converge to x and y respectively, and if α, β are scalars, then $\sum_1^\infty (\alpha x_n + \beta y_n)$ converges to $\alpha x + \beta y$.

4W Adapt 4.21 to prove that a linear subspace M of a normed space X is closed iff M contains the limit of each absolutely convergent series of elements of M, whenever this limit exists.

Equivalent norms, products, quotients

4X Use the criterion for equivalent metrics at the end of 1.2 to derive the criterion for equivalent norms given in 4.23.

4Y Prove directly that the ℓ_p^n norms on $\mathbf{R}^n$ $(1 \leqslant p \leqslant \infty)$ are all equivalent and that they all give the usual topology on $\mathbf{R}^n$.

4Z Let X and Y be normed spaces and let $1 < p < \infty$. Prove that $(x, y) \mapsto (\|x\|^p + \|y\|^p)^{1/p}$ is a norm on $X \times Y$ equivalent to the norms defined in 4.23. Denoting $X \times Y$, equipped with this norm, by $X \times_p Y$, show that $(\ell_p^n) \times_p (\ell_p^m)$ is congruent to ℓ_p^{m+n}.

4A′ Let M be a closed linear subspace of a normed space X. Show that the quotient map Q of X onto X/M is continuous and also open (i.e. it maps open sets in X to open sets in X/M).

Show also that if M and X/M are complete then X is complete.

Miscellaneous problems

4B′ Sketch the unit ball of (real) $(\ell_p^2) \times_q \mathbf{R}$ for various combinations of $p, q = 1, 2, \infty$ (see 4.2).

4C′ Let $X = \mathcal{C}[0,1]$ with the usual norm, M the subspace of X consisting of the constant functions. Find a simple description of the norm on X/M when the scalars are (a) real (b) complex.

4D′ Let M and Y be closed subspaces of a normed space X with $M \subset Y$. Clearly Y/M is a linear subspace of X/M. Show it is closed in X/M and prove the 'cancellation theorem':

$$(X/M)/(Y/M) \text{ is congruent to } X/Y.$$

4E′ Is it possible in any reasonable sense to say that $(X/M) \times M$ can be identified with X; or that $(X/M)/(X/Y)$ can be identified with Y/M?

Convex sets in normed spaces

A fixed normed space X is under consideration.

4F′ Show that $\operatorname{co}(A + B) = \operatorname{co} A + \operatorname{co} B$ for any $A, B \subset X$. Deduce that $\overline{\operatorname{co}}\,(A + B) \subset \overline{\operatorname{co}}\,A + \overline{\operatorname{co}}\,B$, with equality if X is complete and A is compact.

4G′ Caratheodory's Lemma If X is of finite dimension n, $A \subset X$ and $x \in \operatorname{co} A$ then (see 3.20) not more than $(n + 1)$ terms are needed to express x as a convex combination of points of A.

Deduce that the convex hull of a closed bounded set in $\boldsymbol{R}^n$ is closed. Can 'bounded' be omitted?

4H′ Let C be as in Problem 4M and let C^0 denote its interior. Show that:

$$C^0 = \bigcup (\alpha C \colon 0 < \alpha < 1);$$

C and C^0 have the same closure;

C and C^- have the same interior.

4I′ Let C be a closed bounded convex symmetric set in a real Banach space and let $L = \operatorname{lin} C$. Show that there is a unique norm on L whose closed unit ball is C, and that L with this norm is a Banach space.

Two important Banach spaces (used later in the text)

4J′ $\mathcal{C}^n[a, b]$ denotes the space of scalar functions f on the real interval $[a, b]$ which are n times differentiable and such that the nth derivative $f^{(n)}$ is continuous, with the usual linear operations. Prove:

(i) If functions f_n with continuous derivatives are such that $\{f_n\}$ and $\{f_n{}'\}$ both converge uniformly, to f, g respectively, then $f' = g$.

(ii) $\mathcal{C}^n[a, b]$ is a Banach space under the norm

$$\|f\| = \max\{\|f\|_\infty, \|f'\|_\infty, \ldots, \|f^{(n)}\|_\infty\}.$$

(iii) An equivalent norm is

$$|f| = \sum_{r=0}^{n} \frac{\|f^{(r)}\|_\infty}{r!}$$

and this has the additional property $|fg| \leqslant |f| . |g|$.

4K′ Let X be a Banach space and T a compact metric (or topological) space. Show that

(i) The space $\mathcal{C}(T, X)$ of all continuous mappings of T into X becomes a Banach space under the obvious pointwise operations and the *generalised sup-norm* $\|f\|_\infty = \sup\{\|f(t)\| : t \in T\}$.

(ii) $\mathcal{C}(T, \ell_\infty^n)$ is congruent in a natural way with $\mathcal{C}(T) \times_\infty \cdots \times_\infty \mathcal{C}(T)$ in the notation of 4Z.

5. The $\mathbf{L}_p$ and ℓ_p spaces

5.1 Suppose p is a real number with $p \geqslant 1$. The set of all everywhere finite measurable functions f on a measure space $(S, \mathbf{S}, \mu)$ which are such that $|f|^p$ is integrable is called $\mathcal{L}_p$, or $\mathcal{L}_p(S, \mathbf{S}, \mu)$ when it is necessary to specify the measure space concerned. Trivially $\mathcal{L}_p$ is closed under scalar multiplication, and also under addition in view of the inequality $|f + g|^p \leqslant 2^p(|f|^p \vee |g|^p)$. We define the **$\mathcal{L}_p$-norm** of an arbitrary measurable function, denoted by $\|f\|_p$, to be

$$\|f\|_p = \left(\int |f|^p\right)^{1/p}$$

which is finite if $f \in \mathcal{L}_p$ and $+\infty$ otherwise. Since $\|f - g\|_p = 0$ if (and only if) $f = g$ almost everywhere, we count functions which are equal a.e. as identical. The result (formally stated, it is the quotient $\mathcal{L}/\mathcal{N}$ where $\mathcal{N} = \{f : f = 0 \text{ a.e.}\}$) is **Lebesgue's space of index p**, denoted L_p or $L_p(S, \mathbf{S}, \mu)$. We aim to show that $\|f\|_p$ makes L_p into a Banach space. The L_p spaces provide a storehouse of examples in many branches of analysis, and more particularly the spaces L_1 and L_2 are of great importance in applications. Before we go on to the proofs we give some of the important special cases which will, we hope, help to fix ideas in the reader's mind.

5.2 Example 1 It is not immediately obvious that $\|f\|_p$ satisfies the triangle inequality, but this *is* clear in the case $p = 1$, when

$$\|f\|_1 = \int |f|.$$

Clearly L_1 is just the space $\mathfrak{L}$ of integrable functions, counting functions which are equal a.e. as identical.

Example 2 The most important L_p-spaces, along with those of the next two examples, are those where μ is Lebesgue measure λ on an interval on the real line. As is customary, we shall write $L_p[0,1]$ instead of $L_p([0,1]$, Lebesgue measurable sets, $\lambda)$ and similarly $L_p[0,\infty)$, $L_p(\mathbf{R})]$ and so on.

Example 3 The spaces ℓ_p^n (4.2 Example 3) are the special cases where S is the set of natural numbers $\{1,\dots,n\}$ and μ is counting measure. Canonically identifying functions f on S with n-tuples of scalars $(f_1,\dots,f_n)$ one obtains $\|f\|_p = (\Sigma|f_j|^p)^{1/p}$ – see 2.23. In this case all the measure theory of course is trivial: the only thing that really needs proving is the triangle inequality for the norm.

Example 4 The **sequence space** ℓ_p is defined to be the set of all infinite scalar sequences $x = (x_1, x_2, \dots)$ such that $\Sigma|x_n|^p$ converges, with the norm $\|x\|_p = (\sum_1^\infty |x_n|^p)^{1/p}$. It is (i.e. is canonically identified with) the special case of L_p where $S = \mathbf{N}$ and μ is counting measure (see 2.23). Here the measure theory is not entirely trivial.

The ℓ_p spaces are among the most tractable infinite-dimensional Banach spaces, and for this reason are often used for the construction of simple examples, for instance of linear operators with specified properties.

The analysis of this section blossoms out from the following simple inequality about real numbers:

5.3 Lemma Let $\alpha,\beta > 0$ and $\alpha + \beta = 1$. Let $a, b \geqslant 0$. Then

$$a^\alpha b^\beta \leqslant \alpha a + \beta b$$

with equality iff $a = b$.

Of the many proofs of this fact we shall choose one that depends on the properties of convex functions.

Proof There is nothing to prove if $a = 0$ or $b = 0$, so let $a, b > 0$ and set $x = \log a$, $y = \log b$. Then the inequality reduces to

$$e^{\alpha x + \beta y} \leqslant \alpha e^x + \beta e^y$$

with equality iff $x = y$; and this is nothing more than the statement that the function $x \mapsto e^x$ is strictly convex (see Problem 5A).

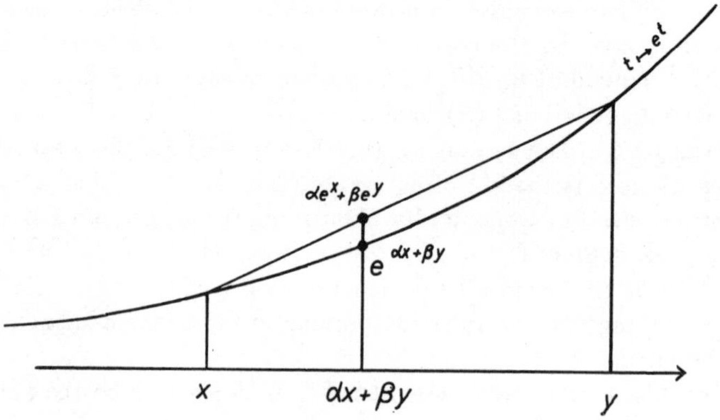

Diagram 6.

5.4 Theorem (Hölder's inequality) Let $p > 1$ and let q be the number such that

$$\frac{1}{p} + \frac{1}{q} = 1.$$

(Note that q is also > 1).
Let $f \in \mathfrak{L}_p$ and $g \in \mathfrak{L}_q$. Then $fg \in \mathfrak{L}_1$, and

$$\int |fg| \leqslant \|f\|_p \, \|g\|_q,$$

with equality iff there exist constants a, b, not both zero, such that $a|f|^p = b|g|^q$ a.e.

Proof $\|f\|_p$ and $\|g\|_q$ are both finite, and if either of them vanishes it is clear that the corresponding function is zero a.e., making the inequality, and the equality condition, trivial, so we may assume they are both strictly positive. Now apply the Lemma, taking

$$\alpha = \frac{1}{p}, \beta = \frac{1}{q}, a = \left(\frac{|f(s)|}{\|f\|_p}\right)^p, \ b = \left(\frac{|g(s)|}{\|g\|_q}\right)^q$$

to obtain for each s in the measure space S

$$\frac{|f(s)\,g(s)|}{\|f\|_p\,\|g\|_q} \leqslant \frac{|f(s)|^p}{p(\|f\|_p)^p} + \frac{|g(s)|^q}{q(\|g\|_q)^q}, \tag{1}$$

in other words

$$\frac{|fg|}{\|f\|_p\,\|g\|_q} \leqslant \frac{|f|^p}{p\int|f|^p} + \frac{|g|^q}{q\int|g|^q}.$$

Integrating over S gives

$$\frac{\int|fg|}{\|f\|_p\,\|g\|_q} \leqslant \frac{1}{p} + \frac{1}{q} = 1 \tag{2}$$

showing that fg is integrable and proving the result.

Since the situation $F \leqslant G$, $\int F = \int G$ implies $F = G$ a.e., equality holds in (2) iff it holds in (1) for almost $s \in S$. By the Lemma this occurs iff

$$\left(\frac{|f(s)|}{\|f\|_p}\right)^p = \left(\frac{|g(s)|}{\|g\|_q}\right)^q \quad \text{a.e.;}$$

and this is easily seen to be equivalent to having $a\,|f|^p = b\,|g|^q$ a.e., where $a = \int|g|^q$ and $b = \int|f|^p$.

It is worth mentioning that this elegant argument is not an isolated trick, but a case of a standard technique for turning number inequalities into integral (or sum, or series) ones. This is an area every budding analyst should delve in – see for instance Hardy, Littlewood and Polya (**9**), a classic and still one of the best books on the subject.

Numbers $p, q > 1$ such that $p^{-1} + q^{-1} = 1$ are called **conjugate indices.** Note that $p = 2$ is the only **self-conjugate** index.

5.5 Proposition (Converse to Hölder's inequality) Let p, q be conjugate indices and c a real number $\geqslant 0$. If $f \in \mathcal{L}_p$ and $|\int fg| \leqslant c$ for all $g \in \mathcal{L}_q$ such that $\|g\|_q \leqslant 1$ then $\|f\|_p \leqslant c$.

Proof If $\|f\|_p = 0$ there is nothing to prove, so suppose $\int|f|^p > 0$. Then the function

$$g = \overline{\operatorname{sgn}}\, f.\, (|f|^p / \textstyle\int|f|^p)^{1/q}$$

is easily seen to have the properties $\|g\|_q = 1$, $fg \geqslant 0$ and

$|f|^p = \lambda |g|^q$ for some $\lambda > 0$. By the equality case of Hölder's inequality, fg is integrable and

$$\|f\|_p = \|f\|_p \|g\|_q = \int |fg| = |\int fg| \leqslant c.$$

5.6 Theorem (Minkowski's inequality) Let $1 \leqslant p < \infty$ and let $f, g \in \mathcal{L}_p$. Then

$$\|f + g\|_p \leqslant \|f\|_p + \|g\|_p.$$

Proof For $p = 1$ the inequality is just the obvious $\int |f + g| \leqslant \int |f| + \int |g|$ so suppose $p > 1$ and let q be the conjugate index. Hölder's inequality implies that for all $h \in \mathcal{L}_q$ with $\|h\|_q \leqslant 1$,

$$|\int (f + g)h| \leqslant \int |fh| + \int |gh| \leqslant \|f\|_p + \|g\|_p$$

and the last Proposition gives the result.

The rather technical result below, a strengthened converse to Hölder's inequality, is basic to the study of continuous linear functionals on the L_p spaces. The reader may if he wishes ignore it till this topic is reached in §14. If σ-finiteness is left out the result is false in the case where $\|f\|_p = +\infty$. Recall that $\mathcal{L}_0$ denotes the space of finite simple functions vanishing outside a set of finite μ-measure, or equivalently the linear span of $\{\chi_A : A \in \mathbf{S}, \mu(A) < \infty\}$.

5.7 Theorem Let p, q be conjugate indices and c a real number $\geqslant 0$. Suppose the measure space is σ-finite and let f be a finite-valued measurable function such that, for all $g \in \mathcal{L}_0$ with $\|g\|_q \leqslant 1$,

$$fg \in \mathcal{L}_1 \text{ and } |\int fg| \leqslant c.$$

Then $f \in \mathcal{L}_p$ and $\|f\|_p \leqslant c$.

Proof First, suppose $\|f\|_p$ is known to be finite. If it is zero there is nothing to prove, so suppose that $0 < \int |f|^p < \infty$ and let g be the function constructed in Proposition 5.5, such that $\|g\|_q = 1$, and $|\int fg| = \|f\|_p$. By Proposition 2.19 there is a sequence $\{g_n\}$ in $\mathcal{L}_0$ such that $|g_n| \leqslant |g|$ and $g_n \to g$ pointwise. Clearly $\|g_n\|_q \leqslant \|g\|_q = 1$, and $\{fg_n\}$ is dominated by fg, which is assumed to be integrable, so the Dominated Convergence Theorem gives $\|f\|_p = |\int fg| = \lim_n |\int fg_n|$. But $|\int fg_n| \leqslant c$ for each n by the hypothesis; therefore $\|f\|_p \leqslant c$.

Suppose $\|f\|_p$ is not known to be finite. Using σ-finiteness one easily constructs an increasing sequence of measurable sets S_n

of finite measure covering the whole space S, such that f is bounded on each S . Then $f\chi_{S_n} \in \mathfrak{L}_p$, $|f\chi_{S_n}|^p \nearrow |f|^p$, and therefore by the Monotone Convergence Theorem, $\|f\chi_{S_n}\|_p \to \|f\|_p$. For any $g \in \mathfrak{L}_0$ with $\|g\|_q \leqslant 1$, $\chi_{S_n} g$ is also in $\mathfrak{L}_0$ and $\|\chi_{S_n} g\|_q \leqslant 1$, and one has $|\int f\chi_{S_n} g| \leqslant c$ by hypothesis. By the first part of the proof it follows that $\|f\chi_{S_n}\|_p \leqslant c$, and letting $n \to \infty$ gives $\|f\|_p \leqslant c$, as required.

We now construct the space L_p out of the space $\mathfrak{L}_p$ by the identification process sketched at the beginning of the section. Recall that two functions in $\mathfrak{L}_p$ are **μ-equivalent** if they are equal μ-almost everywhere. The equivalence class determined by a function $f \in \mathfrak{L}_p$ under this equivalence relation, often denoted by $[f]$, is the set $\{g: g = f\ \mu\text{-a.e.}\}$. The **Lebesgue space** L_p is the set of all such equivalence classes with the obvious definitions $[f] + [g] = [f+g]$, $\alpha[f] = [\alpha f]$, and $\|[f]\|_p = \|f\|_p$.

5.8 Theorem L_p is a normed linear space.

Proof That L_p is a linear space and that the above formula for the norm is well defined, are easily verified. The triangle inequality for the norm is just Minkowski's inequality. If $\|[f]\| = 0$ then $\int |f|^p = 0$, so that $f = 0$ a.e. and therefore $[f]$ is the zero element of L_p. The other properties 4.1 (1,3) in the definition of a norm are obvious.

Although the elements of L_p are actually equivalence classes of functions in $\mathfrak{L}_p$ the usual practice is to denote both a function and its equivalence class by the same symbol f, so as to simplify notation, and we shall do so. Thus 'the element f of L_p' will be short for 'the equivalence class represented by a function f in $\mathfrak{L}_p$' and so on.

We now prove the main result of this section, the

5.9 Riesz–Fischer Theorem L_p is complete and so is a Banach space.

Proof By Theorem 4.21 it is sufficient to show that every absolutely convergent series in L_p has a limit in L_p . Accordingly let $\{f_n\}$ be a sequence in L_p such that $\sum_{n=1}^{\infty} \|f_n\|_p = c < \infty$. Since a function and its modulus have the same L_p-norm, Minkowski's inequality gives for each n

$$\left\| \sum_{j=1}^{n} |f_j| \right\|_p \leqslant \sum_{1}^{n} \|f_j\|_p \leqslant c$$

and so

$$\int \Big(\sum_{j=1}^{n} |f_j|\Big)^p \leqslant c^p \text{ for each } n.$$

As $n \to \infty$ the integrand in this expression converges monotonely to the function g^p, where $g(s) = \sum_{n=1}^{\infty} |f_j(s)|$, $(s \in S)$. The Monotone Convergence Theorem gives $\int g^p \leqslant c^p < \infty$ so that g^p, and hence g, must be finite valued outside some set N of measure zero. In other words $\Sigma f_j(s)$ is absolutely convergent for $s \notin N$, and if we define f by

$$f(s) = \begin{cases} 0 & \text{for } s \in N \\ \sum_{j=1}^{\infty} f_j(s) & \text{otherwise} \end{cases}$$

then f is measurable and finite-valued, $|f| \leqslant g$ and therefore $\int |f|^p \leqslant \int g^p \leqslant \infty$ so that $f \in L_p$. Clearly $f(s) - \sum_{j=1}^{n} f(s)$, and therefore $|f(s) - \sum_1^n f_j(s)|^p$, converges to 0 a.e. But $|f(s) - \sum_1^n f_j(s)| = |\sum_{j=n+1}^{\infty} f(s)|^p \leqslant \sum_{j=n+1}^{\infty} |f_j(s)|^p \leqslant g(s)^p$ a.e. and since $\int g^p < \infty$ we can apply the Dominated Convergence Theorem to obtain

$$\lim_n \int |f - \sum_1^n f_j|^p = \int \lim_n |f - \sum_1^n f_j|^p = 0.$$

In other words $\|f - \sum_1^n f_j\|_p \to 0$ so that Σf_n converges to the limit f in the L_p norm.

5.10 Corollary Let $\sum_1^{\infty} f_n$ be an absolutely convergent series in L_p. Then the partial sums converge pointwise a.e. to their L_p limit function.

Proof Contained in the previous proof.

5.11 Corollary Every sequence $\{f_n\}$ which is Cauchy in L_p has a subsequent pointwise convergent to the L_p limit function.

Proof Choose an increasing sequence of indices $\{n_k\}$ such that $m, n \geqslant n_k \Rightarrow \|f_m - f_n\| < 2^{-k}$. Let f be the L_p limit-function of $\{f_n\}$ guaranteed by the Riesz–Fischer theorem. It follows as in Theorem 4.21 that $\{f_{n_k}\}$ is the sequence of partial sums of the absolutely convergent series $f_{n_1} + (f_{n_2} - f_{n_1}) + (f_{n_3} - f_{n_2}) + \ldots$ and therefore converges pointwise a.e. to the L_p limit of $\{f_{n_k}\}$, which is f.

The student should understand clearly that not every sequence which converges in L_p converges pointwise a.e. (see Problem 5J); nor does every sequence of L_p functions which converges pointwise to an L_p function necessarily converge in L_p (Problem 5K). The second corollary is useful in giving an explicit construction for a pointwise a.e. convergent subsequence – roughly, one that is sufficiently rapidly Cauchy (see Problem 5I).

The space L_∞

5.12 The reader may have noticed that the index $p = 1$ has (so far) no conjugate index q. It is clear from the formula $p^{-1} + q^{-1} = 1$ that $q \to \infty$ as $p \to 1$, and we remarked on p. 47 that in the case of the finite dimensional spaces $\mathbf{R}^n$, $\mathbf{C}^n$ the ℓ_q norm $(\Sigma|x_j|^p)^{1/p}$ converges to the limit $\|x\|_\infty = \max_{1 \leqslant j \leqslant n} |x_j|$ as $q \to \infty$. The space L_∞ is a generalization of this, and bears the same relation to L_1 in many respects as L_q does to L_p.

We say that a real number α is an **essential upper bound** for a non-negative measurable function f if $f(s) \leqslant \alpha$ a.e.

Example The Lebesgue-measurable function on $[0, 1]$ defined by $f(t) = 1/t$ when t is a nonzero rational, and t otherwise, has 1 for an essential upper bound but is not bounded.

5.13 Lemma If f has an essential upper bound it has a least one, denoted by ess sup f and called the **essential supremum** of f.

Proof This amounts to saying that if $\beta = \inf\{\alpha\colon f(s) \leqslant \alpha \text{ a.e.}\}$ then $f(s) \leqslant \beta$ a.e. Choose a sequence $\{\alpha_n\}$ with $f(s) \leqslant \alpha_n$ a.e., $\alpha_n \to \beta$. Then $\{s\colon f(s) > \beta\}$ is the union of the null sets $\{s\colon f(s) > \alpha_n\}$ $(n = 1, 2, \ldots)$ and so is null.

The space L_∞ is the space of (equivalence classes of) all measurable functions f such that ess sup $|f|$ exists, often called the space of **essentially bounded** measurable functions, with the norm

$$\|f\|_\infty = \text{ess sup } |f|.$$

It is clear that equivalent functions have the same value of ess sup $|f|$, so the above norm is well-defined. It is also easy to see that ess sup $|f|$ is the least value of sup $|g|$ taken over all functions g equivalent to f. From these remarks it follows easily that $\| \ \|_\infty$ is indeed a norm on L_∞.

5.14 Theorem L_∞ is a Banach space, under the norm defined above.

Proof Let $\{f_n\}$ be a Cauchy sequence in L_∞. Then there is a null set $N \subset S$ (the union of countable many null sets, one for each pair (m, n)), such that $|f_n(s) - f_m(s)| \leqslant \|f_n - f_m\|_\infty$ for all $s \notin N$ and $m, n = 1, 2, \ldots$. Then the sequence $\{f_n\}$ is uniformly convergent outside N, and we let

$$f(s) = \begin{cases} \lim f_n(s) & s \notin N \\ 0 & s \in N. \end{cases}$$

It follows that f is measurable and it is easily seen that $\|f_n - f\|_\infty \to 0$. Hence L_∞ is complete.

Convergence in L_∞ can be described roughly as uniform convergence outside a set of measure zero. In particular if the only set of measure zero is the empty set then $\| \ \|_\infty$ is just sup $|f|$ and convergence is just uniform convergence. This occurs in particular with the sequence space ℓ_∞.

We define $q = \infty$ to be the **conjugate index** to $p = 1$.

5.15 Theorem Hölder's inequality 5.4, Proposition 5.5 and Theorem 5.7 remain valid when $p = 1$, $q = \infty$ and vice versa subject to the condition of σ-finiteness in 5.5 when $p = \infty$, $q = 1$.

Proof Hölder's inequality is a triviality in this case. When $p = 1$ and $q = \infty$, the proofs of 5.5, 5.7 are almost unchanged, the function g defined there being replaced by the function $g = \overline{\operatorname{sgn} f}$.

Let $p = \infty$, $q = 1$ and the space be σ-finite. To prove 5.5 note that if $\|f\|_\infty > c$ there is a set E of nonzero *finite* measure such that $|f(s)| > c$ $(s \in E)$. The function

$$g = \overline{\operatorname{sgn} f} . (\chi_E / \textstyle\int \chi_E)$$

is then easily seen to have the properties $\|g\|_1 = 1$ but $|\int fg| > c$. The proof of 5.7 is extended to this case by making use of the same idea: this is left as an exercise.

Concluding remarks

All the general results about L_p spaces include corresponding results about the ℓ_p sequence space as a special case. But the student should not feel he has mastered the key theorems of this section unless he is able to translate their notation into direct, concise proofs of their ℓ_p versions (see Problems 5D, 5E).

It may well happen that a function is in L_p for many different values of p – for instance if f is in L_0. The following result may help the reader to get a feel for the behaviour of L_p spaces. The proof is left as Problem 5N.

5.16 Proposition If $\mu(S) < \infty$ then $L_{p_2} \subset L_{p_1}$ whenever $1 \leqslant p_1 < p_2 \leqslant \infty$. On the other hand $\ell_{p_2} \supset \ell_{p_1}$ whenever $1 \leqslant p_1 < p_2 \leqslant \infty$. For a general measure space no such inclusions hold.

Problems

5A (To Lemma 5.3) Let I be an interval in $\boldsymbol{R}$. A function $f: I \to \boldsymbol{R}$ is **convex** if $f(\alpha x + (1-\alpha)y) \leqslant \alpha f(x) + (1-\alpha) f(y)$ whenever $x, y \in I$ and $0 \leqslant \alpha \leqslant 1$, and **strictly convex** if strict inequality holds whenever $x \neq y$ and $0 < \alpha < 1$. Let f be differentiable. Show that

(i) If f' is increasing then f is convex;
(ii) If f' is strictly increasing then f is strictly convex.

Deduce that the function e^t is strictly convex on $\boldsymbol{R}$. [Apply the mean value theorem to the intervals $[x, z]$ and $[z, y]$ where $z = \alpha x + (1-\alpha)y$, assuming $x < y$.]

5B Young's inequality Let $\phi: [0, \infty) \to [0, \infty)$ be continuous and strictly increasing with $\phi(0) = 0$ and $\phi(t) \to \infty$ as $t \to \infty$. Clearly ϕ is one-to-one and its inverse ψ (that is $\phi(u) = v$ iff $\psi(v) = u$) is a function of the same sort. Define

$$\Phi(u) = \int_0^u \phi(t)\,dt \qquad \text{and} \qquad \Psi(v) = \int_0^v \psi(t)\,dt.$$

From a picture of the graph of ϕ deduce the inequality

$$uv \leqslant \Phi(u) + \Psi(v)$$

with equality iff $\phi(u) = v$. Setting $\phi(u) = u^{p-1}$, deduce Lemma 5.3.

5C Hölder's inequality for the case $p = 2$ is called the Cauchy–Schwarz inequality. Prove it by using the fact that the quantity $\int(|f| + t|g|)^2$ is a non-negative quadratic function of the real variable t.

5D State and prove Hölder's inequality and its converse for the space ℓ_p, $(1 < p < \infty)$, making appropriate changes in the notation and (where possible) simplifications in the proofs.

5E Adapt the notation and simplify the proof of Theorem 5.9 to give a proof of the completeness of $\ell_p (1 < p < \infty)$ avoiding any mention of measure and integration.

5F Denote by ℓ_0 the space of scalar sequences $y = (y_1, y_2, \ldots$ with only finitely many nonzero terms, called *finitely nonzero* sequences.

Theorem 5.7 for the ℓ_p spaces takes the form: Let p, q be conjugate indices and $x = (x_1, x_2, \ldots)$ a scalar sequence such that

$$|\sum_1^\infty x_n y_n| \leqslant c \text{ whenever } y \in \ell_0 \text{ and } \|y\|_q \leqslant 1.$$

Then $x \in \ell_p$ and $\|x\|_p \leqslant c$.

Prove this by elementary analysis.

5G Show that ℓ_0 is dense in ℓ_p if $1 \leqslant p < \infty$, but not dense in ℓ_p ; and that ℓ_p is separable if $1 \leqslant p < \infty$, but ℓ_∞ is not separable. [For ℓ_∞ consider the set of all sequences consisting entirely of 0's and 1's.]

5H Let $f \in L_p (1 < p < \infty)$ and let q be the conjugate index. show that Hölder's inequality and its converse together are equivalent to either of the following:

(i) $\|f\|_p$ is the smallest number c such that $|\int fg| \leqslant c \|g\|_q$ for all g in L_q.

(ii) $\|f\|_p = \sup \{|\int fg|: g \in L_q, \|g\|_q \leqslant 1\}$.

5I Let $\{f_n\}$ be a sequence of L_p functions converging in the L_p norm to a function f. Show that $f_n \to f$ pointwise a.e. if $\sum_n \|f - f_n\|$ converges.

5J In $L_p[0,1]$ where $1 \leqslant p < \infty$, consider all the functions χ_I where I is an interval of the form $[(j-1)/k, j/k]$ with $k = 1,2, \ldots$; $j = 1, 2, \ldots, k$. Show that when enumerated as a sequence in any way, these converge to the function 0 in the L_p norm but do not converge pointwise a.e.

5K Show that $\{n\chi_{[0,1/n]}\}$ is a sequence of functions converging to 0 pointwise a.e. but not converging in the norm of $L_p [0,1]$ for any $p \geqslant 1$.

5L A **σ-finite** set means a countable union of measurable sets of finite measure. Show that if $f \in L_p$ and $1 \leqslant p < \infty$ then f vanishes outside some σ-finite set; deduce that for such a function Proposition 2.19 (i) holds without the condition of σ-finiteness of the measure.

5M Fill in the details that $\| \ \|_\infty$ is indeed a norm on L_∞. (See before Theorem 5.14.)

Miscellaneous problems

5N Prove Proposition 5.16.

5O Let u, v be bounded Lebesgue-measurable real functions on $[0,1]$ such that $|\int uf|^2 + |\int vf|^2 \leqslant 1$ whenever $f \in L_1$ and $\|f\|_1 = 1$. Show that $u^2 + v^2 \leqslant 1$ almost everywhere. Does this generalize to complex-valued functions?

(Consider $|\int(uf \cos \alpha + vf \sin \alpha)|$ for $\alpha \in \boldsymbol{R}$.)

5P If $f \in L_{p_1}$ and $f \in L_{p_2}$ then $f \in L_p$ for all p between p_1 and p_2.

[One can choose conjugate indices r, s such that $p_1/r + p_2/s = p$. Then $f^{p_1/r} \in L_r$ and $f^{p_1/s} \in L_s$. Now apply Hölder.]

The space $\mathcal{C}(T)$

The Banach space $\mathcal{C}(T)$ of continuous real or complex functions on a compact topological space T occupies a commanding position in analysis. For one reason, as was mentioned on p. 56, there is a sense in which the class of subspaces of $\mathcal{C}(T)$'s includes all normed spaces whatsoever. For another, $\mathcal{C}(T)$ has the extra structure of *multiplication*, with the help of which one is able to solve many problems far more satisfactorily than if one had to rely solely on the linear structure. We shall not make use of the multiplication till p. 86 where we develop the Stone–Weierstrass approximation theorems, but in view of its importance it is worth starting the chapter with the basic properties of the multiplicative structure.

$\mathcal{C}(T)$ *as a Banach algebra*

6.1 A **linear associative algebra**, or simply an **algebra**, is a linear space A which also possesses a multiplication such that the product xy of two elements of A is an element of A, and which obeys the rules

(1) $(xy)z = x(yz)$

(2) $(\lambda x)y = \lambda(xy) = x(\lambda y)$

(3) $x(y + z) = xy + xz, \ (y + z)x = yx + zx$

where $x, y, z \in A$ and λ is a scalar. Roughly then, an algebra is a ring which is also a vector space. If also

(4) $xy = yx$

it is **commutative** and if A is a normed space and

(5) $\|xy\| \leqslant \|x\| \, \|y\|$

holds, then it is called a **normed algebra,** or a **Banach algebra** if A is also complete under the norm. Finally if there is an element $1 \in A$ called the **unit** such that

(6) $1x = x1 = x$ for all x
(7) $\|1\| = 1.$

then A is called **unital.** It should be immediately clear to the reader that if fg denotes the usual pointwise product and 1 is the function with constant value unity, then $\mathcal{C}(T)$ is a commutative, unital Banach algebra.

Perhaps property (5) needs checking:

$$\|fg\| = \sup_{t \in T} |f(t)\, g(t)| \leqslant \sup_{t \in T} |f(t)| \sup_{t \in T} |g(t)| = \|f\| \, \|g\|.$$

6.2 A **subalgebra** of $\mathcal{C}(T)$ is a linear subspace B which is an algebra in its own right, in other words one such that $f, g \in B \Rightarrow fg \in B$.

Example 1 Let a, b be any two points of T. Then $\{f \in \mathcal{C}(T): f(a) = 0\}$ and $\{f \in \mathcal{C}(T); f(a) = f(b)\}$ are easily seen to be subalgebras.

Example 2 In $\mathcal{C}[0,1]$ the subspace $\mathcal{C}'[0,1]$, of functions with a continuous derivative, is a subalgebra, and so is the subspace $\mathcal{P}[0,1]$ of polynomial functions.

It is clear the the definition of a subalgebra, as well as most of the following discussion, applies to any normed algebra, but we state it in terms of $\mathcal{C}(T)$ so as to help the reader fix his ideas.

6.3 Multiplication in $\mathcal{C}(T)$ is jointly continuous, in other words $f_n \to f$, $g_n \to g$ (in norm) imply $f_n g_n \to fg$. For 6.1 property 5 and the triangle inequality give

$$\begin{aligned} \|f_n g_n - fg\| &= \| f_n(g_n - g) + (f_n - f) g \| \\ &\leqslant \|f_n\| \, \|g_n - g\| + \|f_n - f\| \, \|g\| \to 0. \end{aligned}$$

A host of properties follow from this fact by elementary arguments, such as that $\lim_n f_n = f$ implies $\lim_n (f_n)^k = f^k$ (in norm) for $k = 2, 3, \ldots$ (see Problem 6A).

6.4 A **closed subalgebra** is of course one that is closed in the norm of $\mathcal{C}(T)$. The subalgebras of Example 1 are easily seen to be closed whereas those of Example 2 are not and, as we shall see, are in fact dense proper subsets of $\mathcal{C}[0,1]$. Any intersection of (closed) subalgebras is again a (closed) subalgebra, and therefore the argument that was applied in §4 to linear spans and convex hulls shows that for any family F of functions in $\mathcal{C}(T)$ there is a unique smallest subalgebra containing F and a unique smallest closed subalgebra containing F. These are called the subalgebra and the closed subalgebra **generated** by F and denoted by $\operatorname{alg} F$, $\overline{\operatorname{alg}}\, F$ respectively. The student should have no trouble in proving (cf. Proposition 4.9) that the continuity of multiplication implies:

6.5 Lemma The closure of any subalgebra is a subalgebra; and therefore (cf. Lemma 4.11):

6.6 Lemma $\overline{\operatorname{alg}}\, F$ is the closure of $\operatorname{alg} F$.

6.7 Can one describe $\operatorname{alg} F$ explicitly, in the same way as one could describe $\operatorname{lin} F$ and $\operatorname{co} F$ in terms of linear or convex combinations? In case F consists of 1 and a single other function it is fairly clear that $\operatorname{alg} F$ consists of all *polynomials* in f, that is elements of the form $\alpha_0 1 + \alpha_1 f + \ldots + \alpha_n f^n$.

Definition The **polynomial mapping** $\tilde{p}$ associated with an ordinary polynomial function $p(x) = \sum_0^n \alpha_r x^r$ of the scalar variable x is the mapping that associates to each $f \in \mathcal{C}(T)$ the element $\tilde{p}(f) = \sum_0^n \alpha_r f^r$ of $\mathcal{C}(T)$.

Having made the distinction between p and $\tilde{p}$ we immediately slur it over, and in future we shall write $p(f)$ instead of $\tilde{p}(f)$, as is customary. Thus we have

$$\operatorname{alg}\{1, f\} = \{p(f) \colon p \text{ a polynomial function}\}.$$

It is possible to express the subalgebra generated by an arbitrary family of functions in a similar way using polynomials in n variables, but we shall not need such a representation.

The preceding discussion of polynomials in *one* variable applies to any algebra, but there is one property which is peculiar to algebras whose elements are functions with the operations defined pointwise:

6.8 Lemma For $f \in \mathcal{C}(T)$, p a polynomial function,

$$p(f)(t) = p(f(t)) \quad (t \in T)$$

or more concisely: $p(f) = p \circ f$.

Proof $p(f)(t) = (\sum_0^n a_r f^r)(t) = \sum_0^n a_r f^r(t) = \sum_0^n a_r \{f(t)\}^r = p(f(t))$.

Compactness in $\mathcal{C}(T)$

We now leave the algebraic structure of $\mathcal{C}(T)$ temporarily on one side and turn to the question: How does one recognise compact subsets in the metric space $\mathcal{C}(T)$? Or, to put it in a more down-to-earth way, using the Bolzano–Weierstrass description (1.14) of compactness in metric spaces: what families F of continuous functions on T have the property that every sequence of functions in F has a uniformly convergent subsequence?

6.9 To avoid unneeded generality we assume throughout this section that T is a compact *metric* space. The notion of equicontinuity of F provides the key. Recall that each $f \in \mathcal{C}(T)$ is uniformly continuous in the sense that given $\epsilon > 0$ there exists $\delta > 0$ such that $|f(s) - f(t)| \leq \epsilon$ whenever $s, t \in T$ and $d(s, t) \leq \delta$. An **equicontinuous** family F of functions is one such that the same δ will do for every f, in other words such that given $\epsilon > 0$ there exists $\delta > 0$ such that $|f(s) - f(t)| \leq \epsilon$ whenever $f \in F$ and $s, t \in T$ with $d(s, t) \leq \delta$.

Example 1 In $\mathcal{C}[0,1]$, let F consist of differentiable functions, and suppose their derivatives are uniformly bounded, so that there is $c > 0$ such that $|f'(t)| \leq c \quad (f \in F;\ t \in [0,1])$. Then F is equicontinuous, for it follows from the mean-value theorem that $|f(s) - f(t)| \leq c\,|s - t| \quad (f \in F;\ s, t \in [0, 1])$. Thus given $\epsilon > 0$, choose $\delta = \epsilon/c$ to obtain $|f(s) - f(t)| \leq \epsilon$ for $f \in F$ and $s, t \in [0,1]$ with $|s - t| < \delta$.

Example 2 Let ϕ be a continuous function on the square $0 \leq s \leq 1,\ 0 \leq t \leq 1$. Then the family $\{f_s : 0 \leq s \leq 1\}$ of 'sections' of ϕ, where f_s is defined by

$$f_s(t) = \phi(s, t), \quad t \in [0,1],$$

is equicontinuous. For since the square is a compact metric space, ϕ is uniformly continuous; and the equicontinuity condition

$$|f_s(t_1) - f_s(t_2)| \leq \epsilon \text{ if } d(t_1, t_1) \leq \delta, \quad \text{for every } f_s,$$

is the special case, when $s_1 = s_2 = s$, of the condition

$$|\phi(s_1, t_1) - \phi(s_2, t_2)| \leqslant \epsilon \text{ if } d\{(s_1, t_1), (s_2, t_2)\} \leqslant \delta.$$

The idea in this example, namely that of regarding a function of two variables as a family of functions of one variable, is probably familiar to the reader from elementary analysis; the variable s being termed a 'parameter'. We now reverse this idea. Let F be a family of functions in $\mathcal{C}(T)$ and regard F as a metric space in its own right under the metric $\|f - g\|$. The operation of 'working out f at the point t' can be regarded as a scalar function ϕ on $F \times T$ defined by

$$\phi(f, t) = f(t).$$

6.10 Lemma ϕ is continuous on $F \times T$.

Proof This amounts to saying that $t_n \to t$ in T, and $f_n \to f$ in F, imply $f_n(t_n) \to f(t)$. We have

$$|f_n(t_n) - f(t)| \leqslant |f_n(t_n) - f(t_n)| + |f(t_n) - f(t)|.$$

The first term is $\leqslant \|f_n - f\|$ and the second tends to zero by the continuity of f, so the result follows.

There is no reason why F here should not be a large family, even the whole of $\mathcal{C}(T)$, but it probably helps the reader's intuition if he thinks of F as fairly small so that its members can be 'laid out in a row' like the f_s's in Example 2.

We shall use this idea in the next proof, not because it is vital to the proof but because it illustrates excellently the way a shift of viewpoint can save the mathematician an acre of epsilons and deltas, and because 'continuous functions on the product of compact metric spaces' is really what equicontinuous families are all about – a theme we ask the reader to explore in Problem 6F.

The next result, also called Arzela's theorem, gives a neat and often easily verified characterization of compact sets in $\mathcal{C}(T)$.

6.11 Ascoli's Theorem If T is a compact metric space then a subset of $\mathcal{C}(T)$ is compact iff it is closed, bounded and equicontinuous.

Proof First, let F be a compact subset of $\mathcal{C}(T)$. Then F and T are both compact metric spaces, and therefore so is $F \times T$ under the product topology and metric $\rho\{(f, s), (g, t)\} = \|f - g\| +$

$d(s, t)$. Since the function $\phi: (f, s) \mapsto f(s)$ is continuous on $F \times T$, it is uniformly continuous. Hence given $\epsilon > 0$ we can choose $\delta > 0$ such that

$$|f(s) - g(t)| = |\phi(f, s) - \phi(g, t)| \leqslant \epsilon$$

whenever $f, g \in F$, $s, t \in T$ and $\|f - g\| + d(s, t) \leqslant \delta$. The equicontinuity of F is the special case when $f = g$. That F is closed and bounded follows from the remark after Theorem 1.14.

Conversely suppose F is closed, bounded and equicontinuous. Since $\mathcal{C}(T)$ is complete, so also is F, and by Theorem 1.14, compactness will follow at once if we prove F is totally bounded. To do this, choose $\epsilon > 0$, and using the equicontinuity of F choose $\delta > 0$ such that $|f(s) - f(t)| < \epsilon/3$ when $f \in F$ and $d(s, t) < \delta$. Since T is compact and therefore totally bounded there exists a finite set of points $t_1, \dots, t_m$ forming a δ-net for T. Since F is bounded there is a number $c \geqslant 0$ such that $|f(t)| \leqslant c$ for all $f \in F$, $t \in T$. The set of scalars $\{\lambda: |\lambda| \leqslant c|$ is totally bounded and therefore can be partitioned into finitely many, disjoint, portions $E_1, \dots, E_n$ each of diameter $< \epsilon/3$. (These may be small half-open intervals or 'half-open squares' according as the scalars are $\mathbf{R}$ or $\mathbf{C}$.) We now define two functions f, g in F to be *equivalent*, and write $f \sim g$, if $f(t_j)$ lies in the same E-set as $g(t_j)$ for each $j = 1, \dots, m$. It is obvious that $\sim$ is an equivalence relation, and it only has finitely many equivalence classes (at most n^m in fact). Choosing one f from each equivalence class we get a finite set of functions $f_1, \dots, f_N$. Choose any $g \in F$. Then g is equivalent to some f_r, and since any $t \in T$ lies within δ of some t_j we have

$$|g(t) - f_r(t)| \leqslant |g(t) - g(t_j)| + |g(t_j) - f_r(t_j)| + |f_r(t_j) - f_r(t)|,$$

where the outer terms are $< \epsilon/3$ by the definition of δ and the middle one is $< \epsilon/3$ because $f_r(t_j)$ and $g(t_j)$ lie in the same set E_i. Hence $|g(t) - f_r(t)| < \epsilon \quad (t \in T)$, $\|g - f_r\| \leqslant \epsilon$, and since g was any element of F the $f_1, \dots, f_N$ form an ϵ-net for F. Thus F is totally bounded and the theorem follows.

There is a minor variant of Ascoli's theorem which is worth stating separately.

6.12 Corollary A subset F of $\mathcal{C}(T)$ has compact closure iff it is bounded and equicontinuous.

Proof Necessity is clear. Conversely the second part of the last theorem shows, in effect, that if F is bounded and equicontinuous then F is totally bounded, and the result follows from the corollary to Theorem 1.14.

The following fact, though we have no use for it, is striking and should give the student an enlightening insight into the general behaviour of equicontinuous families.

6.13 Proposition A pointwise convergent, equicontinuous sequence of functions in $\mathcal{C}(T)$ converges uniformly.

The reader should have no trouble in constructing a proof, either directly from the definition of equicontinuity or by first proving that the sequence is bounded (which, be it noted, was not assumed) and using the Corollary to Ascoli's theorem. Note in particular the interesting fact that the pointwise limit of an equicontinuous sequence is necessarily continuous.

† *Application: Montel's theorem*

For arbitrary families of continuous functions there is of course no strong connexion between uniform boundedness and equicontinuity. It is one of the features of *analytic* functions that for them the two notions are intimately, related. Here we describe the main idea of one of the most important applications of Ascoli's theorem: Montel's theorem on normal families of analytic functions. We assume the following facts from complex analysis: Cauchy's integral formula for a circular contour C

$$f(z) = \frac{1}{2\pi i} \int_C \frac{f(w)\,dw}{w - z} \quad (z \text{ inside } C);$$

that contour integrals obey the inequality $|\int_C g(z)\,dz| \leqslant \sup \{|g(z)|: z \in C\} \times$ length of C; and that the uniform limit of analytic functions is analytic. The reader should preferably have some idea of the subtleties of the 'diagonal process' – see 16.12 where it is used in the main part of the text.

Let D be a nonempty open set in the plane and $\mathcal{A}(D)$ the linear space of all functions analytic in D. An f in $\mathcal{A}(D)$ will in general not be bounded on D but certainly, being continuous, will be bounded on any compact $K \subset D$.

6.14 Montel's theorem If F is a family of functions in $\mathcal{A}(D)$, uniformly bounded on each compact $K \subset D$ – in other words if $\sup\{|f(z)| : z \in K, f \in F\} = m(K) < \infty$ for each such K – then any sequence in F has a subsequence converging, uniformly on each compact $K \subset D$, to a function in $\mathcal{A}(D)$.

It is clear that this is about a compactness property of some kind. In fact it is possible to put a metric – not a norm – on $\mathcal{A}(D)$ so that metric convergence means precisely uniform convergence on compact subsets of D. A family satisfying the conditions of the theorem is called *normal*, and Montel's theorem is usually stated in the form

$$F \text{ has compact closure in } \mathcal{A}(D) \Leftrightarrow F \text{ is normal.}$$

We start with two lemmas, one containing mostly complex analysis and the other mostly topology.

Lemma 1 Let the closed disc $\Delta = \{z : |z - \alpha| \leqslant r\}$ be such that the concentric closed disc of radius $2r$ lies in D. Let F be a normal family of functions. Then F is equicontinuous on Δ (in other words $\{f|_\Delta : f \in F\}$ is equicontinuous).

Proof Let C be the circle $|z - \alpha| = 2r$. Then C, together with the disc it encloses, lie in D, and the latter disc contains Δ. Thus

$$f(z) = \frac{1}{2\pi i}\int_C \frac{f(w)\,dw}{w - z} \quad (z \in \Delta, f \in F).$$

For any $\zeta, \eta \in \Delta$, since

$$\frac{1}{w - \zeta} - \frac{1}{w - \eta} = \frac{\zeta - \eta}{(w - \zeta)(w - \eta)}$$

we have

$$f(\zeta) - f(\eta) = \left|\frac{(\zeta - \eta)}{2\pi i}\int_C \frac{f(w)\,dw}{(w - \zeta)(w - \eta)}\right| \quad (f \in F).$$

But $|w - z| \geqslant r (w \in C, z \in \Delta)$ and $|f(w)| \leqslant m(C)\ (w \in C, f \in F)$ and the length of C is $4\pi r$, so

$$|f(\zeta) - f(\eta)| \leqslant \frac{|\zeta - \eta|}{2\pi}\cdot\frac{m(C)}{r^2}\cdot 4\pi r = \frac{2m(C)}{r}$$

$$(f \in F; \quad \zeta, \eta \in \Delta),$$

which implies F is equi-continuous on Δ.

Lemma 2 There exists a sequence of closed discs Δ_n such that: (i) for each n the concentric closed disc of twice the radius lies in D; (ii) $\bigcup_n \text{int}(\Delta_n) = D$.

Proof Let $\{a_n\}$ be a countable dense subset of D, let $r_n = \frac{1}{3}\min\{1, \text{dist}(a_n, C \sim D)\}$ and let $\Delta_n = \{z: |z - a_n| \leqslant r_n\}$. Clearly the Δ_n have property (i). Choose any $z \in D$, let $\rho = \min\{1, \text{dist}(z, C \sim D)\}$ and choose n so that $|z - a_n| < \frac{1}{4}\rho$. Then $\text{dist}(a_n, C \sim D) > \rho - \frac{1}{4}\rho = \frac{3}{4}\rho$ and $1 > \frac{3}{4}\rho$ so that $r_n > \frac{1}{3}\cdot\frac{3}{4}\rho = \frac{1}{4}\rho$ (see Diagram 7).

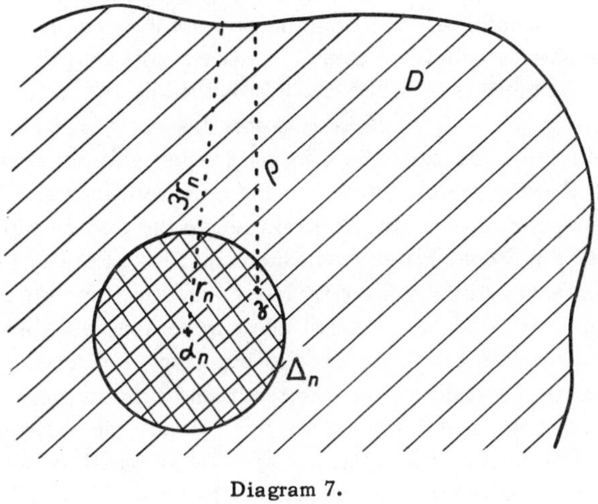

Diagram 7.

Hence $|z - a_n| < r_n$ which shows $z \in \text{int}\,\Delta_n$ and, since z was arbitrary, proves property (ii) and completes the proof.

6.15 Proof of theorem Let F be as stated in the theorem, and let $\{f_m\}$ be a sequence in F. Let $\{\Delta_n\}$ be a sequence of discs in D with the properties in Lemma 2. By Lemma 1, and the hypothesis of the theorem, F is both equicontinuous and bounded on Δ_1, and therefore by Ascoli's theorem, $\{f_m\}$ has a subsequence which is uniformly convergent on Δ_1, and which we denote by $\{f_m^1\}$. For the same reason, $\{f_m^1\}$ has a subsequence $\{f_m^2\}$ uniformly convergent on Δ_2; since it is already uniformly convergent on Δ_1 it follows that $\{f_m^2\}$ converges uniformly on $\Delta_1 \cup \Delta_2$. Continuing in this way gives for each k a subsequence $\{f_m^k\}$ of the previous subsequence $\{f_m^{k-1}\}$, converging uniformly on $\Delta_1 \cup \ldots \cup \Delta_k$.

Choosing the first element of each subsequence gives a sequence $\{f_1^k\}$ which is, apart from a finite number of terms, a subsequence of $\{f_m^k\}$ for each fixed k, and therefore converges uniformly on $\Delta_1 \cup \ldots \cup \Delta_k$ for each k (though not, be it noted, uniformly on $\bigcup_1^\infty \Delta_n$). Since by construction the interiors of the Δ_n form an open cover for D, any compact subset k of D is covered by $\Delta_1, \ldots, \Delta_k$ for some k and it follows that our subsequence converges uniformly on every compact $K \subset D$.

Uniform approximation: the Stone–Weierstrass theorems

The origin of the results of this section is the famous theorem of Weierstrass (1885) which states that any continuous function on a closed real interval $[a, b]$ can be *uniformly approximated* by polynomials: by which we mean, in modern notation, that given $f \in \mathcal{C}[a, b]$ there is a sequence of polynomials p_n on $[a, b]$ such that $\|f - p_n\|_\infty \to 0$, or in other words the set $\mathcal{P}[a, b]$ of polynomial functions is dense in $\mathcal{C}[a, b]$. It was realized about 1940 that the property which makes the theorem work is that $\mathcal{P}[a, b]$ is a *subalgebra* of $\mathcal{C}[a, b]$, and this led to the modern forms, known as the Stone–Weierstrass theorems, which are of immense use in modern analysis. We only have space for the half of the story which tells us which functions *can* be approximated by functions of a certain type: equally important, for numerical analysis, is the study of the existence and construction of *best* approximations; in which we seek for instance the polynomial p, of degree at most k, which makes $\|f - p\|$ as small as possible. This part of the theory was begun by Chebyshev around 1854.

Since the real and complex spaces $\mathcal{C}(T)$ behave somewhat differently we shall take care to emphasise the distinction by denoting them by $\mathcal{C}(T, \mathbf{R})$ and $\mathcal{C}(T, \mathbf{C})$ throughout the section.

6.16 The theorems rest on a simple but vital definition (of which we shall see more in §16). A family F of real or complex functions on a set S **separates the points** of S if for any two distinct $s, t \in S$ there exists $f \in F$ such that $f(s) \neq f(t)$.

Example The single function $u: t \mapsto t$ separates the points of $[0, 2\pi]$, but the functions $\{\sin, \cos\}$ fail to do so – they 'cannot tell $t = 0$ from $t = 2\pi$'.

Our basic result, whose proof occupies the next few pages, is:

6.17 Real Stone–Weierstrass Theorem Let T be a compact topological space, and A be a closed subalgebra of $\mathcal{C}(T, \mathbf{R})$ which contains 1 and separates the points of T. Then $A = \mathcal{C}(T, \mathbf{R})$.

Some simple remarks are in order here to bring out the significance of this theorem. By Lemmas 6.5, 6.6 the closure of a subalgebra in $\mathcal{C}(T, \mathbf{R})$ is again a subalgebra, and the closure of the subalgebra generated by any family $F \subset \mathcal{C}(T, \mathbf{R})$ is just the closed subalgebra generated by F. This allows us to re-state the Theorem in the forms in which it is mostly applied (the reader should note that the Complex Stone–Weierstrass Theorem 6.24 has similar reformulations, equally useful, which he should state and prove):

6.18 Corollary

(i) Let A be a subalgebra of $\mathcal{C}(T, \mathbf{R})$ which contains 1 and separates the points of T. Then A is dense in $\mathcal{C}(T, \mathbf{R})$.

(ii) Let F be a subset of $\mathcal{C}(T, \mathbf{R})$ which separates the points of T. Then the subalgebra $\text{alg}(1, F)$ generated by 1 and F is dense in $\mathcal{C}(T, \mathbf{R})$.

(iii) Let F be a subset of $\mathcal{C}(T, \mathbf{R})$ which separates the points of T. Then every $f \in \mathcal{C}(T, \mathbf{R})$ can be uniformly approximated by polynomials in the elements of F.

Proof (i) A^- is clearly a closed subalgebra which (since $A^- \supset A$) contains 1 and separates the points of T. Hence $A^- = \mathcal{C}(T, \mathbf{R})$, or A is dense in $\mathcal{C}(T, \mathbf{R})$. (ii) follows by taking $A = \text{alg}(1, F)$ in (i), and (iii) is just a restatement of (ii).

The idea is then, that by starting with quite a small set F of continuous functions (in the case $T = [a, b]$, F can consist of the single function $u: t \mapsto t$) and by applying repeatedly the operations of addition, multiplication and scalar multiplication, one can get close to *any* continuous function whatever.

6.19 The proof of the Theorem uses a rather surprising link between the algebra structure of $\mathcal{C}(T, \mathbf{R})$ and the pointwise ordering $f \leqslant g$ for real functions: one shows that a closed subalgebra A is a **lattice**, in the sense that it contains the max and min, $f \vee g$ and $f \wedge g$, of any two of its functions. A fairly easy finite open cover argument, using the compactness of T and the fact that A separates points, then enables us to approximate any $f \in \mathcal{C}(T, \mathbf{R})$ uniformly. We split the proof into several lemmas. The first is one of several ways of proving the lattice property, and illustrates an *iterative process* – a favourite technique of the numerical analyst, and one that has found many applications to more abstract analysis. An alternative proof, based on the binomial series for $(1 + x)^{\frac{1}{2}}$, is sketched in Problem 6S.

6.20 Lemma Define a sequence of polynomials p_n on $[-1, 1]$ by the rule $p_0 = 0$, $p_{n+1}(x) = \frac{1}{2}x^2 + p_n(x) - \frac{1}{2}p_n(x)^2$. Then $p_n(x) \to |x|$ uniformly on $[-1, 1]$ as $n \to \infty$.

Proof Fix $x \in [-1, 1]$ and define the function $F(u) = \frac{1}{2}x^2 + u - \frac{1}{2}u^2$. Simple algebra or calculus (see Diagram 8)

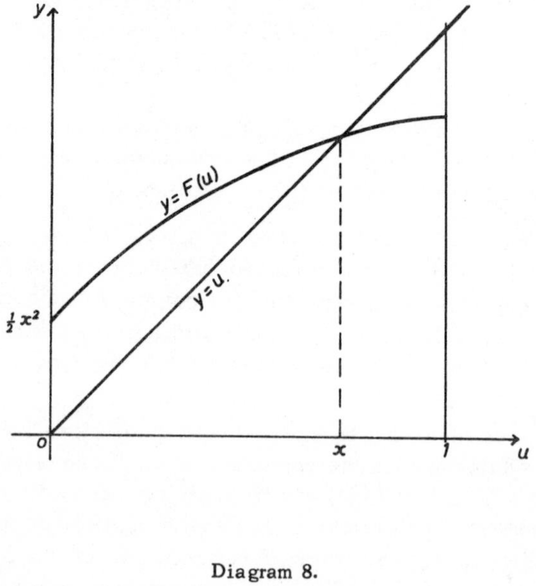

Diagram 8.

shows that

$$0 \leqslant u \leqslant |x| \leqslant 1 \Rightarrow u \leqslant F(u) \leqslant |x| \tag{1}$$

and that in the latter inequality, $u = F(u)$ iff $u = |x|$. Clearly the sequence of values $p_0(x), p_1(x), p_2(x) \ldots$ is just the sequence

$\{u_n\}$ where $u_0 = 0$ and

$$u_{n+1} = F(u_n). \tag{2}$$

From (1), $\{u_n\}$ is a positive, monotone, bounded sequence and therefore converges to some u. Letting $n \to \infty$ in (2) gives $u = F(u)$ and therefore $u = |x|$. In other words we have shown that for any $x \in [-1, 1]$, $p_n(x)$ converges *monotonely* to $|x|$. Since the limit-function $|x|$ is continuous, we deduce by Dini's Theorem 1.30 that convergence is uniform, completing the proof.

6.21 Lemma A closed subalgebra A of $\mathcal{C}(T, \mathbf{R})$ is a lattice.

Proof It is simple to verify that for any real numbers x, y

$$\max\{x, y\} = \left|\frac{x+y}{2}\right| + \left|\frac{x-y}{2}\right|, \ \min\{x, y\} = \left|\frac{x+y}{2}\right| - \left|\frac{x-y}{2}\right|$$

and to deduce that for real functions f, g

$$f \vee g = \left|\frac{f+g}{2}\right| + \left|\frac{f-g}{2}\right|, \ f \wedge g = \left|\frac{f+g}{2}\right| - \left|\frac{f-g}{2}\right|,$$

so that it suffices to show that $|f| \in A$ whenever $f \in A$. Again, since $|\alpha f| = \alpha|f|$ for $\alpha \geqslant 0$ there is no loss in assuming that $\|f\| \leqslant 1$. Thus, given $f \in A$ with $\|f\| \leqslant 1$, define a sequence $\{g_n\}$ by

$$g_0 = 0, \ g_{n+1} = \tfrac{1}{2}f^2 + g_n - \tfrac{1}{2}g_n^2.$$

By induction we see that: first, $g_n \in A$ for all n because A is a subalgebra; second, in terms of the p_n of the last lemma, $g_n = p_n \circ f$. Now

$$\begin{aligned}\|g_n - |f|\| &= \sup\{|p_n(f(t)) - |f(t)||: t \in T\} \\ &\leqslant \sup\{|p_n(x) - |x||: x \in [-1,1]\} \quad \text{since } f(T) \subset [-1,1] \\ &\to 0.\end{aligned}$$

Since A is closed, $|f| = \lim g_n \in A$, as required.

6.22 Proof of Theorem 6.17 Assume A satisfies the conditions of the theorem, and fix $f \in \mathcal{C}(T, \mathbf{R})$. We note that A has the **two-point property**, namely that for any distinct $s, t \in T$ and any $\alpha, \beta \in \mathbf{R}$ there is $g \in A$ with $g(s) = \alpha$ and $g(t) = \beta$: for since A separates points we can find $h \in A$ with $h(s) \neq h(t)$, and it is now only a matter of suitably adjusting λ, μ in the formula $g = \lambda h + \mu 1$.

Stage 1 Given $\epsilon > 0$, $t_0 \in T$ there is $g \in A$ with $g(t_0) = f(t_0)$ and $g < f + \epsilon 1$. Proof: for each $s \in T$ there is a function, say g_s, in A which equals f at both s and t_0 by the two-point property. Define $U_s = \{t \in T : g_s(t) < f(t) + \epsilon\}$, then U_s is open. Since $s \in U_s$ the U_s cover T as s varies, so by compactness there is a finite subcover, say $U_{s_1}, \ldots, U_{s_n}$. Set $g = g_{s_1} \wedge \ldots \wedge g_{s_n}$, which lies in A by lemma 6.21. Each $t \in T$ lies in one of the U_{s_i}, giving $g(t) \leqslant g_{s_i}(t) < f(t) + \epsilon$; since t is arbitrary this implies $g < f + \epsilon 1$. On the other hand $g_{s_i}(t_0) = f(t_0)$. for each i by construction, so $g(t_0) = f(t_0)$. This proves stage 1 and we pass to:

Stage 2 There is $h \in A$ with $f - \epsilon 1 < h < f + \epsilon 1$. Proof: By Stage 1, for each $t \in T$ there is a function $h_t \in A$ which equals f at t and lies below $f + \epsilon 1$. Another open cover argument, this time applied to the sets $V_t = \{u \in T : h_t(u) > f(t) - \epsilon\}$, provides a function $h = h_{t_1} \vee \ldots \vee h_{t_m}$ in A, such that $h > f - \epsilon 1$. (The student should supply the details.) But since $h_{t_j} < f + \epsilon 1$ for each j, we have $h < f + \epsilon 1$. Hence, $f - \epsilon 1 < h < f + \epsilon 1$, that is $\|f - h\| < \epsilon$. Since ϵ is arbitrary, $f \in A^- = A$; since f is arbitrary, $A = \mathcal{C}(T, \mathbf{R})$, completing the proof of Theorem 6.17 and hence of its equivalent formulations 6.18.

6.23 What about algebras of complex-valued functions? Not only does Theorem 6.17 fail in the complex case but the fact that it does so has significant and far-reaching connections with complex variable theory, which we have not the space to explore. For the reader who knows some complex analysis we note that the classic example is the **disc algebra.** Let D be the disc $|z| \leqslant 1$ in $\mathbf{C}$, and let A be the set of all continuous complex functions on D which are analytic on $|z| < 1$. Then A is a subalgebra of $\mathcal{C}(D, \mathbf{C})$; A is closed since the uniform limit of analytic functions is analytic; A contains 1, and separates points of D because it contains the function $u : z \mapsto z$. But $A \neq \mathcal{C}(D, \mathbf{C})$, because for instance the function $\bar{u} : z \mapsto \bar{z}$ is not analytic, and so is not in A.

It turns out that it is precisely this fact – that the disc algebra doesn't contain complex conjugates of its functions – that allows it to be strictly smaller than $\mathcal{C}(D, \mathbf{C})$. We say that a subalgebra A of $\mathcal{C}(T, \mathbf{C})$ (where T is a compact space) is **self-adjoint** if $f \in A \Rightarrow \bar{f} \in A$. Arguing in a sequence with which the reader should be thoroughly familiar we note that: the whole of $\mathcal{C}(T, \mathbf{C})$, the intersection of any family of self-adjoint subalgebras, and the closure of any self-adjoint subalgebra, are clearly self-adjoint subalgebras. This makes sense of defining the **self-adjoint subalgebra generated by** a given subset F of $\mathcal{C}(T, \mathbf{C})$ to be the intersection of all self-adjoint subalgebras containing F. The **closed self-adjoint subalgebra generated by** F is defined similarly. The closure of the former is the latter, as the reader should verify.

6.24 Complex Stone–Weierstrass Theorem Let T be a compact space, and A be a closed self-adjoint subalgebra of $\mathcal{C}(T, \mathbf{C})$ which contains 1 and separates the points of T. Then $A = \mathcal{C}(T, \mathbf{C})$.

Proof We reduce to the Real Stone–Weierstrass Theorem. Restricting the scalars to be real makes $\mathcal{C}(T, \boldsymbol{R})$ a (real) subalgebra of $\mathcal{C}(T, \mathbf{C})$ and therefore the set $A_r = A \cap \mathcal{C}(T, \boldsymbol{R})$ – in other words the *real* functions in A – is a closed (real) subalgebra of $\mathcal{C}(T, \boldsymbol{R})$.

Clearly $1 \in A_r$. By hypothesis, if s, t are distinct points of T there is f in A such that $f(s) \neq f(t)$. Set $g = f - f(s)1$, then $g, \bar{g} \in A$ and therefore the positive function $g\bar{g}$, which is zero at t, and nonzero at s, is in A_r. Hence A_r separates points and so is all of $\mathcal{C}(T, \boldsymbol{R})$ by Theorem 6.17. Since every $f \in \mathcal{C}(T, \mathbf{C})$ can be written as $f_1 + if_2$ where $f_1, f_2 \in \mathcal{C}(T, \boldsymbol{R}) = A_r \subset A$ it follows at once that $A = \mathcal{C}(T, \mathbf{C})$.

It is a result of complex analysis that a function which is analytic and real valued on an open disc must be constant there, so we see that if A is the disc algebra – a very non-self-adjoint algebra – then the preceding proof breaks down rather spectacularly, since A_r consists only of constant functions.

As a simple – but very important – application we prove the two classical approximation theorems of Weierstrass. Here and for the rest of the section we revert to using $\mathcal{C}[a, b]$ and so on to denote either the real or the complex continuous functions on $[a, b]$, specifying the scalars verbally when it matters.

6.25 Polynomial Approximation Theorem Any real or complex continuous function on a compact interval $[a, b]$ in $\boldsymbol{R}$ can be uniformly approximated by polynomials.

Proof Apply Corollary 6.18 or its corresponding complex version (note that complex $\mathcal{P}[a, b]$ is self-adjoint) to the subalgebra $\mathcal{P}[a, b]$ of $\mathcal{C}[a, b]$.

6.26 It should be clear that the complex form of this theorem follows rather trivially from the real case, roughly because the algebra of polynomials is generated by a *real*-valued function $u: t \mapsto t\ (t \in [a, b])$. In the next application it is much more convenient to deal with complex scalars throughout. A function f on $\mathbf{R}$ is **periodic with period** 2π or **periodic** (2π) for short, if $f(t + 2\pi) = f(t)$ for all $t \in \mathbf{R}$. A **trigonometric polynomial** on $\mathbf{R}$ is a function p of the form

$$p(t) = \tfrac{1}{2}a_0 + a_1 \cos t + b_1 \sin t + \dots + a_n \cos nt + b_n \sin nt \quad (1)$$

with complex coefficients a_r, b_r. The relations $\exp(\pm irt) = \cos rt \pm i \sin rt$ show that an equivalent representation is

$$p(t) = \sum_{r=-n}^{n} c_r\, e^{irt} \quad (2)$$

where the a_r, b_r are related to the c_r by the formulae $c_r + c_{-r} = a_r$, $i(c_r - c_{-r}) = b_r$. (The $\tfrac{1}{2}$ in formula (1) is to make this work when $r = 0$). Clearly every trigonometric polynomial is a periodic (2π) continuous complex function on $\mathbf{R}$ and we aim to prove:

6.27 Trigonometric Polynomial Approximation Theorem Every periodic (2π) continuous complex function g on $\mathbf{R}$ can be uniformly approximated by trigonometric polynomials.

Proof The substitution $e^{it} = z$ expresses the trigonometric polynomial (2) in the form $q(e^{it})$ where q is the function on the unit circle $\Gamma = \{z : |z| = 1\}$ defined by

$$q(z) = \sum_{r=-n}^{n} c_r z^r \quad (z \in \Gamma) \quad (3)$$

It is clear that, similarly, *every* periodic (2π) function g on $\mathbf{R}$ is of the form $g(t) = f(e^{it})$ for some function f on Γ, and we leave as an exercise the intuitively obvious but slightly tricky fact that, if g is continuous, so if f; (that is, a continuous periodic (2π) function 'rolls up' into a continuous function on the circle).

The set A of all q of the form (3) is clearly a subalgebra of (complex) $\mathcal{C}(\Gamma)$ containing 1; since A contains the function $u(z) = z$, it separates the points of Γ. Since Γ is just the set of points in $\mathbf{C}$ such that $\bar{z} = z^{-1}$, the complex conjugate of a function of the form (3) is of the same form (with $\bar{c}_{-r}$ replacing c_r) so that A is self-adjoint. It follows that A is dense in $\mathcal{C}(\Gamma)$. Thus, given $\epsilon > 0$ we can pick $q \in A$ such that $\|q - f\|_\infty < \epsilon$, where f is the function such that $f(e^{it}) = g(t)$. The function $p(t) = q(e^{it})$ is a trigonometric polynomial, and

$$|g(t) - p(t)| = |f(e^{it}) - q(e^{it})| < \epsilon \quad (t \in \mathbf{R})$$

which proves the theorem.

We remark that this result of Weierstrass is the most proved theorem of the book, being a byproduct of three other, independent, proofs, two in §13 and one in Problem 16N.

The reader should note that the trick of regarding periodic functions as 'really' being functions on the circle is an important and frequently used one – see Problem 6M where the correspondence is made precise.

An equivalent form of the theorem is:

6.28 Corollary A function $f \in \mathcal{C}[-\pi, \pi]$ can be approximated by trigonometric polynomials, uniformly on $[-\pi, \pi]$, if and only if $f(-\pi) = f(\pi)$.

This follows from the theorem by noting that the functions satisfying the stated condition are just the ones which are restrictions of continuous periodic (2π) functions on $\mathbf{R}$, and implies it because by periodicity a sequence of trigonometric polynomials uniformly convergent on $[-\pi, \pi]$ necessarily converges uniformly on $\mathbf{R}$.

Problems

Where $\mathcal{C}(T)$ is mentioned it is assumed that T is a compact metric or topological space.

Algebraic Structure

6A Prove that for any polynomial p the map $f \mapsto p(f)$ of $\mathcal{C}(T)$ into itself is continuous.

6B Let A be a normed algebra with a unit 1 such that $\|1\| \neq 1$. Show that the formula $|a| = \sup\{\|ax\| : \|x\| \leqslant 1\}$ defines a norm on A equivalent to the original one and making it into a unital normed algebra.

6C Represent ℓ_1 as the set of sequences $x = (x_0, x_1, x_2, \ldots)$ with $\|x\| = \sum_0^\infty |x_n| < \infty$. Show that ℓ_1 is a unital Banach algebra if one defines $x.y = z$ where $z_n = x_0 y_n + x_1 y_{n-1} + \ldots + x_n y_0$.

Ascoli's Theorem

6D Show that a sequence $\{f_n\}$ in $\mathcal{C}(T)$ is uniformly convergent iff it is equicontinuous and pointwise convergent. (A slightly strengthened form of Proposition 6.13.)

6E A family $F \subset \mathcal{C}(T)$ is called **equicontinuous at a point** t_0 of T if given $\epsilon > 0$ there exists $\delta > 0$ such that $|f(s) - f(t_0)| < \epsilon$ whenever $f \in F$ and $d(s, t_0) < \delta$. Show that a family which is equicontinuous at each point of T is equicontinuous.

6F Let S and T be compact metric spaces and ϕ a continuous function on $S \times T$. Let A be the set of sections f_s of ϕ defined by $f_s(t) = \phi(s, t)$. Show that:

(i) A is a bounded equicontinuous subset of $\mathcal{C}(T)$ and the map $s \mapsto f_s$ of S into $\mathcal{C}(T)$ is continuous.

(ii) Conversely if $s \mapsto f_s$ is a continuous map of S into $\mathcal{C}(T)$ then the function ϕ on $S \times T$ defined by $\phi(s, t) = f_s(t)$ is continuous.

(iii) Deduce that the Banach spaces $\mathcal{C}(S, \mathcal{C}(T))$, $\mathcal{C}(T, \mathcal{C}(S))$ and $\mathcal{C}(S \times T)$ are congruent in a natural way (notation of 4K$'$).

Other compactness theorems

6G Prove the following criteria for compactness of a closed, bounded set A in the Banach spaces indicated:

(i) In c_0: for any $\epsilon > 0$ there exists n_0 such that $|x_n| \leqslant \epsilon$ for all $n \geqslant n_0$ and all $x = (x_1, x_2, \ldots)$ in A.

(ii) In ℓ_p $(1 \leqslant p < \infty)$: for any $\epsilon > 0$ there exists n_0 such that $\sum_{n=n_0}^{\infty} |x_n|^p < \epsilon$ for all x in A.

(iii) In $\mathcal{C}^1[a, b]$: the set $\{f' : f \in A\}$ is equicontinuous.

6H (The general Ascoli theorem) The space $\mathcal{C}(T, X)$ of continuous functions from a compact metric space T into a Banach space X becomes a Banach space with the norm $\|f\| = \sup_{t \in T} \|f(t)\|$, by Problem 4K$'$. Equicontinuity of a subset A of $\mathcal{C}(T, X)$ is defined exactly as for scalar-valued functions, with norm replacing absolute value. Show that A is compact iff it is closed and equicontinuous and the subset $\{f(t) : f \in A,\ t \in T\}$ of X is totally bounded. Verify that when $X = \mathcal{C}(S)$ this agrees with what one would expect in view of Problem 6F (iii).

6I Let ϕ be a continuous scalar function on the unit square $0 \leqslant s \leqslant 1$, $0 \leqslant t \leqslant 1$; thus for each s in $[0,1]$ the section f_s: $t \mapsto (s,t)$ of ϕ belongs to $\mathcal{C}[0,1]$. Show that, for any $g \in \mathcal{C}[0,1]$, the function f on $[0,1]$ defined by

$$f(t) = \int_0^1 \phi(s,t)\, g(s)\, ds$$

is in the closed linear span of the functions f_s. [Take Riemann sums.]

The Stone–Weierstrass Theorems

6J (To Lemma 6.20) The binomial series for $(1+x)^{1/2}$ is

$$1 + \tfrac{1}{2}x + \frac{1.(-1)}{2.4}x^2 + \frac{1.(-1).(-3)}{2.4.6}x^3 + \ldots = \sum_0^\infty c_n x^n, \text{ say.}$$

(i) Show that $\Sigma |c_n|$ converges.

(ii) Deduce that the polynomials $p_n(x) = \sum_{r=0}^{n} c_r (x^2-1)^r$ converge uniformly on $[-1,1]$ to $|x|$.

6K Let f be the function $t \mapsto e^{2\pi i t}$ in $\mathcal{C}[0,1]$. Show that $\bar{f} \notin \overline{\text{alg}(1,f)}$.

6L A function f on $\boldsymbol{R}$ is **even** if $f(-t) = f(t)$ for all $t \in \boldsymbol{R}$.

(i) Assuming Theorem 6.27, show that every continuous, even, periodic (2π) complex function on $\boldsymbol{R}$ can be uniformly approximated by trigonometric polynomials containing only cosine terms.

(ii) Show that $\cos nx$ is a polynomial (the nth **Chebyshev polynomial**) in $\cos x$ and by considering $f(\cos x)$, where $f \in \mathcal{C}[-1,1]$, deduce Theorem 6.25 from Theorem 6.27.

6M (To Corollary 6.28) Let Γ denote the unit circle in the complex plane and define

$$\mathcal{C}_p[-\pi,\pi] = \{f \in \mathcal{C}[-\pi,\pi]: f(-\pi) = f(\pi)\},$$

$$\mathcal{C}_{2\pi}(\boldsymbol{R}) = \{\text{the continuous periodic } (2\pi) \text{ scalar functions on } \boldsymbol{R}\}.$$

Show that each of these, with the sup norm and pointwise operations, is a Banach algebra. Let α be the map $t \mapsto e^{it}$ $(t \in \boldsymbol{R})$. Verify that the maps $f \mapsto f \circ \alpha$ and $f \mapsto f \circ \alpha|_{[\pi,\pi]}$ establish isometric algebra isomorphisms (that is, maps that preserve linear operations, multiplication and norm) of $\mathcal{C}(\Gamma)$ with $\mathcal{C}_{2\pi}(\boldsymbol{R})$ and $\mathcal{C}_p[-\pi,\pi]$ respectively.

In view of this it is customary to think of $\mathcal{C}(\Gamma)$, $\mathcal{C}_{2\pi}(\boldsymbol{R})$ and $\mathcal{C}_p[-\pi,\pi]$ as being identical – at least for the purpose of studying their algebraic structure.

Two important applications

6N A simple version of the iterated integral theorem Let Q be a rectangle $\{(s,t): a \leqslant s \leqslant b, c \leqslant t \leqslant d\}$ in the plane. Show by direct calculation that for functions of the form $f(s,t) = s^m t^n$ the formula

$$\int_a^b \left(\int_c^d f(s,t)\, dt \right) ds = \int_c^d \left(\int_a^b f(s,t)\, ds \right) dt \qquad (*)$$

holds, and deduce by the Stone–Weierstrass Theorem that (*) holds for all f in $\mathcal{C}(Q)$.

6O Urysohn's lemma and Tietze's theorem Any compact topological space T such that the Stone–Weierstrass theorems can be applied to $\mathcal{C}(T)$ must clearly have:

Property (S): $\mathcal{C}(T)$ separates the points of T.

Two other topological properties which T may (or may not) have are

Property (T): Every continuous scalar function f on a closed subset X of T has a continuous extension to T, that is there exists g in $\mathcal{C}(T)$ with $g|_X = f$.

Property (U): Given disjoint closed sets X, Y in T there exists $f \in \mathcal{C}(T)$ with $f = 0$ on X and $f = 1$ on Y.

A fundamental result (see any topology text) is **Urysohn's Lemma,** which says that every compact Hausdorff space has (U); from which, since points are closed sets, (S) follows.

(i) Show that (S) is trivial for compact metric spaces.
(ii) Assuming property (S) deduce property (T). [Let A denote the set of $f \in \mathcal{C}(X)$ having an extension to a $g \in \mathcal{C}(T)$. First show that one may suppose $\|g\| = \|f\|$. Deduce, using Problem 4W or otherwise, that A is a closed subalgebra separating the points of X.]
(iii) Show that (U) is a special case of (T) and hence that (S) implies (U).

In short: compact metric spaces have all three properties (S), (T), (U); and so, assuming Urysohn's Lemma, do compact Hausdorff spaces. Property (T) is called **Tietze's Theorem.**

6P The identification of $\mathcal{C}(\Gamma)$ with $\mathcal{C}_p[-\pi, \pi]$ (Problem 6M) leads one to suspect that subalgebras of $\mathcal{C}[a, b]$ are associated in some way with spaces $\mathcal{C}(T)$ where T is obtained by gluing together some of the points of $[a, b]$. The two examples below explore this idea.

(i) Let T be a figure-of-eight curve in the plane, e.g. the curve parametrized by $x = \sin 2t$, $y = \sin t$. Show that $\mathcal{C}(T)$ is isometrically algebra-isomorphic to the subalgebra $\{f: f(0) = f(1), f(\frac{3}{4}) = f(\frac{1}{4})\}$ of $\mathcal{C}[0,1]$.

(ii) More pathologically, show that $\mathcal{C}[0,1]$ contains a subalgebra isometrically algebra-isomorphic to $\mathcal{C}(Q)$ where Q is the unit square $[0,1] \times [0,1]$.

3

BASIC THEORY OF OPERATORS AND FUNCTIONALS

7. Linear maps and functionals on normed spaces

7.1 When studying normed spaces it is natural to consider those maps from a normed space into itself, or into some other normed space, that are both linear and continuous; such maps have many useful properties and are the commonest in applications. We shall have a lot to say about the simplest linear maps on X, the functionals. A **linear functional**, or just a **functional**, on a linear space X is a linear map from X into the scalar field, in other worc a scalar valued function f on X with the property that

$$f(\alpha x + \beta y) = \alpha f(x) + \beta f(y).$$

In Molière's play (**10**) there is a character who finds to his amazement that Prose is what he has been talking all his life. So for us, for we have been using linear maps and functionals since our schooldays: indeed this is the reason for inventing a subject called linear functional analysis at all. Roughly speaking, whenever we evaluate a function at a point; or integrate it from a to b; or extract the coefficient of x^k from its Taylor expansion – we are defining a linear functional on a space of functions. Whenever we form the derivative, or the indefinite integral, of a function; or its Laplace transform $\bar{f}(s) = \int_0^\infty e^{-st} f(t)\, dt$ – we are defining a linear mapping on a space of functions to some other space of functions. We shall give examples shortly to make these vague statements precise.

Continuous linear maps
Let X and Y be normed spaces over the same scalars, and, to avoid trivial cases, assume that X is not the trivial space $\{0\}$. Let T be a linear map of X into Y. To say T is continuous means of course that

$$x_n \to x \text{ in the norm of } X \Rightarrow T(x_n) \to T(x) \text{ in the norm of } Y.$$

In this section we shall derive a number of more useful conditions equivalent to the continuity of T, and show that the set of *all* continuous linear maps of X into Y becomes a normed linear space in a natural way.

Recall that a set in a metric space is *bounded* if it lies in some ball (of finite radius), and that B_X, B_Y denote the closed unit balls in X and Y.

7.2 Theorem For a linear map $T: X \to Y$ the following conditions are equivalent

(i) T is continuous;
(ii) T is continuous at 0; that is $x_n \to 0 \Rightarrow T(x_n) \to 0$;
(iii) There exists a real number $k \geqslant 0$ such that $\|T(x)\| \leqslant k\|x\|$ for all x in X;
(iv) $T(B_X)$ is bounded in Y.

Proof (i) $\Rightarrow$ (ii) is trivial.

(ii) $\Rightarrow$(iii). Suppose (iii) fails, so there is no k such that $\|Tx\| \leqslant k\|x\|$ for all x. Then for each n there is an x_n in X with $\|Tx_n\| \geqslant n\|x_n\|$. If we let $y_n = x_n/(n\|x_n\|)$ then it is easy to see that $y_n \to 0$ but $Ty_n \not\to 0$ contradicting (ii). Thus if (ii) holds then so must (iii).

(iii) $\Rightarrow$ (iv). If (iii) holds then clearly $T(B_X)$ lies in the ball kB_Y.

(iv) $\Rightarrow$ (i). Assume (iv), let x be any point of X, and let $\epsilon > 0$. It is easy to see that a set lies in some ball iff it lies in some ball round 0, so by (iv), $T(B_X) \subset kB_Y$ for some $k > 0$. If we choose $\delta = k^{-1}\epsilon$ then $T(x + \delta B_X) = T(x) + \delta T(B_X) \subset T(x) + \delta k B_Y = T(x) + \epsilon B_Y$.

That is, the δ-ball round x maps into the ϵ-ball round $T(x)$. Since x, ϵ were arbitrary this shows T is continuous.

A mapping T from one normed space to another which has the property $\|Tx\| \leqslant k\|x\|$ is called **bounded**, the number k being a **bound** for T: thus for linear maps we can, and do, use the terms 'bounded' and 'continuous' interchangeably. We state this separately for emphasis:

7.3 Theorem A linear map from one normed space to another is continuous $\Leftrightarrow$ it is bounded.

We define the **norm** of a bounded linear map by the formula

$$\|T\| = \sup\{\|T(x)\|: \|x\| \leqslant 1\}.$$

This is clearly equivalent to

$$\|T\| = \sup\{\|T(x)\|: \|x\| = 1\}.$$

The proof of Theorem 7.2 shows that k is a bound for $T \Leftrightarrow T(B_X) \subset kB_Y$. Since $\|T\|$ is clearly the smallest k such that $T(B_X) \subset kB_Y$ this gives a further expression for the norm:

$$\|T\| = \inf\{k: \|T(x)\| \leqslant k\,\|x\| \text{ for all } x\}.$$

In other words, $\|T\|$ is the **smallest bound** for T.

We now illustrate these ideas by some examples.

7.4 Example 1 Let $\mathcal{P} = \mathcal{P}[0, 1]$ denote the set of polynomial functions on the interval $[0, 1]$ and let D be the differentiation operator:

$$D(p) = p'.$$

Then D is clearly a linear map of $\mathcal{P}$, into itself (the derivative of a polynomial is a polynomial). Consider two norms on $\mathcal{P}$ – the one as a subspace of $\mathcal{C}[a, b]$, namely the usual sup norm $\|p\| = \|p\|_\infty$, and the other as a subspace of $\mathcal{C}^1[a, b]$, namely $\|p\|_1 = \max\{\|p\|_\infty, \|p'\|_\infty\}$. D is *unbounded* as a map from $(\mathcal{P}, \|\cdot\|)$ to $(\mathcal{P}, \|\cdot\|)$ because for the functions $p_n(t) = t^n$ $(0 \leqslant t \leqslant 1)$ one has $\|p_n\| = 1$, but $\|D(p_n)\| = n$. D is *bounded* as a map from $(\mathcal{P}, \|\cdot\|_1)$ to $(\mathcal{P}, \|\cdot\|)$ because $\|D(p)\| = \|p'\| = \|p'\|_\infty \leqslant \max\{\|p\|, \|p'\|_\infty\} = \|p\|_1$. This shows that $\|D\| \leqslant 1$ in this case. (As an exercise, show $\|D\| = 1$.) This points up the obvious fact that the continuity of a mapping depends on the norm one is using.

Example 2 Let $X = \mathcal{C}[a, b]$ with the sup-norm and let $T: X \to X$ be defined by $T(f) = g$ where $g(t) = \int_a^t f(u)\,du$. Then

$$|g(t)| \leqslant \int_a^t |f(u)|\,du \leqslant \int_a^b \|f\|\,du = (b-a)\|f\|,$$

so that $\|T(f)\| \leqslant (b-a)\|f\|$, $\|T\| \leqslant b-a$. In fact by taking $f = 1$ we see that $\|T\|$ actually is equal to $b-a$.

This is an example of a **Volterra integral operator** (Problem 70).

Example 3 Let X, Y be $\boldsymbol{R}^n$ and $\boldsymbol{R}^m$ respectively and let $A = (a_{ij})$ be a real $m \times n$ matrix. The map

$$T_A : x \mapsto Ax,$$

where Ax denotes the matrix product of A with x regarded as a column vector, is a linear map from $\boldsymbol{R}^n$ into $\boldsymbol{R}^m$. We recall from linear algebra that *every* linear map of $\boldsymbol{R}^n$ into $\boldsymbol{R}^m$ is represented by a matrix in this way.

Now let $\boldsymbol{R}^n$ and $\boldsymbol{R}^m$ have the Euclidean (ℓ_2) norm. Then T_A is bounded, and in fact it is not hard to see (Problem 7D) that $\sum_{i=1}^{m} \sum_{j=1}^{n} |a_{ij}|$ is a bound for T_A. Rather surprisingly perhaps, there is no explicit purely algebraic formula for $\| T_A \|$, although (as will follow from the results of §18) it is known that $\| T_A \|^2$ is the largest root of the polynomial equation

$$p(\lambda) \equiv \det(A^t A - \lambda I) = 0$$

where A^t means the transpose of A.

Topological isomorphisms

7.5 A linear map T from a normed space X into a normed space Y is a **topological isomorphism** if it is a homeomorphism of X with Y, that is if T is one-to-one and onto, and both T and T^{-1} are bounded. (Recall that if T^{-1} exists it is clearly also linear.) Normed spaces X, Y for which such a T exists are **topologically isomorphic**, or **equivalent.**

7.6 Proposition T is a topological isomorphism $\Leftrightarrow T$ is onto and there exist constants a, $b > 0$ such that $a\|x\| \leqslant \|T(x)\| \leqslant b\|x\|$ for all x.

Proof The condition implies that T is bounded; also that T is one-to-one, for $T(x) = 0$ implies $\|x\| \leqslant a^{-1}\|T(x)\| = 0$. Thus T^{-1} exists and, since $a\|x\| \leqslant \|T(x)\|$ is equivalent to $\|T^{-1}(y)\| \leqslant a^{-1}\|y\|$ where $y = T(x)$, we see T^{-1} is bounded. Thus T is a topological isomorphism. The converse is similar.

7.7 Corollary Two norms $\|\cdot\|$ and $|\cdot|$ on a linear space X are equivalent $\Leftrightarrow$ there exist a, $b > 0$ such that $a\|x\| \leqslant |x| \leqslant b\|x\|$ for all x.

Proof Apply the proposition to the identity map $x \mapsto x$ from $(X, \|\cdot\|)$ to $(X, |\cdot|)$.

7.8 Corollary If X and Y are equivalent normed spaces, and either is a Banach space, then so is the other.

Proof Left as an exercise, with the remark that the proposition implies that a topological isomorphism from X onto Y necessarily maps a Cauchy sequence in X to one in Y.

These results indicate that, for most purposes, equivalences are just as useful as congruences. It is worth pointing out that a homeomorphism between arbitrary metric spaces does *not* always preserve completeness – a simple example is the map $x \mapsto x/(1+|x|)$ which is a homeomorphism of $\mathbf{R}$ (complete) onto $(-1, 1)$ (incomplete). So this result is not quite as obvious as it may look.

Finite dimensional spaces
This is a natural place to insert a number of facts about norms, linear maps and linear functionals on finite dimensional spaces, which all follow from the theorem below.

7.9 Theorem All norms on a finite dimensional linear space are equivalent.

Proof Let X be an n-dimensional linear space and denote the scalar field by $\mathbf{F}$ $(=\mathbf{R}$ or $\mathbf{C})$. Choose a basis $e_1, \ldots, e_n$ for X (this basis is quite arbitrary but will remain fixed throughout). It is easy to see that the function $\|x\|_0 = \sum_1^n |\alpha_j|$, where $x = \alpha_1 e_1 + \ldots + \alpha_n e_n$, is a norm on X.

Now let $\|\cdot\|$ be any other norm on X. By the continuity of the algebraic operations on the normed space $(X, \|\cdot\|)$ – plus induction – the map $(\alpha_1, \ldots, \alpha_n) \to \alpha_1 e_1 + \ldots + \alpha_n e_n$ of $\mathbf{F}^n$ onto X is continuous, and therefore the set

$$K = \{\alpha_1 e_1 + \ldots + \alpha_n e_n : |\alpha_1| + \ldots + |\alpha_n| = 1\},$$

being the continuous image of a closed bounded set in $\mathbf{F}^n$, is compact in $(X, \|\cdot\|)$. Since $\|x\|$ is a continuous function it attains its maximum and minimum values – say M, m – on K. Moreover $m > 0$, since $\|\alpha_1 e_1 + \ldots + \alpha_n e_n\| = 0 \Rightarrow \alpha_1 e_1 + \ldots + \alpha_n e_n = 0 \Rightarrow \alpha_1 = \ldots = \alpha_n = 0$. Thus

$$\sum |\alpha_j| = 1 \Rightarrow 0 < m \leqslant \|\sum \alpha_j e_j\| \leqslant M < \infty.$$

It follows easily from the homogeneity of the norm that

$$m \sum |\alpha_j| \leqslant \|\sum \alpha_j e_j\| \leqslant M \sum |\alpha_j| \text{ for all } \alpha_1, \ldots, \alpha_n,$$

in other words $m\|x\|_0 \leqslant \|x\| \leqslant M\|x\|_0$, $(x \in X)$. Hence, by

Corollary 7.7 *any* norm on X is equivalent to the norm $\| \|_0$ which was fixed at the beginning, and the result follows.

7.10 Corollary Any linear map from a finite dimensional normed space into any other normed space, not necessarily finite dimensional, is continuous.

Proof Let $T: X \to Y$ be linear, where X, Y are normed spaces and $\dim X < \infty$. If we define

$$\|x\|' = \|x\| + \|T(x)\| \quad (x \in X)$$

then $\| \|'$ is obviously a norm on X, and T is bounded with respect to it since $\|T(x)\| \leq \|x\|'$. By the theorem, $\| \|'$ is equivalent to $\| \|$, and the result follows.

7.11 Corollary

(i) Any linear functional on a finite dimensional normed space is continuous.

(ii) Any linear isomorphism between finite dimensional normed spaces is a topological isomorphism.

Proof Obvious.

A metric space is **locally compact** if every point has a compact neighbourhood.

7.12 Corollary Any finite-dimensional normed space X is complete and locally compact.

Proof X is linearly isomorphic, and therefore topologically isomorphic, to $\mathbf{F}^n$ ($= \mathbf{R}^n$ or $\mathbf{C}^n$) for suitable n, with the Euclidean norm. $\mathbf{F}^n$ is complete and locally compact, and topological isomorphism preserves completeness (by Corollary 7.8) and local compactness (trivially). The result follows.

7.13 Corollary Any finite dimensional subspace M of a normed space X is closed.

Proof With the restriction of the norm of X, M is a finite dimensional normed space and so is complete. Hence M is a complete subset of the metric space X and so is closed.

A rough summary of all these results would be to say that all finite dimensional normed spaces, and this includes of course all finite dimensional subspaces of normed spaces, are really just $\mathbf{R}^n$ or $\mathbf{C}^n$ in disguise.

The next theorem shows that the converse of the local compactness property of Corollary 7.12 holds. We shall not use it till §19, so the reader may omit the proof till then, but it is a useful piece of general knowledge about normed spaces.

7.14 Theorem A normed space is locally compact $\Leftrightarrow$ it is finite dimensional.

Proof Implication $\Leftarrow$ has been proved already, so suppose X is a locally compact normed space. Then some ball rB round 0 is compact, and since multiplying by r^{-1} is continuous, B, the unit ball, is compact. It follows from 4.14 that there exists a *finite* set of points, say $F = \{x_1, \ldots, x_n\}$, with

$$B \subset F + \tfrac{1}{2}B$$

so that if M is the finite dimensional subspace spanned by F, then

$$B \subset M + \tfrac{1}{2}B$$

Using repeatedly the computation rules of 3.23 we have $\frac{1}{2}B \subset \frac{1}{2}M + \frac{1}{4}B = M + \frac{1}{4}B$, so $B \subset M + (M + \frac{1}{4}B) = M + \frac{1}{4}B$. Again, $\frac{1}{4}B \subset M + \frac{1}{8}B$ which leads to $B \subset M + \frac{1}{8}B$. Continuing in this way we find $B \subset M + 2^{-n}B$ for all n, so that

$$\begin{aligned} B &\subset \bigcap_{n=1}^{\infty} (M + 2^{-n}B) \\ &= M^{-} \quad \text{by Proposition 4.8} \\ &= M \quad \text{by Corollary 7.13} \end{aligned}$$

Hence the unit ball of X lies inside a finite dimensional subspace, so clearly X is finite dimensional.

7.15 A bounded, hence continuous, linear map from a normed space X to a normed space Y will often be called an **operator from X to Y**. The set of all operators from X to Y will be denoted by $\mathcal{B}(X, Y)$, the letter $\mathcal{B}$ standing from 'bounded'. We shorten $\mathcal{B}(X, X)$ to $\mathcal{B}(X)$ and call its members **operators on X**. The next theorem shows that $\mathcal{B}(X, Y)$ has a natural structure as a normed space, indeed a Banach space if Y is complete. In particular $\mathcal{B}(X)$ is a Banach space if X is a Banach space.

7.16 Theorem With the linear operations defined pointwise, namely

$$\begin{aligned} (S + T)(x) &= S(x) + T(x) \\ (\lambda T)(x) &= \lambda \cdot T(x) \end{aligned}$$

and the **operator norm** $\|T\|$ defined above, $\mathcal{B}(X, Y)$ becomes a normed linear space. If Y is complete, so is $\mathcal{B}(X, Y)$.

Proof To show $\mathcal{B}(X, Y)$ is a linear space the main point is to show it is closed under the addition and scalar multiplication operations, since once this is done it is a trivial matter to verify the linear space axioms. That is, if S, T are linear and bounded, is $S+T$ linear and bounded? This follows from

$$\begin{aligned}(S+T)(\alpha x+\beta y) &= S(\alpha x+\beta y)+T(\alpha x+\beta y) \text{ (definition of } S+T)\\ &= \alpha S(x)+\beta S(y)+\alpha T(x)+\beta T(y) \text{ (linearity of } S, T)\\ &= \alpha(S(x)+T(x))+\beta(S(y)+T(y))\\ &= \alpha\cdot(S+T)(x)+\beta\cdot(S+T)(y) \text{ (definition of } S+T),\end{aligned}$$

and $\|(S+T)(x)\| \leqslant \|S(x)\| + \|T(x)\| \leqslant \|S\|\ \|x\| + \|T\|\ \|x\|$, which also shows that $\|S+T\| \leqslant \|S\| + \|T\|$.

Similarly, for any λ, λT is linear and bounded with

$$\|(\lambda T)(x)\| = |\lambda|\ \|T(x)\|$$

which also shows that $\|\lambda T\| = |\lambda|\ \|T\|$.

It is trivial that $\|T\| \geqslant 0$, and $\|T\| = 0$ iff T is the zero element of $\mathcal{B}(X, Y)$. Hence, $\mathcal{B}(X, Y)$ is a linear space and $\|T\|$ defines a norm on it.

Now assume Y is complete. To show $\mathcal{B}(X, Y)$ is complete we need to show that, given a Cauchy sequence $\{T_n\}$ in $\mathcal{B}(X, Y)$, there is an element T of $\mathcal{B}(X, Y)$ such that $\|T_n - T\| \to 0$. Any Cauchy sequence in a metric space is bounded (why?) so there exists M such that $\|T_n\| < M$ for all n.

Fix a point $x \in X$. The sequence of points $\{T_n(x)\}$ is then a Cauchy sequence in Y, since $\|T_n(x) - T_m(x)\| \leqslant \|x\|\ \|T_m - T_n\|$, and $\{T_n\}$ is Cauchy. Since Y is complete, $\{T_n(x)\}$ converges, and we may therefore define a new map $T: X \to Y$ by setting

$$T(x) = \lim_n T_n(x).$$

By the continuity of the linear operations in Y, T is linear, for we see that $\lim_n T_n(\alpha x+\beta y) = \alpha \lim_n T_n(x) + \beta \lim_n T_n(y)$. Since $\|T(x)\| = \lim_n \|T_n(x)\| \leqslant \sup_n \|T_n(x)\| \leqslant \sup_n \|T_n\|\ \|x\| \leqslant M\|x\|$, it follows that T is bounded, and so is in $\mathcal{B}(X, Y)$. The proof is not yet complete, however, for we must show that $\|T_n - T\| \to 0$. Given $\epsilon > 0$ choose n_0 with $\|T_m - T_n\| \leqslant \epsilon$ for $m, n \geqslant n_0$. Then $\|T_m(x) - T_n(x)\| \leqslant \epsilon\|x\|$ for $m, n \geqslant n_0$. Letting $n \to \infty$

gives $\|T_m(x) - T(x)\| \leqslant \epsilon \|x\|$ for $m \geqslant n_0$, and therefore $\|T_m - T\| \leqslant \epsilon$ for $m \geqslant n_0$, completing the proof.

Problems

Linear maps and their bounds

7A Add to the list of equivalent statements in Theorem 7.2:

(a) T is uniformly continuous;
(b) The sequence $\{\|T(x_n)\|\}$ is bounded for each sequence $\{x_n\}$ converging to zero in X.

7B (See after Theorem 7.3) Show that the norm of a linear operator is also given by $\|T\| = \sup\{\|T(x)\| : \|x\| < 1\}$.

7C Let $\{x_n\}$ be a sequence in a normed space X and T a bounded linear map from X into some other normed space Y. Show that $\sum_{n=1}^{\infty} T(x_n) = T(\sum_{n=1}^{\infty} x_n)$ whenever the series on the right hand side converges.

7D Let T be a linear map of $\boldsymbol{R}^n$ into $\boldsymbol{R}^m$ with matrix (a_{ij}), and let $p, r \in [1, \infty]$. Show that the norm of T considered as an element of $\mathcal{B}(\ell_p^n, \ell_r^m)$ is at most $\sum_i \sum_j |a_{ij}|$. [Write $T = \sum_i \sum_j T_{ij}$ where the matrix of T_{ij} consists of zeros except in the i, j position.] Finer bounds on $\|T\|$ are considered in Problem 7N.

Topological isomorphisms

7E Let X and Y be Banach spaces, let $T \in \mathcal{B}(X, Y)$ and define $\delta_T = \inf\{\|Tx\| : x \in X, \|x\| = 1\}$. Suppose that $\delta_T > 0$ and im T is dense in Y. Show that T is a topological isomorphism of X and Y.

7F (To Corollary 7.8) A one-to-one map f of a metric space X onto a metric space Y is a **uniform equivalence** if f and f^{-1} are uniformly continuous. One says that Y is uniformly equivalent to X under f.

(i) Show that if two metric spaces are uniformly equivalent and one is complete then so is the other.
(ii) Show that a topological isomorphism of normed spaces is a uniform equivalence in the metric space sense.
(iii) Verify that the map described after Corollary 7.8 is a homeomorphism but not a uniform equivalence of $\boldsymbol{R}$ with $(-1, 1)$.

Finite-dimensional spaces

7G The rational plane $X = \mathbf{Q}^2$ is a linear space over the rationals $\mathbf{Q}$. Defining a *norm* on X to be a $\mathbf{R}$-valued function on X satisfying the norm properties 1 to 4 in 4.1 (with α rational in property 3), show that $\|(r, s)\| = |r - s\sqrt{2}|$ is a norm on X not equivalent to any of the usual (e.g. Euclidean) norms. This shows that some compactness notion is essential in Theorem 7.9.

(Harder!) Show that a norm on X which only takes rational values is necessarily equivalent to the usual norms.

7H Let M and N be closed linear subspaces of a normed space X with N finite-dimensional. Show that the subspace $M + N$ is closed.

The space $\mathcal{B}(X, Y)$

7I Let X_1 and Y_1 be normed spaces and let X_2 and Y_2 denote the same linear spaces endowed with different but equivalent norms. Show that $\mathcal{B}(X_1, Y_1)$ and $\mathcal{B}(X_2, Y_2)$ consist of the same operators and are identical as linear spaces, and that the norms on them (though generally different) are equivalent.

Miscellaneous problems

7J Riesz' Lemma

(i) Let M be a proper closed linear subspace of a normed space X. Show that for any $\epsilon > 0$ there exists x in X with $\|x\| = 1$ and $d(x, M) > 1 - \epsilon$. If $\dim X < \infty$ one can actually take $d(x, M) = 1$.

(ii) Use Riesz' Lemma to give an alternative proof that a locally compact normed space is finite-dimensional.

(iii) Let Q be the quotient map of X onto X/M, and let U, V be the open unit balls in X and X/M respectively. Show that Riesz' Lemma is equivalent to the statement $Q(U) = V$.

7K Factorization Theorem Let X and Y be normed spaces and $T \in \mathcal{B}(X, Y)$; let $M = \ker T$ and let Q be the quotient map of X onto X/M. Show

(i) There is a unique $S \in \mathcal{B}(X/M, Y)$ such that $T = S \circ Q$.

(ii) S is one-to-one and $\|S\| = \|T\|$.

7L Extension by continuity

(i) Let X be a normed space, Y a Banach space, X_0 a dense linear subspace of X and $T_0 \in \mathcal{B}(X_0, Y)$. Then T_0 has a unique extension to an operator $T \in \mathcal{B}(X, Y)$; moreover, $\|T\| = \|T_0\|$.

[Given $x \in X$, let $\{x_n\}$ be a sequence in X_0 converging to x. Show that $\{T_0(x_n)\}$ converges to a limit $T(x)$ that is independent of the sequence chosen; that the resulting map $T : X \to Y$ is linear and bounded and agrees with T_0 on X_0. Note that by definition, if B is the unit ball of X, then $\|T\|$ and $\|T_0\|$ are respectively the supremum of $\|T(x)\|$ on B and on $B \cap X_0$.]

This essentially simple but very useful result, a special case of a similar extension theorem for uniformly continuous maps on metric spaces, is called the principle of *extension by continuity*.

(ii) Show in particular that if X is complete then each $T_0 \in \mathcal{B}(X_0)$ has a unique extension to an operator $T \in \mathcal{B}(X)$.

(iii) Show that if X and Y are Banach spaces with dense subspaces X_0, Y_0, and if X_0 is congruent [equivalent] to Y_0 then X is congruent [equivalent] to Y.

Norms of operators

7M Multiplication operators

(i) Let $(S, \mathbf{S}, \mu)$ be a measure space; write L_p for $L_p(S, \mathbf{S}, \mu)$. Given $p, r \in [1, \infty]$ with $p \geqslant r$, and a finite measurable function g, show that the map $T : f \mapsto gf$ is in $\mathcal{B}(L_p, L_r)$ iff $g \in L_s$ where $1/s = 1/r - 1/p$, and that $\|T\|$ is then equal to $\|g\|_s$. (Certain σ-finiteness restrictions may be needed if $p = \infty$.)

(ii) By considering the identity map from ℓ_p to ℓ_r show that a different situation arises when $p < r$ and investigate it.

7N Let $\boldsymbol{F}$ denote $\boldsymbol{R}$ or $\boldsymbol{C}$ and let A be a linear map from $\boldsymbol{F}^n$ to $\boldsymbol{F}^m$ with matrix (a_{ij}). Establish the following formulae for the norm of A relative to different norms on the domain and range spaces:

As a member of $\mathcal{B}(\ell_1^n, \ell_\infty^m)$, $\quad \|A\| = \max_{i,j} |a_{ij}|$;

$\mathcal{B}(\ell_1^n, \ell_1^m)$, $\quad \|A\| = \max_j (\sum_i |a_{ij}|)$;

$\mathcal{B}(\ell_\infty^n, \ell_\infty^m)$, $\quad \|A\| = \max_i (\sum_j |a_{ij}|)$.

Integral operators

In the next two problems we write Tf instead of $T(f)$ (where T is a linear operator on a space of functions) to simplify notation, as is customary. Thus $Tf(t)$ means $g(t)$ where $g = Tf$.

7O

(i) Let Q be the square $a \leqslant s \leqslant b$, $a \leqslant t \leqslant b$ in $\mathbf{R}^2$ and let $K \in \mathcal{C}(Q)$. Show that the mapping $T: f \mapsto Tf$ where

$$Tf(s) = \int_a^b K(s, t) f(t)\, dt \quad (a \leqslant s \leqslant b)$$

belongs to $\mathcal{B}(\mathcal{C}[a, b])$ and that $\|T\| \leqslant (b-a) \sup\{|K(s, t)|: (s, t) \in Q\}$. T is a **Fredholm integral operator**; K is the **kernel** of T.

(ii) Let Δ be the triangle $a \leqslant s \leqslant b$, $a \leqslant t \leqslant s$ in $\mathbf{R}^2$ and let $K \in \mathcal{C}(\Delta)$. Show that the mapping $T: f \mapsto Tf$ where

$$Tf(s) = \int_a^s K(s, t) f(t)\, dt \quad (a \leqslant s \leqslant b)$$

belongs to $\mathcal{B}(\mathcal{C}[a, b])$ and that $\|T\| \leqslant (b-a) \sup\{|K(s, t)|: (s, t) \in \Delta\}$. T is a **Volterra integral operator** (with kernel K).

7P This illustrates the trick (put to good use in Problem 17N) of **renorming** a space to make a given operator have a smaller norm. Let T be a Volterra integral operator on $\mathcal{C}[0, b]$, with kernel K, and let $\|K\|_\infty$ denote the supremum of $|K(s, t)|$ on its domain of definition. Prove:

(i) $\|T\|$ can be as great as $b\|K\|_\infty$.

(ii) Given $\epsilon > 0$ there is a norm $\|\ \|$ on $\mathcal{C}[0, b]$ equivalent to the usual norm $\|\ \|_\infty$ such that the norm of T as an operator on $(X, \|\ \|)$ is less than ϵ. [Let $m > 0$ and define $\|f\| = \sup\limits_{0 \leqslant t \leqslant b} |f(t)e^{-mt}|$. Then $\|\ \|$ and $\|\ \|_\infty$ are equivalent. One has $|Tf(s)| \leqslant \|K\|_\infty \|f\| e^{ms}/m$ so that $\|T\| \leqslant \|K\|_\infty/m$.]

Matrix operators

7Q Let $A = (a_{ij})_{i,j \in N}$ be a **double sequence** (or **infinite matrix**) of scalars and suppose that $1 \leqslant p < \infty$ and

$$\sup_j \left(\sum_i |a_{ij}|^p\right)^{1/p} = c < \infty.$$

Show that matrix multiplication by A (that is $Ax = y$ where $y_i = \sum\limits_j a_{ij} x_j$) defines an operator in $\mathcal{B}(\ell_1, \ell_p)$, whose norm is exactly c.

By considering the images of the unit vectors e_j, show that every operator in $\mathcal{B}(\ell_1, \ell_p)$ is of the above form.

8. Continuous linear functionals

8.1 The **dual space** of a normed space X, denoted X^*, is the space of all bounded, and therefore continuous, linear functionals on X. In other words $X^* = B(X, \boldsymbol{F})$ where $\boldsymbol{F}$ is the scalar field of X, and the members of X^* are all the scalar functions $f, g, \ldots$ on X which obey the rule $f(\alpha x + \beta y) = \alpha f(x) + \beta f(y)$ and are such that the **dual norm**

$$\|f\| = \sup\{|f(x)| : \|x\| \leqslant 1\}$$

is finite, with the linear operations defined in the obvious way by

$$(f+g)(x) = f(x) + g(x), \ (\lambda f)(x) = \lambda \cdot f(x).$$

Since $\boldsymbol{R}$ and $\boldsymbol{C}$ are complete, Theorem 7.12 implies:

8.2 Theorem X^* is a Banach space with the norm and linear operations defined above.

Before we go on to study X^* in detail we pause to give a few concrete examples of functionals on an infinite dimensional (and therefore not entirely trivial!) normed space.

8.3 Example Let $\mathcal{P} = \mathcal{P}[0, 1]$ be the space of polynomial functions on $[0, 1]$ that we looked at in 7.4 Example 1, and let the norm on $\mathcal{P}$ be the sup norm $\|p\|_\infty$. Then the following are linear functionals on $\mathcal{P}$:

$f(p) = p(t_0)$ for any given $t_0 \in [0, 1]$;

$g(p) = p'(t_0)$ for any given $t_0 \in [0, 1]$;

$h(p) = \int_a^b p(t)\,dt$ for any interval $[a, b] \subset [0, 1]$;

$k(p) = c_n$, the coefficient of t^n in the expansion of $p(t)$ in powers of t, where n is a fixed integer $\geqslant 1$.

That f is linear follows directly from the definition of the linear operations in $\mathcal{P}$. That g and h are linear expresses properties of the derivative and of the integral. That k is linear, and indeed well-defined at all, arises from the fact that each polynomial can be *uniquely* expressed in the form $c_0 + c_1 t + \ldots + c_n t^n$.

The functionals f and h are *bounded*, and therefore elements of $\mathcal{P}^*$, for it is immediate that $|f(p)| \leqslant \|p\|_\infty$ and

$$|h(p)| \leqslant \int_a^b |p(t)|\,dt \leqslant (b-a)\|p\|_\infty.$$

On the other hand g and k are *unbounded.* The unboundedness of g is easy to see when $t_0 = 1$ – take the sequence of polynomials $p_n(t) = t^n$, for which $\|p_n\| = 1$, $|g(p_n)| = n$. For a general t_0, it needs a bit more work as does the unboundedness of k (Problem 8C).

These simple examples illustrate the commonest way in which functionals arise in practice: by **evaluation**, by **integration** of which summation is a special case, and as **coefficient functionals.** There is always one element of X^*, the rather uninteresting **zero functional** 0 defined by $0(x) = 0$ for all x. As yet there is no guarantee that X^* contains any other elements at all. We shall soon see, however, that X is richly supplied with bounded linear functionals, enough of them in fact to separate the points of X. For a general, infinite dimensional, space this is a consequence of the Hahn–Banach Theorem 8.9. In the finite-dimensional case the situation is simpler because by Corollary 7.11 all linear functionals are automatically continuous, and we look at this case first. Because we get continuity for free, it is really a purely algebraic result.

8.4 Theorem When X is of finite dimension n, X^* is also of dimension n. Moreover to each basis $\{e_1, \ldots, e_n\}$ in X corresponds a unique set of functionals $\{f_1, \ldots, f_n\}$ called the **coefficient functionals** of the basis $\{e_j\}$ such that

$$f_i(e_j) = \delta_{ij} \qquad (1 \leqslant i \leqslant n,\ 1 \leqslant j \leqslant n)$$

(where $\delta_{ij} = 1$ if $i = j$ and 0 otherwise). The $\{f_i\}$ form a basis for X^* called the **dual basis** to $\{e_j\}$.

Proof Each x in X has a unique representation $x = \lambda_1 e_1 + \ldots + \lambda_n e_n$, and the map which extracts the ith coefficient from this expression, that is the map $f_i : x \mapsto \lambda_i$, is therefore well-defined and easily seen to be linear. Since when $x = e_j$ all the λ's are zero except for λ_j, which is 1, it is clear that $f_i(e_j) = \delta_{ij}$.

That the f_i are the *unique* functionals with this property we leave as an exercise.

To show the f_i are linearly independent, suppose $\alpha_1 f_1 + \ldots + \alpha_n f_n = 0$, that is suppose that $\alpha_1 f_1(x) + \ldots + \alpha_n f_n(x) = 0$ for all x. By taking x to be each e_j in turn and using the fact that $f_i(e_j) = 0$ for $i \neq j$ we obtain $\alpha_j = \alpha_j f_j(e_j) = 0$ for each j, proving linear independence.

To show the f_i span X^*, let f be any functional in X^* and *define*

$$\alpha_j = f(e_j).$$

We shall show $f = \alpha_1 f_1 + \ldots + \alpha_n f_n$. Evaluating at e_j gives (again since $f_i(e_j) = 0$ when $i \neq j$)

$$(\alpha_1 f_1 + \ldots + \alpha_n f_n)(e_j) = \alpha_j f_j(e_j) = \alpha_j = f(e_j)$$

Hence the functionals f and $\alpha_1 f_1 + \ldots + \alpha_n f_n$ agree on the spanning set $\{e_j : 1 \leqslant j \leqslant n\}$ and so are equal by Lemma 3.18. Hence the f_i form a basis for X^*. In particular, $\dim X^* = n$, and this completes the proof.

Hyperplanes and the Hahn–Banach Theorems

8.5 We now develop a geometric way of looking at functionals which has proved to be a very useful tool in functional analysis. A **hyperplane** in a linear space X is a proper linear subspace which is not contained in any strictly larger linear subspace except X itself. (The word hyperplane is also used to denote a coset of such a subspace, but we shall not do so.) The case $X = \boldsymbol{R}^3$ may help to fix ideas in the reader's mind: here the hyperplanes are the planes through the origin, that is the sets determined by equations $ax + by + cz = 0$ where a, b, c are not all zero. With this in mind the reader should not be surprised by the next result.

8.6 Proposition

(i) The following properties of a linear subspace M are equivalent:

(a) M is a hyperplane;

(b) The quotient space X/M is one-dimensional;

(c) $M = \ker f$ for some nonzero linear functional f.

(ii) If X is a normed space and $M = \ker f$ as in (i), then M is closed if and only if f is continuous.

Proof (i) ($a \Rightarrow b$). Let M be a hyperplane, and choose a fixed e not in M. Then $\text{lin}(e, M) = X$ by definition, so that every x in X can be written in the form $x = \lambda e + m$ where $m \in M$. This amounts to saying that every coset $x + M$ is of the form $\lambda(e + M)$, so that X/M is spanned by the single nonzero element $e + M$ and so is one-dimensional.

($b \Rightarrow c$). Let $\dim(X/M) = 1$ and choose a fixed nonzero coset $e + M$ in X/M. Then $\{e + M\}$ is a basis for X/M, so for each $x \in X$, there is a unique scalar λ such that $x + M = \lambda(e + M)$.

The mapping f that associates to x the scalar λ is obviously a nonzero linear functional; it vanishes iff $x + M$ is the zero coset, that is iff $x \in M$, so that $M = \ker f$.

($c \Rightarrow a$) Let $M = \ker f$ where $f \neq 0$. Clearly $M \neq X$, and to show M is a hyperplane we only need show that if x is any vector not in M then $\text{lin}(x, M) = X$. For any other vector y we see that

$$f(y - \frac{f(y)}{f(x)} x) = f(y) - \frac{f(y)}{f(x)} f(x) = 0$$

so that $y - \dfrac{f(y)}{f(x)} x \in M$, and hence $y \in \text{lin}(x, M)$ which shows that $\text{lin}(x, M) = X$, as required.

(ii) If f is continuous then $M = \ker f = f^{-1}\{0\}$ is closed because $\{0\}$ is a closed set in $\boldsymbol{R}$ or $\mathbf{C}$. For the converse, we shall assume f is *not* continuous and prove M is not closed. Let u be any vector with $f(u) = \alpha \neq 0$. There exist vectors $x_n \in X$ with $\|x_n\| = 1$ and $f(x_n) = \lambda_n$ where $|\lambda_n| \to \infty$. It is easy to verify that $u - \alpha\lambda_n^{-1}x_n \in \ker f = M$, but $\lim_n (u - \alpha\lambda_n^{-1}x_n) = u \notin M$, so that M is not closed.

The basic result of this section, the Hyperplane Theorem, is remarkable for being essentially a geometrical fact – and even an intuitively obvious one – yet having deep algebraic and topological consequences, as well as applications to areas as diverse as linear programming and the theory of additive set functions. We start with a lemma that contains the main geometric idea.

8.7 Lemma Let U be a convex open set in a real normed space X, such that $0 \notin U$. Let N be a two-dimensional subspace of X. Then for some nonzero $x \in N$ the line through x – that is the one-dimensional subspace $L_x = \{\lambda x : \lambda \in \boldsymbol{R}\}$ – misses U.

Proof Since any such line L_x is contained in N it is clearly sufficient that L_x should miss the set $V = U \cap N$, which is a convex open set in the two dimensional normed space N, such that $0 \notin V$. Thus we may restrict attention to N and forget about the rest of X. Note that by Corollary 7.11 we can, for the purpose of visualizing the proof, think of N as being the ordinary Euclidean plane (Diagram 9).

Assume there is no nonzero $x \in N$ such that L_x misses V. Then for every such x, $\lambda x \in V$ for some real λ. Since $0 \notin V$, λ cannot be 0, so we can write this as $x \in \mu U$ where $\mu = \lambda^{-1} \neq 0$

and deduce that

$$N \sim \{0\} = W \cup W'$$

where $W = \{x \in N : x \in \mu U \text{ for some } \mu > 0\}$
$W' = \{x \in N : x \in \mu U \text{ for some } \mu < 0\}$.

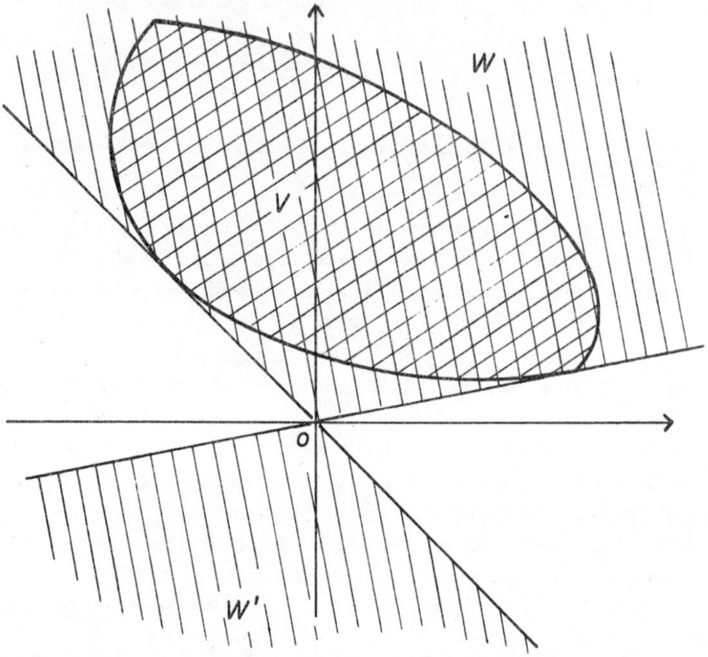

Diagram 9.

Since $x \in W \Leftrightarrow -x \in W'$ and at least one of the sets is nonempty it is clear that both are. Since W is the union of the open sets μU for all $\mu > 0$, W is open; similarly W' is open. Also $W \cap W' = \emptyset$, for if $x \in W \cap W'$ then there exist $\alpha, \beta > 0$ and $u, v \in V$ with $x = \alpha u = -\beta v$. But then the convexity of V and the formula

$$0 = \frac{\alpha}{\alpha + \beta} u + \frac{\beta}{\alpha + \beta} v$$

give the contradiction $0 \in V$.

We thus have disjoint nonempty open sets W, W' covering $X \quad \{0\}$. Since the plane with one point removed is connected, this is absurd; hence there must be a point satisfying the conditions of the lemma.

The student who dislikes the connectedness argument above may like to know that there are more algebraic proofs of the next theorem which appeal to the completeness of $\boldsymbol{R}$ instead of to connectedness; we sketch one such approach in Problem 8P.

8.8 Hyperplane Theorem Let C be a nonempty open convex set in a real normed space X and let M_0 be a linear subspace not meeting C. Then there is a hyperplane M, which contains M_0 and does not meet C. Moreover any such hyperplane is necessarily closed.

Proof First we show that if M is any linear subspace which does not meet C and is *not* a hyperplane, then there must be a subspace strictly larger than M which also does not meet C. To show this, note by Proposition 8.6 the quotient space X/M must have dimension at least two, so we may choose linearly independent $u + M$, $v + M$ in X/M. It is then easy to see that u, v span a two-dimensional subspace N of X such that $N \cap M = \{0\}$. Now the set $C - M$ is open since C is open, convex since C and M are convex, and does not contain 0 since $C \cap M = \emptyset$. By the Lemma there exists a one-dimensional subspace $L_x \subset N$ such that $L_x \cap (C - M) = \emptyset$; equivalently $(L_x + M) \cap C = \emptyset$, and since $L_x \cap M \subset N \cap M = \{0\}$, the subspace $L_x + M$ must be strictly larger than M and so has the required property.

Now form the class $\mathfrak{M}$ of all linear subspaces of X which contains M_0 and do not meet C, and partially order $\mathfrak{M}$ by letting $K \geqslant L$ mean $K \supset L$. If $\{L_i\}$ is any chain in $\mathfrak{M}$ it is clear that $L = \bigcup_i L_i$ is also a linear subspace of X, containing M_0 and not meeting C, and thus L is an upper bound for $\{L_i\}$ in $\mathfrak{M}$. Hence $\mathfrak{M}$ meets the conditions of Zorn's Lemma (see Appendix, page 312), and therefore has a maximal member M_1 with respect to the ordering $\geqslant$. If M_1 were not a hyperplane the first part of the proof would show M_1 to be contained in some strictly larger member of $\mathfrak{M}$, contradicting maximality. Thus M_1 is a hyperplane which contains M_0 and does not meet C, and the first assertion of the theorem is proved. To show that any such M_1 is closed note that the closed subspace M_1^- does not meet C because C is open, and so M_1^- cannot be all of X. Hence $M_1 \subset M_1^- \neq X$, so $M_1 = M_1^-$ by the definition of a hyperplane.

When X is *finite dimensional* the Zorn's lemma part of the proof can be replaced by a simple induction argument; the same thing, though less simply, can be done when X is *separable* (see Problem 8I).

The next theorem is the one that is usually called the **Hahn–Banach Theorem.**

8.9 Extension Theorem Let f_0 be a bounded linear functional defined on a linear subspace M of a real or complex normed space X. Then f_0 has an extension to a bounded linear functional f on X such that $\|f\| = \|f_0\|$. (Note that $\|f_0\|$ means $\sup\{|f_0(x)|: x \in M, \|x\| \leqslant 1\}$.)

Proof (Real case) We give a geometric proof using the product space $X \times \mathbf{R}$ which is a normed space under either of the norms defined in 4.23 – it does not matter which. The reader should try to visualize what is going on in the case when X is 2-dimensional and so can be thought of as the xy plane in $X \times \mathbf{R} = \mathbf{R}^3$ (Diagram 10).

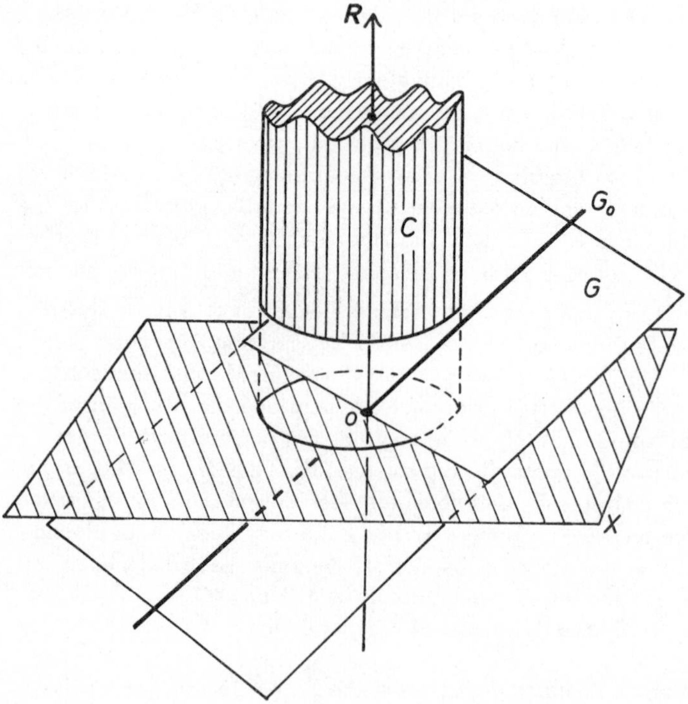

Diagram 10.

The graph G_0 of f_0 – that is the set $\{(x, f_0(x)) : x \in M\}$ – is easily seen to be a linear subspace of $X \times \boldsymbol{R}$. Let $a = \|f_0\|$. Then, for $(x, t) = (x, f_0(x))$ in G_0 it is impossible simultaneously to have $\|x\| < 1$ and $t > a$; in other words the set $C = \{(x, t) \in X \times R : \|x\| < 1, t > a\}$, which is clearly open and convex, misses G_0. By the last theorem there is a hyperplane G of $X \times \boldsymbol{R}$ such that $G \supset G_0$ and C misses G. If we grant for the moment that G is the graph $\{(x, f(x)) : x \in X\}$ of some linear functional f on X then the same argument as above shows that $\|x\| < 1 \Rightarrow f(x) \leqslant a$. Replacing x by $-x$ gives $-f(x) = f(-x) \leqslant a$, and so $|f(x)| \leqslant a$. Hence $\|f\| = \sup\{|f(x)| : \|x\| < 1\} \leqslant a = \|f_0\|$ (using Problem 7B). On the other hand since f is an extension of f_0 it is obvious that $\|f\| \geqslant \|f_0\|$, and so $\|f\| = \|f_0\|$.

To show that G is indeed the graph of an f, we note that $(0, 1) \notin G$, for otherwise $(0, a+1) = (a+1)\cdot(0, 1)$ would be a vector common to G and C, contrary to construction. The definition of hyperplane now implies that for each vector in $X \times \boldsymbol{R}$, in particular for each $(y, 0)$, $y \in X$, there is a unique λ, which we denote by $f(y)$, such that $(y, 0) + \lambda(0, 1) \in G$, in other words $(y, \lambda) \in G$. Then G is the graph of f, and it is clear that f is linear since G is a linear subspace.

Proof (Complex case) If we restrict scalar multiplication to be by real scalars only, then the complex normed space X becomes a real normed space X_r, and the subspace M becomes a linear subspace M_r of X_r. Of course $X = X_r$, $M = M_r$ as sets, but the linear structure is significantly different. If f is a bounded complex-linear functional on X, its real part $(\operatorname{Re} f)(x) = \operatorname{Re}(f(x))$ is clearly a bounded real-linear functional on X_r. More surprisingly, if f_r is a real-linear functional on X_r, the complex function f defined by

$$f(x) = f_r(x) + i f_r(-ix)$$

can easily be verified to be a complex-linear functional on X having f_r for its real part. (The reason for this rather improbable formula can be seen if one inspects the identity:

$$z = \operatorname{Re} z + i \operatorname{Re}(-iz) \qquad (z \in \mathbf{C}).)$$

We next show that the norm of f as an element of X^* equals that of f_r as an element of X_r^*. Clearly $|f_r(x)| \leqslant |f(x)|$, so $\|f_r\| \leqslant \|f\|$.

On the other hand, given $x \in X$, choose α such that $|\alpha| = 1$, $\alpha f(x) = |f(x)|$. Then

$$\begin{aligned} |f(x)| &= f(\alpha x) = f_r(\alpha x) \qquad \text{(since } f(\alpha x) \text{ is real)} \\ &\leqslant \|f_r\| \, \|\alpha x\| = \|f_r\| \, \|x\|, \end{aligned}$$

so that $\|f\| \leqslant \|f_r\|$ and hence $\|f\| = \|f_r\|$.

Applying the real case of the theorem we obtain an extension of Re f_0 to a real-linear f_r on X_r with the same norm, and f_r is the real part of a complex-linear f on X. It is simple to verify that f is an extension of the original complex f_0, and $\|f\| = \|f_r\| = \|\text{Re } f_0\| = \|f_0\|$, completing the proof.

It is interesting that the complex version of the theorem, due to Bohnenblust and Sobczyk, was discovered as much as eleven years after Hahn proved the real version in 1927.

A consequence of the Extension Theorem, important for the algebraic and topological aspects of functional analysis, is that X^* has enough elements to distinguish points of X (Corollary 8.11 below) and to determine the norm in X (Corollary 8.12).

8.10 Theorem Let X be a real or complex normed space. Then for each nonzero $x_0 \in X$ there exists $f \in X^*$ with $\|f\| = 1$, $f(x_0) = \|x_0\|$.

Proof Define f_0 on the one-dimensional subspace $M = \text{lin}\{x_0\}$ by $f_0(\lambda x_0) = \lambda \|x_0\|$. Then $|f_0(\lambda x_0)| = |\lambda| \, \|x_0\| = \|\lambda x_0\|$, so $\|f_0\| = 1$. The Extension Theorem now guarantees that there is an f in X^* extending f_0 and having the required properties.

8.11 Corollary Given distinct $u, v \in X$ there exists $f \in X^*$ with $f(u) \neq f(v)$.

Proof By the Theorem there exists $f \in X^*$ with $f(u - v) = \|u - v\| \neq 0$.

8.12 Corollary For any $x \in X$, $\|x\| = \sup\{|f(x)| : f \in X^*, \|f\| \leqslant 1\}$.

Proof Trivially $\geqslant$ holds, and the Theorem shows that $\leqslant$ holds. Note the symmetry between this and the formula

$$\|f\| = \sup\{|f(x)| : x \in X, \|x\| \leqslant 1\}.$$

8.13 Theorem Let M be a closed linear subspace of a normed space X and x_0 a point of $X \sim M$. Then there exists $f \in X^*$ such that $f = 0$ on M but $f(x_0) \neq 0$.

Proof By hypothesis the open convex set $V = \{x : \|x - x_0\| < r\}$ misses M for some $r > 0$. By Theorem 8.8 there is a closed hyperplane $M_1 \supset M$ such that $M_1 \cap V = \emptyset$, and by Proposition 8.6, M_1 is the null space of the required f.

The geometrical consequences of the Hyperplane Theorem are various **separation theorems.** There are several slightly different versions of these, the basic one being the

8.14 Theorem Let A, B be disjoint nonempty convex sets in a real normed space X and let B be open. Then there exist $f \in X^*$, $\alpha \in R$ such that

$$f(x) \geqslant \alpha \quad (x \in A), \quad f(x) < \alpha \quad (x \in B).$$

Proof The set $C = A - B$ is open, convex and does not contain 0, so there exists a closed hyperplane M missing C by Theorem 8.8. By Proposition 8.6, $M = \ker f$ for some $f \in X^*$, and we may suppose, multiplying f by a suitable scalar, that $f(u) = 1$ for some $u \in C$. It follows that $f > 0$ on C, for if $f(v) = -\alpha$, $\alpha > 0$, for some $v \in C$ then by convexity the point

$$w = \frac{\alpha u + v}{\alpha + 1}$$

would be in C, whereas $f(w) = 0$, $w \in \ker f = M$, a contradiction. Since $C = A - B$, this implies $f(a - b) > 0$ $(a \in A,\ b \in B)$ so that

$$f(a) > f(b) \qquad (a \in A,\ b \in B)$$

Letting $\alpha = \sup\{f(b) : b \in B\}$, holding a fixed and varying b, we obtain

$$f(a) \geqslant \alpha \qquad (a \in A)$$

while clearly

$$f(b) \leqslant \alpha \qquad (b \in B). \qquad (*)$$

Since B is open, given $b \in B$ there exists $\epsilon > 0$ such that $b + \epsilon u \in B$, (where u is the point with $f(u) = 1$ chosen above). Thus $f(b) + \epsilon = f(b + \epsilon u) \leqslant \alpha$, strengthening (*) to

$$f(b) < \alpha \qquad (b \in B)$$

as required. The proof is complete.

8.15 Corollary Let A be a nonempty closed convex set in a real normed space X, and x_0 a point not in A. Then there is $f \in X^*$ such that $f(x\ ) < \inf\{f(a) : a \in A\}$.

Proof Since A is closed there is an open ball $V = \{x : \|x - x_0\| < r\}$ missing A. Apply the Theorem to A and V.

Of course there is an equivalent statement of these results in which $\leqslant$, $>$ and sup replace $\geqslant$, $<$ and inf, and vice versa.

† *An application to complex analysis*

As an example of the Hahn–Banach theorem going to work on a nontrivial problem we shall show how it can be used to prove a famous result due to Runge [*Acta Math*, **6** (1885)] concerning the approximation of analytic functions by polynomials. If f is analytic in an open set containing the closed unit disc, Δ, then the partial sums of the Taylor series for f about 0 form a sequence of polynomials which converge uniformly to f on Δ. This result is so useful that it is natural to try to approximate analytic functions by polynomials on more general compact sets. This cannot always be done (e.g. $f(z) = \frac{1}{z}$ on $\frac{1}{2} \leqslant |z| \leqslant 1$), but Runge's theorem gives a sufficient topological condition on the compact set for polynomial approximation to occur. In fact the condition is also necessary; the proof, which depends on the maximum modulus principle, is left to the reader as an easy exercise in complex analysis.

8.16 Theorem Let K be a compact subset of $\mathbf{C}$ such that $\mathbf{C} \sim K$ is connected. Let D be an open set containing K, and let $f: D \to \mathbf{C}$ be analytic on D. Then given $\epsilon > 0$ there exists a polynomial function p such that $|p(z) - f(z)| < \epsilon$ $(z \in K)$.

To see the kind of situation, look at three possible K's

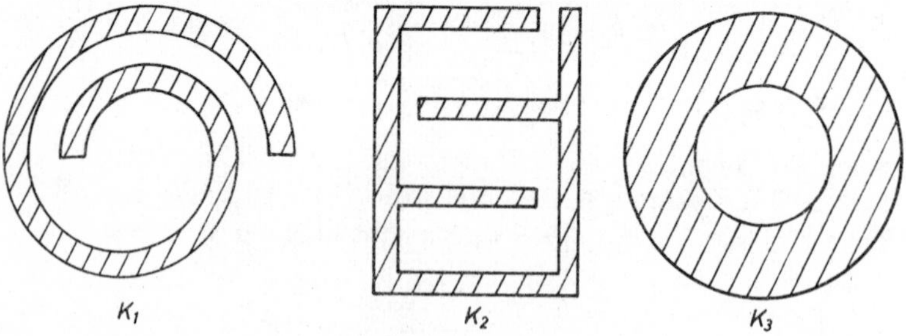

Diagram 11.

sketched in the diagram. There is no need for K to be connected, but the essential point is that K must enclose no 'holes', so the first two sets are all right, but the last is not.

We make use of the following result, which tells us that f can be approximated by rational functions:

8.17 Lemma With the above notation, for each $\epsilon > 0$ there exist $z_1, \ldots, z_q \in D \sim K$ and $\alpha_k \in \mathbf{C}$ such that

$$\left| f(z) - \sum_{k=1}^{q} \frac{\alpha_k}{z_k - z} \right| < \epsilon \qquad (z \in K).$$

Proof This is an easy application of Cauchy's integral formula. Since K is compact, D open and $K \subset D$ we can construct a contour Γ lying in $D \sim K$ consisting of a finite union of oriented simple closed contours such that

$$f(\alpha) = \frac{1}{2\pi i} \int_\Gamma \frac{f(z)\,dz}{z - \alpha} \tag{1}$$

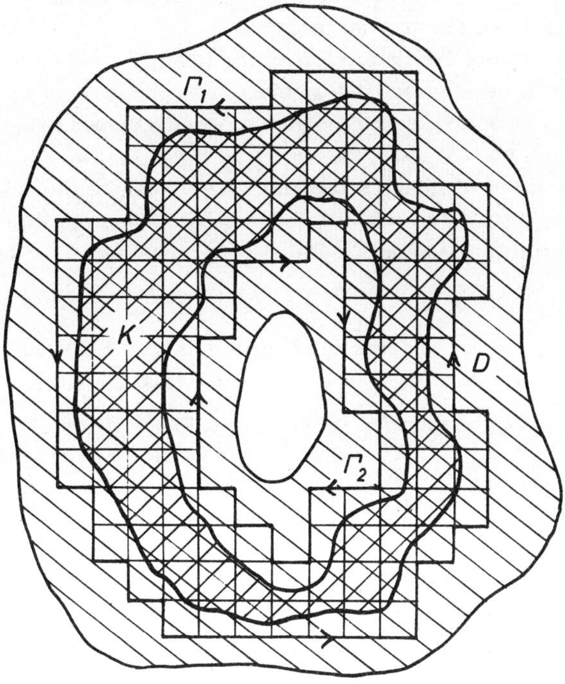

$\Gamma = \Gamma_1 + \Gamma_2$

Diagram 12.

for all α in an open set containing K (Diagram 12). This looks as if it begs a lot of topological questions, but in fact it uses nothing more than a covering of K by a grid of sufficiently small squares, and Cauchy's integral formula for a square. Therefore we leave it for the interested student to justify as an exercise (or see Rudin **11**, p. 254).

The function $(z, \alpha) \mapsto \dfrac{1}{z - \alpha}$ is continuous and therefore uniformly continuous on the compact set $\Gamma \times K$, and it follows (cf. Problem 6I) that we can approximate the integral (1), *uniformly* for $\alpha \in K$, by Riemann sums of the form

$$\frac{1}{2\pi i}\sum_{k=1}^{q}\frac{f(z_k)}{z_k - \alpha}(z_k - z_{k-1})$$

where $z_1, \ldots, z_k \in \Gamma$, and this gives a representation of the required form, proving the lemma.

The proof of the theorem uses the **identity theorem** for analytic functions, namely that if G is open and connected in $\mathbf{C}$, $f: G \to \mathbf{C}$ is analytic, and $f = 0$ on a nonempty open subset of G, then $f = 0$ on G.

8.18 Proof of Runge's theorem Let X denote the complex Banach space $\mathcal{C}(K)$, and let Y denote the subspace $\{p|_K : p$ is a polynomial function on $\mathbf{C}\}$. We wish to prove that

$$\bar{Y} = X.$$

If $\bar{Y} \neq X$ then by Theorem 8.13 there exists $\phi \in X^*$ such that $\phi(f_0) \neq 0$ for some $f_0 \in X$, but $\phi(y) = 0$ $(y \in \bar{Y})$. Given $z \in \mathbf{C} \sim K$, let u_z be the function $\alpha \mapsto \dfrac{1}{\alpha - z}$ $(\alpha \in K)$. Since $z \notin K$ we have $u_z \in \mathcal{C}(K) = X$. Define

$$h(z) = \phi(u_z) \qquad (z \in \mathbf{C} \sim K).$$

We prove that h is *analytic* on $\mathbf{C} \sim K$.

In fact given $z \in \mathbf{C} \sim K$, choose r with $0 < r < \operatorname{dist}(z, K)$. Then

$$|w| \leqslant r \Rightarrow \frac{w}{\alpha - z} \leqslant \frac{r}{\operatorname{dist}(z, K)} < 1 \qquad (\alpha \in K)$$

so $|w| \leqslant r$ and $\alpha \in K$ imply

$$u_{z+w}(\alpha) = \frac{1}{\alpha - z - w} = \frac{1}{\alpha - z}\left(1 - \frac{w}{\alpha - z}\right)^{-1}$$
$$= \frac{1}{\alpha - z} + \frac{w}{(\alpha - z)^2} + \dots$$

the convergence being uniform for $\alpha \in K$. Therefore in the normed space X,

$$u_{z+w} = u_z + wu_z^2 + w^2u_z^3 + \dots \qquad (|w| \leqslant r)$$

the series converging in norm. Since ϕ is linear and continuous, this gives

$$h(z+w) = \phi(u_{z+w}) = \phi(u_z) + \phi(u_z^2)w + \phi(u_z^3)w^2 + \dots \ (|w| \leqslant r).$$

This expresses $h(z + w)$ as a power series in w convergent for $|w| \leqslant r$; since z was arbitrary, h is analytic on $\mathbf{C} \sim K$. We next show $u_z \in \bar{Y}$ for large z. In fact, K is bounded so choose M such that $|\alpha| < M$ $(\alpha \in K)$. Then for $|z| \geqslant 2M$,

$$\frac{1}{\alpha - z} = -\frac{1}{z}\left(1 - \frac{\alpha}{z}\right)^{-1} = \frac{-1}{z} - \frac{\alpha}{z^2} - \dots$$

the series converging uniformly for $\alpha \in K$, which shows $u_z = \lim p_n$ in the norm of X, where $p_n \in Y$ is defined by $p_n(\alpha) = -\sum_0^n \alpha^n/z^{n+1}$.

It follows that $h(z) = \phi(u_z) = 0$ $(|z| \geqslant 2M)$ and therefore by the Identity Theorem and the connectedness of $\mathbf{C} \sim K$,

$$\phi(u_z) = h(z) = 0 \qquad (z \in \mathbf{C} \sim K)$$

Consider the element f_0 of X chosen at the start of the proof. By the Lemma, given $\epsilon > 0$, there exist $z_1, \dots, z_q \in \mathbf{C} \sim K$, $\alpha_1, \dots, \alpha_q \in \mathbf{C}$ such that

$$\|f_0 - \textstyle\sum \alpha_k u_{z_k}\| \leqslant \epsilon$$

But then

$$|\phi(f_0)| = |\phi(f_0 - \textstyle\sum \alpha_k u_{z_k})|, \text{ since } \phi(u_{z_k}) = 0$$
$$\leqslant \|\phi\| \, \|f_0 - \textstyle\sum \alpha_k u_{z_k}\| \leqslant \epsilon \, \|\phi\|.$$

Since ϵ is arbitrary this contradicts the fact that $\phi(f_0) \neq 0$. Hence $\bar{Y} = X$ and the proof is complete.

As this proof shows, the remarkable thing is that the functions $u_z(\alpha) = \dfrac{1}{\alpha - z}$ can be approximated by polynomials. Roughly, we know this to be true for large z, and the construction of the function h together with the Identity Theorem allows us to creep right into all the crevices of K and prove it true for all $z \in \mathbf{C} \sim K$.

Problems

Functionals and their norms

8A Verify that in Example 8.3, $\|f\| = 1$ and $\|h\| = b - a$.

8B Let u be a fixed element of $\mathcal{C}[a, b]$. Clearly

$$\phi(f) = \int_a^b u(t)\, f(t)\, dt \qquad (f \in \mathcal{C}[a, b])$$

defines a bounded linear functional ϕ on $\mathcal{C}[a, b]$. Show that its norm is exactly $\int_a^b |u(t)|\, dt$. [Consider a sequence of continuous functions approximating to $\overline{\operatorname{sgn} u}$, such as $n\bar{u}/(1 + n|u|)$.]

8C Show that the functionals g, k in Example 8.3 are unbounded. [Use the Weierstrass approximation theorem.]

Finite-dimensional spaces

8D Use Theorem 8.4 to show that if M is a linear subspace of a finite-dimensional normed space X and x a point of $X \sim M$, there exists f in X^* such that $f = 0$ on M but $f(x) \neq 0$ (Theorem 8.13).

Continuity of functionals

8E Let f be a nonzero linear functional on a normed space X. Show that f is bounded $\Leftrightarrow$ $\ker f$ is not everywhere dense in X $\Leftrightarrow$ $\ker f$ is nowhere dense in X.

8F Let f be a nonzero bounded linear functional on a normed space X and let $x \in X \sim \ker f$. Show that

$$\|f\| = \frac{|f(x)|}{d(x, \ker f)}.$$

The extension theorems

8G Let M be a linear subspace of a normed space X, and x a point of X with $d(x, M) = \delta > 0$. Show that there exists $f \in X^*$

such that

$$f = 0 \text{ on } M, \ \|f\| = 1, \ f(x) = \delta.$$

[It is interesting to note that $\|f\| \geqslant \delta^{-1}|f(x)|$ for any f which vanishes on M, by the last problem.]

8H A linear functional on a linear space over $\mathbf{Q}$ means (of course) a $\mathbf{Q}$-valued function satisfying $f(\alpha x + \beta y) = \alpha f(x) + \beta f(y)$. Show that there are no nonzero linear functionals on $\mathbf{Q}^2$ that are bounded with respect to the norm $\|(r, s)\| = |r - s\sqrt{2}|$ of Exercise 7G. Thus the completeness of the scalar field is an essential ingredient of the Hahn–Banach theorems.

8I (To Theorem 8.8) A linear space X is of **countably infinite** dimension if it is infinite-dimensional but contains a countable spanning set. Give a proof of the Hyperplane Theorem avoiding Zorn's Lemma

(i) when X has finite or countably infinite dimension;
(ii) when X is separable. [Use Proposition 8.6 (ii) and the extension by continuity theorem, Problem 7L, to deduce (ii) from (i).]

8J Use Problem 8F to deduce Theorem 8.9 (real case) from the Hyperplane Theorem without using the product space $X \times \mathbf{R}$.

The separation theorems

8K Let A and B be disjoint nonempty convex sets in a real normed space X with A closed and B compact. Prove that for some $f \in X^*$,

$$\sup_{x \in A} f(x) < \inf_{x \in B} f(x)$$

[Let U be the open unit ball. For some $r > 0$, $A \cap (B + rU) = \emptyset$.]

8L A **closed half-space** in a real normed space X means a set of the form $\{x \in X\colon f(x) \geqslant \alpha\}$ where $f \in X^*$ and $\alpha \in \mathbf{R}$. Show that a closed convex set in X is the intersection of all the closed half-spaces that contain it.

Miscellaneous problems

8M A functional ϕ on $\mathcal{C}(T)$, where T is a compact space, is called **positive** if $\phi(f) \geqslant 0$ whenever f is a non-negative real-valued member of $\mathcal{C}(T)$. Show that a positive linear functional on $\mathcal{C}(T)$ is bounded, and that any two of the conditions

(a) $\|\phi\| = 1$ (b) $\phi(1) = 1$ (c) ϕ is positive

imply the third. [The case of complex scalars is distinctly trickier – and more important – than the real case.]

8N Let X be the linear space of all scalar sequences $x = (x_1, x_2, \ldots)$ with the usual linear operations and let f_n denote the co-ordinate functional

$$f_n(x) = x_n \qquad (x \in X).$$

Show that there is no norm on X making all the f_n continuous. [If there were, consider the sequence whose nth term is $n\,\|f_n\|$.]

The algebraic extension theorem

A **sublinear functional** on a real linear space X is a real function p on X satisfying

$$p(x + y) \leqslant p(x) + p(y)$$

$$p(\alpha x) = \alpha p(x) \quad \text{whenever } \alpha \geqslant 0.$$

The following theorem is more-or-less equivalent to the Hyperplane Theorem; two proofs are sketched in the next two problems:

Let f_0 be a linear functional on a linear subspace M of X such that

$$f_0(x) \leqslant p(x) \qquad (x \in M). \tag{1}$$

Then f_0 has an extension to a linear functional f_1 on all of X with

$$f_1(x) \leqslant p(x) \tag{2}$$

for all $x \in X$.

8O **Proof 1** Assume (for simplicity – it is not really necessary) the existence of a norm on X such that $p(x) \leqslant \|x\|$ $(x \in X)$. Show that p is norm-continuous [Hint: prove $-p(y - x) \leqslant p(x) - p(y) \leqslant p(x - y)$.] Argue as in Theorem 8.9, letting C be the set

$$\{(x, t) \in X \times \mathbf{R} \colon t > p(x)\}.$$

8P **Proof 2** (independent of the Hyperplane Theorem) Suppose f_0 has an extension f_1 to a subspace Y of X, satisfying (2) for all $x \in Y$.

(i) Given any $e \in X \sim Y$ show that

$$\alpha = \sup_{y \in Y} (f_1(y) - p(y - e)) \leqslant \inf_{x \in Y} (p(x + e) - f_1(x)) = \beta.$$

(ii) Deduce that for any θ between α and β the formula

$$f(y + \lambda e) = f_1(y) + \lambda\theta \qquad (y \in Y, \lambda \in \boldsymbol{R})$$

extends f_1 to a linear functional f on $\text{lin}(e, Y)$ satisfying (2).

(iii) Use Zorn's Lemma to deduce the existence of an extension to all of X.

8Q Let U be an open convex subset of a normed space with $0 \in U$. Show that

$$p(x) = \inf\{\beta: \beta > 0 \text{ and } x \in \beta U\}$$

defines a sublinear functional (the **Minkowski functional** of U).

Use this and 8P to prove the Hyperplane Theorem.

8R Let X and Y be normed spaces and T a congruence of X onto Y. Show that T^*, where $(T^*g)(x) = g(Tx)$, $(g \in Y^*, x \in X)$ is a congruence of Y^* onto X^*.

8S Let f be a real integrable function on a measure space $(S, \mathbf{S}, \mu)$ such that the total mass of μ is 1, and let $F: \boldsymbol{R} \to \boldsymbol{R}$ be continuous and convex. Show that

$$\int F \circ f \cdot d\mu \geqslant F\left(\int f\, d\mu\right).$$

[Use the separation theorem to show that F is the pointwise supremum of all functions of the form $G(t) = at + b$ lying below F.]

8T **Banach limits** It is often useful, especially in the study of Fourier series, to assign a 'generalized limit' to a wider class of sequences of numbers than those that converge in the ordinary sense. Let X be the space ℓ_∞ of all bounded real sequences $x = (x_1, x_2, \ldots)$ with the sup-norm. Let e denote the sequence $(1, 1, 1, \ldots)$ and for x in X let Sx denote the sequence $(x_2, x_3, \ldots)$. Let $M = \{x - Sx: x \in X\}$.

(i) Show that M is a linear subspace and $d(e, M) = 1$.

(ii) Deduce that there exists a functional L in X^* such that $\|L\| = L(e) = 1$ and such that $L(Sx) = L(x)$ for all x. [Use 8G.]

Derive the following properties:

(iii) $L(x) \geqslant 0$ whenever $x_n \geqslant 0$ for all n.

(iv) $\liminf x_n \leqslant L(x) \leqslant \limsup x_n$ for all x in X.

(v) $L(x) = \lim x_n$ whenever the latter exists.

9. Density theorems

Approximation, especially approximation by functions of a fairly simple kind, to more general functions, plays an important role in analysis. Approximation is about dense subsets of a space, or subsets that give rise to dense subsets under appropriate algebraic operations, and in this section we prove some of the helpful properties of such subsets.

9.1 A subset F of a normed space X is **fundamental** if $\overline{\text{lin}}\, F = X$. Thus, to know F is fundamental is to know each of the following equivalent statements:

The set of linear combinations of members of F is dense in X.
Any closed linear subspace containing F equals X.
Any linear subspace whose closure contains F is dense in X.

The third of these may not be quite obvious, but is perhaps the most useful. If M is the subspace in question, then by the definition of $\overline{\text{lin}}\, F$, $M^- \supset F \Rightarrow M^- \supset \overline{\text{lin}}\, F = X \Rightarrow M$ is dense. Proving that a given set is fundamental can involve deep analysis and have important consequences: for instance Weierstrass' Theorem, which can be regarded as saying that $\{1, u, u^2, \ldots\}$, where $u(t) = t$, is fundamental in $\mathcal{C}[a, b]$.

The next lemma is a topological version of Lemma 3.18.

Note It is customary, when T is a linear mapping, to shorten $T(x)$ to Tx wherever this is unlikely to cause confusion. We shall do so henceforward, since it avoids a plethora of brackets in algebraic manipulation.

9.2 Lemma If two bounded linear maps agree on a fundamental set, they are equal. That is, if $S, T \in \mathcal{B}(X, Y)$, $F \subset X$, F is fundamental, and $Sx = Tx$ $(x \in F)$ then $S = T$.

Proof $\ker(S - T)$ is a closed linear subspace of X containing F, and so equals X. Thus $S - T = 0$, $S = T$.

9.3 Theorem Let X and Y be normed spaces, let $T_1, T_2, \ldots$ and $T \in \mathcal{B}(X, Y)$, with $\sup \| T_n \| < \infty$, and suppose $T_n x \to Tx$ for all x in a fundamental subset F of X. Then $T_n x \to Tx$ for all x in X.

Corollary Note that this includes an analogous result about functionals in the case when Y is the scalar field.

Proof Let $M = \{x \in X : T_n x \to Tx\}$. If $T_n x \to Tx$ and $T_n y \to Ty$ it is clear that $T_n(\alpha x + \beta y) = \alpha T_n x + \beta T_n y \to \alpha Tx + \beta Ty = T(\alpha x + \beta y)$ for any α, β, so that M is a linear subspace. Given

$x \in M^-$ and $\epsilon > 0$, choose $y \in M$ with $\|x - y\| < \epsilon/3c$ where $c = 1 + \sup \|T_n\|$, and then choose n_0 so that $\|T_n y - Ty\| < \epsilon/3$ for $n \geqslant n_0$. Then for all $n \geqslant n_0$,

$$\|T_n x - Tx\| = \|T_n(x - y) + (T_n y - Ty) + T(y - x)\|$$

$$\leqslant \|T_n\| \, \|x - y\| + \|T_n y - Ty\| + \|T\| \, \|y - x\|$$

$$< \epsilon/3 + \epsilon/3 + \epsilon/3 = \epsilon,$$

which proves $T_n x \to Tx$, $x \in M$ and therefore M is closed. Thus M is a closed linear subspace containing F, so $M = X$.

Dense and fundamental sets in L_p

9.4 Proposition The set L_0 of finite simple functions vanishing outside a set of finite measure is dense in $L_p = L_p(S, \mathbf{S}, \mu)$ for any measure space $(S, \mathbf{S}, \mu)$ and for $1 \leqslant p < \infty$. If μ is finite then this holds also for $p = \infty$.

Proof Given $f \in L_p$, Proposition 2.19 and Problem 5L guarantee the existence of a sequence $\{f_n\}$ in L_0 such that $f_n \to f$ pointwise, and $|f_n| \leqslant |f|$ for all n. Thus $|f_n - f|^p \leqslant (|f_n| + |f|)^p \leqslant 2^p |f|^p$, and since the last function is integrable and $|f_n - f|^p \to 0$ we obtain $\int |f_n - f|^p \to 0$ by the Dominated Convergence Theorem. Thus f is the limit in the L_p norm of a sequence of functions in L_0, proving the result.

If μ is finite and $f \in L_\infty$ we can assume, by choosing an equivalent function, that f is bounded. By the second part of Proposition 2.19, there is a sequence $\{f_n\}$ in L_0 converging uniformly to f, i.e. $\|f_n - f\|_\infty \to 0$, as required.

9.5 Corollary The set of characteristic functions of measurable sets of finite measure is fundamental in L_p, $1 \leqslant p < \infty$, and in L_∞ if μ is finite.

Proof Immediate from the definition of L_0.

This result is useful but not very deep. We turn our attention next to the case of Lebesgue measure on $\boldsymbol{R}$, and prove some facts which lie rather deeper. For the rest of the section the measure will be Lebesgue measure λ, and the σ-algebra $\mathbf{L}$ will be the Lebesgue-measurable subsets of some fixed interval (or other measurable set) in $\boldsymbol{R}$, in accord with the notation of 2.7. We recall from 2.22 that the measure and topology in $\boldsymbol{R}$ are tied together by the:

Regularity Property Given $E \in \mathbf{L}$ and $\epsilon > 0$ there is an open set $U \supset E$ with $\lambda(U \sim E) < \epsilon$.

From this we deduce a slightly weaker property. The **symmetric difference** of two sets A, B is the set $A \Delta B = (A \sim B) \cup (B \sim A)$.

9.6 Lemma Given $E \in \mathbf{L}$ with $\lambda(E) < \infty$, and $\epsilon > 0$, there is a set J which is a finite union of bounded intervals such that $\lambda(E \Delta J) < \epsilon$.

Proof Choose open $U \supset E$ with $\lambda(U \sim E) < \epsilon/2$. Then also $\lambda(U) < \infty$. Every open set in $\mathbf{R}$ is a countable union of intervals so there is an increasing sequence of sets J_n, each being a finite union of intervals, with $\bigcup J_n = U$. Since $\lambda(U)$ is finite this implies $\lambda(U) = \sup \lambda(J_n)$ by 2.6. Choose n so that $\lambda(J_n) > \lambda(U) - \epsilon/2$, so $\lambda(U \sim J_n) < \epsilon/2$. Then $E \Delta J_n = (E \sim J_n) \cup (J_n \sim E) \subset (U \sim J_n) \cup (U \sim E)$ which shows that $\lambda(E \Delta J_n) < \epsilon/2 + \epsilon/2 = \epsilon$, as required.

9.7 A *step function* on an interval $I \subset \mathbf{R}$ is a scalar function ϕ on I for which there is a partition of I into finitely many subintervals on each of which ϕ is constant. A step function is clearly a particularly simple kind of simple function.

We denote by $St(I)$ the set of (λ-equivalence classes of) step-functions which vanish outside some bounded interval (which of course means all step-functions, if I is bounded). The functions that do not fulfil this requirement are clearly not integrable, which is why we do not consider them.

9.8 Lemma $St(I)$ is a linear subspace of $L_0(I)$, and consists precisely of the linear combinations of characteristic functions of bounded subintervals of I.

Proof Trivial.

9.9 Theorem Let I be an interval in $\mathbf{R}$ (possible unbounded, possibly the whole of $\mathbf{R}$). The following sets are dense linear subspaces of $L_p(I)$ for $1 \leqslant p < \infty$:

(i) $St(I)$;
(ii) $\mathcal{C}(I)$ if I is closed and bounded;
(iii) The class $\mathcal{C}_{00}(I)$ of functions which are continuous on I and vanish outside some closed, bounded subinterval of I, in the general case.

Proof It is clear that in (i) and (ii) the sets referred to are subsets, and in fact linear subspaces, of $L_p(I)$. A continuous function f on I that vanishes outside an interval $[a, b]$ must be bounded, and therefore $\|f\|_p{}^p \leqslant \int_a^b (\|f\|_\infty)^p < \infty$, which shows that $\mathcal{C}_{00}(I)$ is also contained in – and clearly a linear subspace of – $L_p(I)$ for any I.

(i) By Corollary 9.5, $\{\chi_E: E$ a measurable subset of I, $\lambda(E) < \infty\}$ is fundamental in $L_p(I)$, so it is sufficient to show that $\chi_E \in St(I)^-$ for every such χ_E. By Lemma 9.6, given $\epsilon > 0$ we can choose a finite union J of bounded intervals such that $\lambda(E \Delta J) < \epsilon^p$. It is clear that $\chi_{E \Delta J} = |\chi_E - \chi_J|$, so that

$$\|\chi_E - \chi_J\|_p = \left(\int (\chi_{E\Delta J})^p\right)^{1/p} = \left(\int \chi_{E\Delta J}\right)^{1/p}$$
$$= (\lambda(E \Delta J))^{1/p} < \epsilon.$$

Since χ_J is clearly in $St(I)$ it follows that $\chi_E \in St(I)$, proving that $St(I)$ is dense.

(ii) This is a particular case of (iii).

(iii) It follows from (i) that $\{\chi_J: J$ a bounded subinterval of $I\}$ is fundamental in $L_p(I)$ for any I. Thus it is sufficient to show $\mathcal{C}_{00}(I)^-$ contains every such χ_J. Let the endpoints of J be a, b.

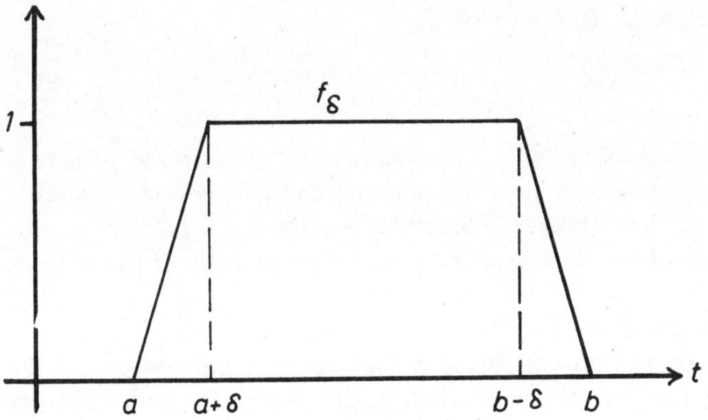

Diagram 13.

Then, for suitable small $\delta > 0$ the function f_δ shown in Diagram 13 (the reader may verify the formula:

$$f_\delta(t) = \begin{cases} (t-a)/\delta & a \leqslant t \leqslant a+\delta \\ 1 & a+\delta \leqslant t \leqslant b-\delta \\ (b-t)/\delta & b-\delta \leqslant t \leqslant b \\ 0 \text{ elsewhere in } I) \end{cases}$$

is in $\mathcal{C}_{00}(I)$, and it is routine to verify that $\|f_\delta - \chi_J\|_p \to 0$ as $\delta \to 0$, which shows that $\chi_J \in \mathcal{C}_{00}(I)^-$.

9.10 Corollary The set of all characteristic functions χ_J, where J is a subinterval of I, is fundamental in $L_p(I)$ for $1 \leqslant p < \infty$.

We note for later use the fact that since, in the case when I is a closed bounded interval $[\alpha, \beta]$, all the f_δ lie in the subspace $\{f: f(\alpha) = f(\beta) = 0\}$ of $\mathcal{C}(I)$, we have in fact shown that this subspace of $\mathcal{C}(I)$ is dense in $L_p(I)$.

The reader should perhaps pause here to observe that the task of approximating a given L_p-function f by continuous functions, explicitly, involves quite a chain of successive approximations. First f is approximated by a simple function ϕ; ϕ is expressed as a combination of characteristic functions χ_{E_i}; the E_i are smoothed out to finite unions of intervals; this gives an approximation to ϕ by a step-function ψ; finally the 'corners' of ψ are smoothed off to give a continuous function g. In fact, there are more efficient ways of doing this smoothing process: for example given $f \in L_p(\boldsymbol{R})$ the function

$$f^\delta(t) = \frac{1}{2\delta} \int_{t-\delta}^{t+\delta} f(u)\, du$$

is continuous and $\|f - f^\delta\|_p \to 0$ as $\delta \to 0$ (Problem 9C) but to prove this one needs to know already that continuous functions are dense. Thus Theorem 9.9 can be regarded as a prototype, with the aid of which one can derive slicker methods of approximation.

† *Application to two classical theorems*

An early stimulus to the study of Lebesgue integration and the L_p-spaces was the wish to put a firm foundation under some of the techniques of applied mathematics, such as the use of *Fourier transform*, *Laplace transform* and *Fourier series*. Here we show the density theorems at work on two important results, the **Riemann–Lebesgue Lemma** and **Lerch's Uniqueness Theorem.**

9.11 Theorem (Riemann–Lebesgue Lemma) Let $f \in L_1(\boldsymbol{R})$. Then

$$\lim_{s \to +\infty} \int_{\boldsymbol{R}} f(t)\, e^{ist}\, dt = 0 \qquad \text{(here } s \text{ is real)}$$

Remark The **Fourier transform** of a function $f \in L_1(\boldsymbol{R})$ is the function $\hat{f}(s) = \int_{\boldsymbol{R}} f(t)\, e^{ist}\, dt \,(s \in \boldsymbol{R})$, so that this says that the Fourier transform of an L_1 function tends to 0 at $+\infty$ (and clearly by a similar argument, at $-\infty$).

Proof of Theorem If the result were false then there would exist $f_0 \in L_1(\boldsymbol{R})$, $\epsilon > 0$ and a sequence $s_n \to +\infty$ such that

$$\left| \int_{\boldsymbol{R}} f_0(t)\, e^{is_n t}\, dt \right| \geqslant \epsilon \text{ for all } n$$

Define

$$\phi_n(f) = \int_{\boldsymbol{R}} f(t)\, e^{is_n t}\, dt \qquad (f \in L_1(\boldsymbol{R}),\ n = 1, 2 \ldots)$$

It is clear that each ϕ_n is a linear functional on L_1 with

$$|\phi_n(f)| \leqslant \int_{\boldsymbol{R}} |f(t)\, e^{is_n t}|\, dt = \int |f| = \|f\|_1$$

so that the $\{\phi_n\}$ are a uniformly bounded sequence in $L_1{}^*$. A simple calculation shows that for any bounded interval J with endpoints a, b,

$$|\phi_n(\chi_J)| = |(e^{is_n b} - e^{is_n a})/is_n| \leqslant 2/s_n \to 0.$$

Since by Corollary 9.10 (applied in the case $I = \boldsymbol{R}$) the χ_J's are a fundamental set in L_1, the corollary to Theorem 9.3 gives $\phi_n(f) \to 0$ ($f \in L_1$) which contradicts the assumption that $|\phi_n(f_0)| \geqslant \epsilon$ for all n. The result follows.

9.12 The set Lap$[0, \infty)$ of **functions which have a Laplace transform** consists, by definition, of those measurable functions f on $[0, \infty)$ such that for some real s the function $e^{-st}f(t)$ is in $L_1[0, \infty)$. Since $e^{-s't} \leqslant e^{-st}$ whenever $s' \geqslant s$, $t \geqslant 0$, it then follows that $e^{-s't} f(t)$ is in $L_1\,[0, \infty)$ for each $s' \geqslant s$. Thus the set of s, for which the **Laplace transform**

$$\tilde{f}(s) = \int_0^\infty e^{-st} f(t)\, dt$$

of such a function is defined, forms an interval – let us call it I_f – in $\boldsymbol{R}$, extending to $+\infty$ on the right. Examples: the functions $t\,e^{37t}$, $\cos t\, e^{-37t}$, e^{-t^2} are in Lap$[0, \infty)$, with $I_f = (37, \infty)$, $(-37, \infty)$ and $(-\infty, \infty)$ respectively; the functions $(t - 99)^{-1}$ and e^{t^2} are not, the first having a nasty 'infinity' at $t = 99$, while the second grows too rapidly as $t \to \infty$.

The importance of the Laplace transform in applied mathematics lies in the fact that difficult equations involving an unknown function f can often be translated into quite simple equations involving the transformed function $\tilde{f}$. In order to recover f from $\tilde{f}$ one needs to know that $\tilde{f}$ more-or-less uniquely determines f. This is the content of

9.13 Lerch's Theorem If $f, g \in \text{Lap}[0, \infty)$ and if there exists s_0 such that $\tilde{f}(s) = \tilde{g}(s)$ for $s \geqslant s_0$, then $f = g$ a.e. on $[0, \infty)$.

We shall need two facts for the proof. The first is that the one-to-one correspondence between functions f on $[0, \infty)$ and functions g on $(0, 1]$ set up by $f \leftrightarrow g$ where $f(t) = g(e^{-t})$, or equivalently $g(x) = f(-\log x)$, maps measurable functions to measurable functions, and null functions to null functions, and that a sequence $\{f_n\}$ converges a.e. on $[0, \infty)$ iff the corresponding $\{g_n\}$ converges a.e. on $(0, 1]$. This is an easy piece of measure theory and will be left as an exercise. The second fact is of interest in its own right:

9.14 Lemma Let g be a bounded measurable function on $[0, 1]$. Then there is a sequence of polynomials $\{p_n\}$ uniformly bounded on $[0, 1]$, such that $p_n \to g$ pointwise a.e. on $[0, 1]$.

Proof For simplicity of notation we take the scalars to be real, the extension to the complex case being obvious. The hypothesis implies that $g \in L_1\,[0, 1]$, so that by Theorem 9.9 there is a sequence of *continuous* functions g_n converging to g in the L_1 norm, and by Corollary 5.11 we may assume $g_n \to g$ pointwise a.e. on $[0, 1]$. The g_n need not be uniformly bounded; however if c is a bound for g, so that $-c1 \leqslant g \leqslant c1$, it is clear that the 'chopped down' functions

$$h_n = (g_n \wedge c1) \vee (-c1)$$

are uniformly bounded, continuous, and converge to g a.e. By Weierstrass' Theorem we may choose, for each n, a polynomial p_n such that $|p_n(t) - h_n(t)| \leqslant 1/n$ $(t \in [0, 1])$. The sequence $\{p_n\}$ clearly has the required property.

Proof of theorem 9.13 In the notation used in the statement of the theorem, $\tilde{f}(s_0)$ and $\tilde{g}(s_0)$ are assumed to exist, so that $e^{-s_0 t} f(t)$ and $e^{-s_0 t}\, g(t)$ are integrable on $[0, \infty)$ and hence so also is the function

$$h(t) = e^{-s_0 t}\,(f(t) - g(t)). \tag{1}$$

By assumption, $\tilde{f}(s) = \tilde{g}(s)$ for $s = s_0, s_0 + 1, s_0 + 2, \ldots$ (rather surprisingly, equality at this discrete set of points is sufficient for the purposes of the proof) so that

$$\int_0^\infty h(t)\, e^{-kt}\, dt = (\tilde{f} - \tilde{g})(s_0 + k) = 0 \qquad (k = 0, 1, 2, \ldots) \tag{2}$$

It follows from (2) by taking linear combinations that

$$\int_0^\infty h(t)\, p(e^{-t})\, dt \;=\; 0 \qquad (3)$$

for all polynomials p, since $p(e^{-t})$ is of the form $c_0 + c_1 e^{-t} + \dots + c_n e^{-nt}$. By the remarks on measurable functions following the statement of the theorem, the function $x \mapsto \overline{\operatorname{sgn}}\, h(-\log x)$ is measurable on $(0, 1]$. By the Lemma we may choose a sequence of polynomials p_n, uniformly bounded on $[0, 1]$ by some constant c, such that $p_n(x) \to \overline{\operatorname{sgn}}\, h(-\log x)$ for almost all $x \in (0, 1]$. Again by the above remarks, it follows on setting $x = e^{-t}$ that $p_n(e^{-t}) \to \overline{\operatorname{sgn}}\, h(t)$ a.e. on $[0, \infty)$. The sequence $\{h(t)\, p_n(e^{-t})\}$ is dominated by the integrable function $c|h|$, so the Dominated Convergence Theorem gives

$$\int_0^\infty |h| \;=\; \int_0^\infty h(t)\, \overline{\operatorname{sgn}}\, h(t)\, dt \;=\; \lim_n \int_0^\infty h(t)\, p_n(e^{-t})\, dt \;=\; 0$$

using (3). Hence $h(t) = 0$ a.e. on $[0, \infty)$, so $f(t) - g(t) = e^{s_0 t}\, h(t) = 0$ a.e. on $[0, \infty)$ and the proof is complete.

Separable normed spaces

The first result of this section will be used once or twice later on. The remaining ones, though we do not use them in the text, are worthwhile general knowledge for anyone using normed spaces, since normed spaces which are separable are in many ways more convenient to work with than nonseparable ones.

9.15 Proposition A normed space X is separable $\Leftrightarrow$ it has a countable fundamental set.

Proof Since a dense subset of X is certainly fundamental the necessity of the condition is clear. Conversely, let $\{x_1, x_2, \dots\}$ be a countable fundamental set in X. We shall assume real scalars, leaving the student to make the modifications, which are simple, for the complex case. Let C be the set of all linear combinations of the x_n with *rational* coefficients. That C is countable should be clear to the reader, but we give a typical proof of this fact for completeness. Each $y \in C$ can be represented, by padding out with zero coefficients, as a finite sum $y = \sum_1^m (p_j/q_j)x_j$ where p_j, q_j are integers and $q_j \geqslant 1$. Let us say the 'height' of such a representation of y is $\max\{|p_1|, \dots |p_m|, q_1, \dots, q_m, m\}$. Clearly the number of y for which there exists a representation of height at most N is finite

($\leqslant (2N+1)^{2N+1}$ in fact); thus C is a countable union of finite sets and so is countable.

Now let $x \in X$, $\epsilon > 0$ be given. Since $\{x_n\}$ is fundamental there is a number m and real coefficients $\lambda_1, \ldots, \lambda_m$ with $\|x - \sum_1^m \lambda_j x_j\| < \epsilon$. Since rationals are dense in $\boldsymbol{R}$ we can choose rationals $r_1, \ldots, r_m$ such that $|r_j - \lambda_j| \, \|x_j\| < \epsilon/2m$ for each j. The vector $z = \sum_1^m r_j x_j$ is in C, and the identity $x - z = (x - \sum \lambda_j x_j) + \sum (\lambda_j - r_j)x_j$ and the triangle inequality show that $\|x - z\| < \epsilon$, proving C is dense in X.

Thus X contains a countable dense subset, as required.

We noted at the beginning of §9 that the Weierstrass approximation theorem implies that $\mathcal{C}[a, b]$ has a countable fundamental set $F = \{1, u, u^2 \ldots\}$ where $u(t) = t$ so we deduce the

9.16 Corollary $\mathcal{C}[a, b]$ is separable.

For instance the polynomials with rational coefficients form a countable dense set, as follows at once from the construction of the set $\mathcal{C}$ in the last theorem.

The set F above consists of the successive powers of the single element u of $\mathcal{C}[a, b]$, which suggests that it would be convenient to have a version of the preceding proposition which made use of the multiplicative structure. Indeed there is one:

9.17 Proposition A normed algebra A is separable $\Leftrightarrow$ there is a countable subset G of A such that $\overline{\text{alg}\, G} = A$.

Proof This is a straightforward elaboration of the preceding Proposition and we leave it as an exercise.

With the help of this fact and the Stone–Weierstrass theorem one can generalize Corollary 9.16, replacing $[a, b]$ by any compact metric space.

9.18 Theorem If T is a compact metric space then $\mathcal{C}(T)$ is separable.

In fact – as we ask the reader to prove in Problem 16D – the converse is also true, namely that T compact Hausdorff, $\mathcal{C}(T)$ separable, implies T is metrizable.

Proof of theorem A compact metric space is separable, so we can choose a countable dense set $\{t_n\}$ in T. Define a sequence of real functions f_n on T by $f_n(t) = d(t, t_n)$. By 1.7 each f_n is in $\mathcal{C}(T)$. Moreover the f_n separate points, for if $s, t \in T$, $s \neq t$ then $d(t, t_n) \neq d(s, t_n)$ whenever n is chosen so that $d(t, t_n) < \frac{1}{2}d(s, t)$. The Stone–Weierstrass Theorem now gives $\overline{\text{alg}}\{f_1, f_2, \ldots\} = \mathcal{C}(T)$ (when the scalars are complex note that since the f_n are real-valued $\overline{\text{alg}}\{f_n\}$ is automatically self-adjoint) and by the last Proposition, $\mathcal{C}(T)$ is separable.

9.19 Theorem Let I be a, possibly unbounded, interval in $\mathbf{R}$. Then $L_p(I)$ is separable for $1 \leqslant p < \infty$, while $L_\infty(I)$ is nonseparable (if I is not a trivial interval).

Proof It is easy to see, using Theorem 9.9, that the set of characteristic functions of intervals with rational endpoints in I, is a countable fundamental set. We leave the second assertion as an exercise.

Problem 9J describes various countable dense sets of L_p spaces on the real line, consisting of polynomials and similar well-behaved functions.

Problems

9A Prove the following strengthening of Theorem 9.3, where $\{T_n\}$ is a uniformly bounded sequence of linear operators from a normed space X to a Banach space Y:

The set $\{x \in X\colon \lim T_n x \text{ exists in } Y\}$ is a closed linear subspace M of X and the map T defined by $Tx = \lim_n T_n x$ belongs to $\mathcal{B}(M, Y)$.

Applications of Theorem 9.3

9B Given $f \in L_p(\mathbf{R})$, $1 \leqslant p < \infty$, and $a \in \mathbf{R}$ let f_a denote the translation of f by a,

$$f_a(t) = f(t-a) \qquad (t \in \mathbf{R}).$$

Show that if $a_n \in \mathbf{R}$ and $a_n \to a$ then $f_{a_n} \to f_a$ in the norm of L_p. (One often expresses this by saying that 'translation is continuous'.) [Clearly the map defined by $T_a f = f_a$ is in $\mathcal{B}(L_p)$ with $\|T_a\| = 1$ so it is sufficient to verify convergence for all f in a suitable fundamental subset.]

9C (This problem assumes the theorem that one may reverse the order of integration in a repeated integral under suitable conditions) Given $f \in L_1(\mathbf{R})$ and $\delta > 0$ define

$$f^\delta(t) = \frac{1}{2\delta}\int_{-\delta}^{\delta} f(t+u)\,du.$$

(i) Show that f^{δ} is continuous and belongs to $L_1(\mathbf{R})$ with $\|f^{\delta}\|_1 \leqslant \|f\|_1$.
(ii) Show that $f^{\delta} \to f$ in the norm of L_1 as $\delta \to 0$.

(The same thing works in L_p but the proof is trickier.)

9D Total sets Let X be a normed space. A subset A of X is **total for** X^* if 0 is the only functional f such that $f(x) = 0$ for all $x \in A$. A subset B of X^* is **total for** X if 0 is the only vector x such that $f(x) = 0$ for all $f \in B$. Show that

(i) A is total for X^* iff A is a fundamental subset of X.
(ii) B is total for X iff B separates the points of X.

9E Show that the set of all functionals of the form

$$\phi(f) = \int_0^c f(t)\,dt, \qquad 0 \leqslant c \leqslant 1,$$

is total for $L_1[0, 1]$.

9F The **moment sequence** of an integrable function f on $[0, 1]$ is the sequence $\{y_n\}$ where

$$y_n = \int_0^1 t^n f(t)\,dt \qquad (n = 1, 2, \ldots).$$

Show that f is uniquely determined by $\{y_n\}$ in the sense that two functions with the same moment sequence are equal a.e.

9G Show that any continuous scalar function on an interval $[a, b]$ can be uniformly approximated by step-functions taking rational values and having their discontinuities at rational points. Hence give an elementary proof that $\mathcal{C}[a, b]$ is separable.

9H Let X be a normed space such that X^* is separable. Show that X is separable.

[Let $\{f_n\}$ be dense in X^*, and choose x_n with $\|x_n\| = 1$, $|f_n(x_n)| > \frac{1}{2}\|f_n\|$. Then $\{x_n\}$ is total for X^* and so fundamental for X.]

9I Uniqueness theorem for the Fourier transform Define a 'generalized trigonometric polynomial' (GTP) to be a function of the form $q(t) = c_1 e^{i\alpha_1 t} + \ldots + c_n e^{i\alpha_n t}$ (c_r complex, α_r real) on $\mathbf{R}$.

(i) Show that every $g \in L_\infty(\mathbf{R})$ is the pointwise a.e. limit of a uniformly bounded sequence of GTP's. [Use the argument of Lemma 9.14 to show that for each $m = 1, 2, \ldots$ there is a GTP q_m (having period $2\pi m$) with $\|(q_m - g)\chi[-\pi m, \pi m]\|_1 < 2^{-m}$, and with $\|q_m\|_\infty$ uniformly bounded.]

(ii) The Fourier transform $\hat{f}$ of a function $f \in L_1(R)$ was defined in 9.11. Show that two functions with the same Fourier transform must be equal almost everywhere.

9J Show that, for $1 \leqslant p < \infty$,

(i) Polynomials are dense in $L_p(I)$ if I is a bounded interval.
(ii) Functions $q(t)\, e^{-t}$, where q is a polynomial, are dense in $L_p[0, \infty)$.
(iii) Functions $q(t) e^{-t^2}$, where q is a polynomial, are dense in $L_p(R)$.

10. The category theorems

There are four basic results – or rather groups of results – in elementary functional analysis; so far we have met the Stone–Weierstrass and Hahn–Banach theorems. In this section we use a basic theorem due to Baire, in which the completeness of the normed space plays an essential part, to derive the remaining two, namely the uniform boundedness principle and the open mapping theorem (and the equivalent closed graph theorem). These four theorems, closely followed by Ascoli's theorem and the completeness of L_p, can be rightly regarded as the foundations of functional analysis in normed spaces.

10.1 A set A in a topological space T is called **nowhere dense** if its closure has empty interior – in particular a closed set is nowhere dense iff its interior is empty. A set which is not nowhere dense is called **somewhere dense.** For the purpose of the next theorem we note that if A is nowhere dense and U is any nonempty open set in T then, since A^- cannot contain U, $U \sim A^-$ must again be a nonempty open set.

10.2 Baire's Theorem It is impossible for a complete metric space T to be covered by a countable family of nowhere dense subsets. Indeed, the complement of the union of such a family must be dense in T.

Remark The terminology that is usually used here is as follows: A set C in a topological space T is said to be **of the first category** in T (or **meagre**) if it can be covered by countably many nowhere dense subsets of T, and to be **of the second category** if this is not possible. Thus Baire's theorem says:

a complete metric space is of the second category in itself. We shall not use these terms except that, by calling the results of this section the Category Theorems, we remind the reader that they rest on a common idea.

Proof of theorem It suffices to prove the second assertion of the theorem. Suppose $A_1, A_2, \ldots$ is a sequence of nowhere dense subsets of T. We have to show that $T \sim \bigcup_n A_n$ is dense in T, or what is the same thing, that given any nonvoid open set U in T, U contains points not in $\bigcup_n A_n$. Since A_1 is nowhere dense, $U \sim A_1^- \neq \emptyset$, so there is an open ball S_1 of radius $< \frac{1}{2}$ contained in U and not meeting A_1. Let F_1 be the concentric closed ball of radius half that of S_1 and consider its interior int F_1. Since A_2 is nowhere dense, int $F_1 \sim A_2^- \neq \emptyset$, so int F_1 contains an open ball S_2 of radius $< \frac{1}{4}$ not meeting A_2. Let F_2 be the concentric closed ball of radius half that of S_2. Since A_2 is nowhere dense, int F_2 contains an open ball S_3 of radius $< \frac{1}{8}$ not meeting A_3.

Continuing in this way gives a decreasing sequence of closed balls F_n in T with diameters tending to zero, each F_n not meeting A_n. By Cantor's intersection theorem there exists one (and only one) point x lying in *all* the sets F_n (and hence in U since $F_1 \subset U$) but in *none* of the sets A_n. Hence $U \sim \bigcup_n A_n \neq \emptyset$ and the proof is complete.

Baire's theorem has applications in many fields outside linear functional analysis and like most good theorems it often crops up in unexpected places. Here is one example, quite unrelated to any of the work in this book, but worth including because it is simple yet illuminating. Every complete metric space T is the union of all its one-point subsets: $T = \bigcup_{t \in T} \{t\}$. A one-point subset $\{t\}$ is necessarily nowhere dense unless t is both open and closed, in which case we say t is an **isolated point** of T. The following fact is now an immediate consequence of Baire's Theorem.

10.3 Proposition A complete metric space T with no isolated points must be uncountable. In particular $\boldsymbol{R}$ is uncountable.

The next lemma is the special case of Baire's Theorem most used in elementary functional analysis.

10.4 Lemma Let X be a Banach space and let C be a closed, convex, symmetric subset such that $\bigcup_{n=1}^{\infty} nC = X$. Then C is a neighbourhood of 0.

Proof Since multiplying by a nonzero scalar is a homeomorphism, all the sets nC are closed and it follows from Baire's theorem that one of them, say $n_1 C$, has nonempty interior. Multiplying by $1/n_1$ we deduce C has nonempty interior, so that

$$C \supset x + rU \text{ for some } x \in C,\ r > 0,$$

U denoting the open unit ball in X. Since C is symmetric, $-x \in C$, and by convexity

$$C \supset \tfrac{1}{2}(-x) + \tfrac{1}{2}(x + rU) = \tfrac{1}{2} rU$$

which proves that C is a neighbourhood of 0.

The next result is also known as the *Banach–Steinhaus Theorem.*

10.5 Uniform Boundedness Theorem Let X be a Banach space, Y a normed space, and let $\mathcal{Q}$ be a subset of $\mathcal{B}(X, Y)$ – that is, a family of bounded linear maps from X into Y. If, for each fixed x in X,

$$\sup_{T \in \mathcal{Q}} \|Tx\| < \infty,$$

then also

$$\sup_{T \in \mathcal{Q}} \|T\| < \infty,$$

Proof Let C be the subset of X defined by

$$C = \{x \in X : \|Tx\| \leqslant 1 \text{ for all } T \in \mathcal{Q}\}.$$

It is trivial to verify that C is convex and symmetric. Further, C is closed, for if x is the limit of a sequence $\{x_n\}$ in C then $\|Tx\| = \lim \|Tx_n\| \leqslant 1$ for all $T \in \mathcal{Q}$, showing that $x \in C$. Finally, for any x we may choose an integer n greater than the finite quantity $\sup_{T \in \mathcal{Q}} \|Tx\|$. We then have

$$\sup_{T \in \mathcal{Q}} \|T(n^{-1}x)\| = n^{-1} \sup_{T \in \mathcal{Q}} \|Tx\| \leqslant 1$$

so that $n^{-1}x \in C$, $x \in nC$, proving $\bigcup_{1}^{\infty} nC = X$. Thus the conditions of the lemma are satisfied, so C is a neighbourhood of 0. That is for some $r > 0$,

$$\|x\| \leqslant r \Rightarrow \|Tx\| \leqslant 1 \text{ for all } T \leqslant \mathcal{Q}$$

which clearly gives

$$\|x\| \leqslant 1 \Rightarrow \|Tx\| \leqslant r^{-1} \text{ for all } T \in \mathcal{A}$$

so that

$$\|T\| = \sup\{\|Tx\| : \|x\| \leqslant 1\} \leqslant r^{-1} \text{ for all } T \in \mathcal{A}$$

and the theorem is proved.

The reader probably agrees that this is a remarkable result, but finds it rather hard to visualize what the general form of the Uniform Boundedness Theorem is saying. The following simpler special case is one that often crops up in applications.

10.6 Corollary Let $\{T_n\}$ be a sequence of bounded linear maps from a Banach space X to a normed space Y such that $Tx = \lim_n T_n x$ exists for *all* x in X. Then $\sup_n \|T_n\|$ is finite and the mapping T thus defined is also a bounded linear map from X to Y.

Proof That T is linear follows trivially from the continuity of the algebraic operations in Y. Since the numbers $\{\|T_n x\| : n = 1, 2, \ldots\}$ form a sequence converging to $\|Tx\|$ they are a bounded set for each $x \in X$. The theorem now implies $\sup_n \|T_n\|$ is finite, let us call it c. Then for each x

$$\|Tx\| = \lim \|T_n x\| \leqslant \sup \|T_n x\| \leqslant \sup \|T_n\| \, \|x\| = c\|x\|,$$

which shows that $\|T\| \leqslant c$.

Another formulation of this fact is that if $T_n \in \mathcal{B}(X, Y)$, with X complete, and if $\|T_n\| \to \infty$ then there is necessarily some point x of X for which $T_n x$ fails to converge (and, in fact, for which $\|T_n x\|$ forms an unbounded sequence). Thus the uniform boundedness theorem gives a powerful technique for proving the *existence* (without actually giving specific examples) of elements of Banach spaces having pathological behaviour. In Problem 13H we ask the reader to fill in the details of one famous counterexample, which we briefly sketch below.

10.7 Let X be the space $\mathcal{C}_{2\pi}(\boldsymbol{R})$ of continuous complex functions on $\boldsymbol{R}$ with period 2π, which is a Banach space under the sup-norm $\|f\|_\infty$. The **Fourier series** of $f \in X$ is the series $\frac{1}{2}a_0 + \sum_1^\infty (a_k \cos kt + b_k \sin kt)$ where

$$a_k = \frac{1}{\pi}\int_{-\pi}^{\pi} f(t) \cos kt \, dt, \quad b_k = \frac{1}{\pi}\int_{-\pi}^{\pi} f(t) \sin kt \, dt.$$

If f is nice and smooth (e.g. twice differentiable) then the series converges to $f(t)$ at each point, and even uniformly. However *there exist $f \in X$ for which the Fourier series fails to converge* at $t = 0$ (this choice of t is purely for convenience: any other point would do equally well). This is established as follows. Let $s_n(t) = \frac{1}{2} a_0 + \sum_1^n (a_k \cos kt + b_k \sin kt)$ denote the nth partial sum of the series. Then it is straightforward to show that the map

$$\phi_n : f \mapsto s_n(0)$$

is a bounded linear functional on X, and, with a bit more work, to show that

$$\|\phi_n\| = \int_{-\pi}^{\pi} \left| \frac{\sin (n + \frac{1}{2}) t}{2\pi \sin \frac{1}{2} t} \right| dt \to \infty$$

Corollary 10.6, in the special case where the T_n are functionals, now shows that there exists some $f \in X$ for which the $\phi_n(f)$ do not converge, in other words whose Fourier series does not converge at $t = 0$.

The next theorem, also of frequent usefulness, can be derived as a corollary of the Uniform Boundedness Theorem but we give a direct proof.

10.8 Theorem Let A be a subset of a normed space X (which need not be complete). Then A is bounded $\Leftrightarrow f(A)$ is a bounded set of scalars for each $f \in X^*$.

Proof Since $\sup_{x \in A} |f(x)| \leqslant \sup_{x \in A} \|f\| \, \|x\|$ it is clear that if A is bounded in X then $f(A)$ is a bounded set of scalars for each $f \in X^*$.

Conversely suppose $f(A)$ is bounded for each $f \in X^*$. Define

$$C = \{f \in X^* : \sup_{x \in A} |f(x)| \leqslant 1\}.$$

Then, just as in the proof of the Uniform Boundedness Theorem, it follows that C is closed, convex, symmetric, and $\bigcup_1^\infty nC = X^*$, so that, since X^* is a Banach space, C contains a ball round 0 by Lemma 10.4. If r is the radius of this ball,

$$f \in X^*, \ \|f\| \leqslant r \Rightarrow |f(x)| \leqslant 1 \quad \text{for all } x \in A$$

so

$$f \in X^*, \ \|f\| \leqslant 1 \Rightarrow |f(x)| \leqslant r^{-1} \text{ for all } x \in A$$

By Corollary 8.12 to the Hahn–Banach theorem, $\|x\| = \sup\{|f(x)|:$

$f \in X^*, \|f\| \leqslant 1\}$, so $\|x\| \leqslant r^{-1}$ for all $x \in A$ and the result follows.

The Open Mapping Theorem

A mapping f between topological spaces X and Y is called **open** if the image $f(U)$ of every open set U in X is open in Y. Thus openness is a sort of reverse property to continuity. The following very easy lemma, which we leave as an exercise because we make no use of it in what follows, explains the name of the main result of the section. As usual B_X, B_Y denote the unit balls in spaces X, Y.

10.9 Lemma Let T be a linear map from a normed space X to a normed space Y. Then $T(B_X)$ is a neighbourhood of 0 in Y $\Leftrightarrow$ T is open.

10.10 Open Mapping Theorem Let X and Y be Banach spaces and T a bounded linear map of X *onto* Y. Then $T(B_X)$ is a neighbourhood of 0 in Y (so that T is open).

Proof Let $V = T(B_X)$. It is clear that V is convex and symmetric, but it need not be closed; however its closure V^- is closed, convex and symmetric since the latter two properties are preserved by taking the closure, and we have

$$\bigcup_n nV^- \supset \bigcup_n nV = \bigcup_n T(nB_X) = T(\bigcup_n nB_X) = T(X) = Y.$$

Thus V^- satisfies the conditions of Lemma 10.4 and we conclude that for some $r > 0$,

$$V^- \supset rB_Y. \tag{1}$$

So far we have only used the completeness of the image space Y, via Lemma 10.4; now we use that of X, to show that

$$V^- \subset 2V. \tag{2}$$

Fix any point y_0 in V^-, so that there exist points of $V = T(B_X)$ arbitrarily close to y_0. Thus we can find a point $x_1 \in B_X$ with $\|y_0 - Tx_1\| \leqslant \frac{1}{2}r$, or in other words

$$y_0 - Tx_1 \in \tfrac{1}{2}rB_Y.$$

Since $\frac{1}{2}rB_Y \subset \frac{1}{2}V^- = (T(\frac{1}{2}B_X))^-$ we can, for the same reason, find $x_2 \in \frac{1}{2}B_X$ with $\|(y_0 - Tx_1) - Tx_2\| \leqslant \frac{1}{4}r$, in other words

$$y_0 - T(x_1 + x_2) \in \tfrac{1}{4}rB_Y.$$

Since $\frac{1}{4}rB_Y \subset (T(\frac{1}{4}B_X))^-$ we can find $x_3 \in \frac{1}{4}B_X$ with $\|(y_0 - Tx_1 - Tx_2) - Tx_3\| \leqslant \frac{1}{8}r$, and so on. Continuing in this way gives a sequence of points x_n in X such that

$$\|y_0 - T(x_1 + \dots + x_n)\| \leqslant 2^{-n} r$$

and $x_n \in 2^{-(n-1)} B_X$, in other words $\|x_n\| \leqslant 2^{-(n-1)}$. The series $\Sigma\, x_n$ is absolutely convergent and, since X is complete, converges to some $x_0 \in X$. Then we have

$$x_1 + \dots + x_n \to x_0,$$

$$T(x_1 + \dots + x_n) \to y_0\,,$$

so that $Tx_0 = y_0$ by the continuity of T. But

$$\|x_0\| \leqslant \sum_1^\infty \|x_n\| \leqslant 1 + \tfrac{1}{2} + \tfrac{1}{4} + \dots = 2$$

so that $x_0 \in 2B_X$, $y_0 \in T(2B_X) = 2V$, and since y_0 was any point of V^-, (2) follows. Combining (1) and (2) gives $T(B_X) = V \supset \frac{1}{2} r B_Y$ and the theorem is proved.

The particular case of the Open Mapping Theorem which is used most frequently in applications is

10.11 Banach's Isomorphism Theorem A one-to-one bounded linear map from one Banach space onto another is a topological isomorphism.

Proof In the notation of the last theorem, suppose T is one-to-one. Then T^{-1} exists, and the assertion $T(B_X) \supset sB_Y$ for some $s > 0$ is easily seen to be equivalent to $T^{-1}(B_Y) \subset s^{-1} B_X$, so that T^{-1} is bounded and the result follows.

Another useful result, which the student should ponder carefully because it gives a good idea of what the Open Mapping Theorem is about, is the next one. Two norms on a linear space, say $\|\ \|_1$ and $\|\ \|_2$, are **comparable** if either $\|x\|_2 \leqslant k\|x\|_1$ for all x, or $\|x\|_1 \leqslant k'\|x\|_2$ for all x, k and k' being suitable constants.

10.12 Corollary Two comparable norms on a linear space X, both making it into a Banach space, must be equivalent.

Proof Suppose without loss that $\|x\|_2 \leqslant k\|x\|_1$ $(x \in X)$. This amounts to saying that the identity map $I : x \mapsto x$ from the Banach space $(X, \|\ \|_1)$ to the Banach space $(X, \|\ \|_2)$ is bounded; since I is one-to-one and onto it is a topological isomorphism by the theorem, and hence the two norms are equivalent.

10.13 The **graph** of a mapping f from a metric space X to a metric space Y is of course the subset $G = \{(x, f(x)) : x \in X\}$ of $X \times Y$, which we regard as carrying the product metric and topology. To say that f has **closed graph** means then, that if a point (x, y) of $X \times Y$ is the limit of a sequence $(x_n, f(x_n))$ in G, then (x, y) lies on G, or in other words (this is the form usually used in applications of the theorem below) that

$$x_n \to x \text{ and } f(x_n) \to y \Rightarrow y = f(x)$$

Since f continuous implies that

$$x_n \to x \Rightarrow f(x_n) \to f(x)$$

it is clear that every continuous function has a closed graph. The next result says that for linear maps the converse is true in suitable circumstances. We derive it as a consequence of the Open Mapping Theorem; but Problem 10 O) the process can be reversed, so that these two apparently very different theorems are equivalent.

10.14 Closed Graph Theorem Let X, Y be Banach spaces and T a linear map from X into Y. Then T is continuous $\Leftrightarrow$ its graph is closed. (Cf. 8.6 for a similar result in the case $Y = F$.)

Proof In view of the last paragraph we only have to prove $\Leftarrow$, so let us suppose the graph $G = \{(x, Tx) : x \in X\}$ is closed in $X \times Y$. Now $X \times Y$ is a Banach space under the norm $\|(x, y)\| = \max\{\|x\|, \|y\|\}$ and G is clearly a closed *linear* subspace, and is therefore a Banach space in its own right under this norm. Consider the 'coordinate projection' $P : G \to X$ defined by $P(x, Tx) = x$. Clearly P is linear, and it is also bounded since $\|P(x, Tx)\| = \|x\| \leqslant \|(x, Tx)\|$. Also, P is one-to-one and maps onto X, the inverse being defined by the formula $P^{-1}x = (x, Tx)$. We conclude by the Isomorphism Theorem 10.11 that P^{-1} is bounded. This gives, for any $x \in X$,

$$\|Tx\| \leqslant \|(x, Tx)\| = \|P^{-1}x\| \leqslant \|P^{-1}\| \, \|x\|,$$

showing that T is continuous, in fact $\|T\| \leqslant \|P^{-1}\|$.

Putting this slightly differently: to show T is continuous we need to show that if x_n converges to x then Tx_n converges to Tx. In the circumstances of the theorem, we are at liberty to assume that Tx_n converges to *something*, and prove that this something is Tx. The next theorem shows what a powerful idea this can be. It is a tricky result to prove without using some form of Baire's theorem. If the reader can produce a direct proof he deserves recognition as a competent classical analyst!

10.15 Theorem Let f be a measurable function on some σ-finite measure space $(S, \mathcal{S}, \mu)$. If $fg \in L_1$ for every $g \in L_p$ then $f \in L_q$. Here, as usual, $1 \leqslant p \leqslant \infty$ and q is the number (or $+\infty$) such that $p^{-1} + q^{-1} = 1$.

Proof The map $T : g \mapsto fg$ is clearly a linear map of L_p into L_1, both of which spaces are complete by the Riesz–Fischer Theorem. We show that T is bounded: in fact by the Closed Graph Theorem it suffices to show that if $f_n \to g$ in L_p and $Tg_n \to h$ in L_1 then $h = Tg$. That is, we assume $\|g_n - g\|_p \to 0$, $\|fg_n - h\|_1 \to 0$. By Corollary 5.11 we can, by choosing a subsequence and then a further subsequence, assume that $g_n(t) \to g(t)$ a.e. and $f(t)\, g_n(t) \to h(t)$ a.e. It follows at once that $h(t) = f(t)\, g(t)$ a.e., so that $h = fg = Tg$, where equality is taken in the L_1 sense. Hence T is bounded, which is to say that for some $k \geqslant 0$,

$$\int |fg| = \|fg\|_1 = \|Tg\| \leqslant k \qquad (g \in L_p,\ \|g\|_p \leqslant 1)$$

which gives $f \in L_q$ with $\|f\|_q \leqslant k$, by Proposition 5.5.

To show that Baire's theorem is at the bottom of this result we give an alternative proof, this time using the Uniform Boundedness principle.

Proof 2 By σ-finiteness and Proposition 2.19 there is a sequence $\{\phi_n\}$ in L_0 (the integrable simple functions) such that $\phi_n \to f$ pointwise, $|\phi_n| \leqslant |f|$ for all n. Since $L_0 \subset L_q$ it follows at once from Hölder's inequality that the maps

$$T_n : g \mapsto \phi_n g$$

are in $\mathcal{B}(L_p, L_1)$ for each n, with $\|T_n\| \leqslant \|\phi_n\|_q$. Defining $T : g \mapsto fg$ as in the first proof we observe that, for each fixed $g \in L_p$, the sequence of functions $|\phi_n g - fg|$ is dominated by the L_1 function $2|fg|$ and tends pointwise to zero, so that the Dominated Convergence Theorem gives

$$\|T_n g - Tg\|_1 = \int |\phi_n g - fg| \to \int 0 = 0.$$

Thus $\{T_n\}$ is a sequence of bounded linear maps on the Banach space L_p such that $\lim T_n g = Tg$ exists (in the norm of the range space!) for each $g \in L_p$. By Corollary 10.6, T is bounded, and as in the first proof it follows that $f \in L_q$, with $\|f\|_q \leqslant \|T\|$.

Baire's Theorem

10A Show that the following is equivalent to Baire's Theorem: If $\{G_n\}$ is a sequence of dense open sets in a complete metric space then $\bigcap_n G_n$ is non-empty (in fact dense).

10B Show that $\mathbf{Q}$ is not expressible as the intersection of a countable family of open subsets of $\mathbf{R}$.

10C (Stronger form of Baire's Theorem) If a complete metric space X is covered by a sequence of closed sets A_n then $\bigcup_n \operatorname{int}(A_n)$ is dense in X. [If some closed ball B is disjoint from each $\operatorname{int}(A_n)$ then the sets $B \cap A_n$ are nowhere dense in B.]

Uniform Boundedness Theorem

10D Show that for a convex symmetric set of C in a real linear space, $\bigcup_{n=1}^{\infty} nC$ and $\bigcup_{\alpha>0}^{\infty} \alpha C$ both coincide with $\operatorname{lin} C$ (see Lemma 10.4). What about the complex case?

10E On the space ℓ_0 (Problem 5F) with the ℓ_1 norm, construct a sequence of bounded linear functionals f_n with $\sup_n |f_n(x)| < \infty$ $(x \in X)$ but $\|f_n\| \to \infty$, thus showing completeness is necessary in Theorem 10.5.

10F Let $\{\alpha_n\}$ be a scalar sequence such that $\sum_n \alpha_n x_n$ is (conditionally) convergent whenever $\{x_n\} \in c_0$. Show that $\{\alpha_n\} \in \ell_1$.

[Consider the functionals f_n on c_0 defined by

$$f_n(x) = \sum_{r=1}^{n} \alpha_r x_r \qquad (x = \{x_n\} \in c_0).]$$

10G Let T be a linear map from a normed space X to a normed space Y such that $f \circ T \in X^*$ for every $f \in Y^*$. Prove T is bounded.

Open Mapping Theorem

10H Prove Lemma 10.9.

10I It is well-known that if $\{\alpha_n\}$ is a sequence of strictly positive real numbers with $\sum \alpha_n < \infty$ then there is a sequence $\{y_n\}$ with $y_n > 0$, $\lim y_n = \infty$ such that $\sum \alpha_n y_n < \infty$. Prove this by the open mapping theorem.

[If no such $\{y_n\}$ exists then the map $T : \ell_\infty \to \ell_1$ defined by

$$Tx = y \text{ where } y_n = \alpha_n x_n,$$

which is clearly one-to-one, would be onto, hence a topological isomorphism.]

10J (i) Prove 10F by the open mapping theorem.

(ii) Let X be a Banach space under a norm $\| \ \|_1$ and let $\| \ \|_2$ be a weaker norm on X (i.e. it gives a smaller topology). Show that either the two norms are equivalent or else the unit ball of the first norm is nowhere dense with respect to the second.

10K Given a sequence of real numbers $c_n > 0$, show that a necessary and sufficient condition for the existence of numbers $b_n > 0$ such that $\sum_n b_n^2 c_n < \infty$ but $\sum_n b_n = \infty$, is that $\sum_n c_n^{-1} = \infty$.

10L Let A be the **disc algebra** of continuous functions on $\Delta = \{z \in C: |z| \leqslant 1\}$ defined in 6.23. A compact subset K of Δ is an **interpolation set** of A if every $f \in \mathcal{C}(K)$ is the restriction to K of some $g \in A$, briefly if $A|_K = \mathcal{C}(K)$. Show that if this occurs there is a constant c depending only on K such that for all $f \in \mathcal{C}(K)$ there exists $g \in A$ with $g|_K = f$ and $\|g\| \leqslant c\|f\|$.

Closed Graph Theorem

10M 'Closed graph' arguments are useful in other contexts: let f be a map from a metric space X to a compact metric space Y. Prove f is continuous iff its graph is closed in $X \times Y$.

10N (cf. Corollary 10.12) Two norms $\| \ \|_1$ and $\| \ \|_2$ on a linear space X are **compatible** if the situation

$$\|x_n - x\|_1 \to 0, \qquad \|x_n - y\|_2 \to 0$$

implies $x = y$. Show that two compatible norms on X, both making X into a Banach space, must be equivalent.

10O Derive Theorem 10.10 from its special case 10.11, and derive the latter from Theorem 10.14, to complete the proof that these three results are essentially equivalent. [Problems 7K, 7J may help for the first part.]

10P Derive Theorem 10.8 from the Closed Graph Theorem. [If $A \subset X$ and $f(A)$ is a bounded set of scalars for each f in X^* then the map T where $Tf = f|_A$ takes X^* into the Banach space $B(A)$ of all bounded scalar functions on A. Show T has closed graph.]

10Q Let Z be a closed linear subspace of $\mathcal{C}[0, 1]$ such that each f in Z is differentiable on $[0, 1]$ with bounded derivative. Show that Z is finite dimensional. [$f \mapsto f'$ is a continuous map of Z into $B[0, 1]$; apply Ascoli to show the unit ball of Z is compact.]

Miscellaneous problems

10R It is important in studying differential equations to know whether a small change in the parameters gives a small change in the solution. A simple case is analysed below. Let p, q be continuous scalar functions on $[a, b]$. Assume (Problem 17N) that for each $f \in \mathcal{C}[a, b]$ and each pair of scalars α, β there is a unique y in $\mathcal{C}^2[a, b]$ satisfying

$$y''(t) + p(t)\,y'(t) + q(t)\,y(t) = f(t), \qquad a \leqslant t \leqslant b \qquad (*)$$

with

$$y(a) = \alpha, \qquad y'(a) = \beta.$$

Prove this solution depends continuously on f, α, β in the sense that if $f_n \to f$ uniformly, $\alpha_n \to \alpha$ and $\beta_n \to \beta$ then the corresponding solutions y_n, y have the property that y_n, y_n' and y_n'' converge uniformly to y, y' and y'' respectively.
[Let $L(y)$ denote the left hand side of (*). Consider the map $y \mapsto (L(y), y(a), y'(a))$ of $\mathcal{C}^2[a, b]$ to $\mathcal{C}[a, b] \times \mathbf{F}^2$.]

Projections

Let X be a normed space. A linear operator $P: X \to X$ is a **projection** if $P^2 = P$ (i.e. $P(Px) = Px$ for all x). Note P is not assumed bounded. X is the **direct sum** of linear subspaces M, N if

$$M + N = X, \qquad M \cap N = \{0\};$$

we write $X = M \oplus N$. A closed linear subspace M is **direct** if $X = M \oplus N$ for some closed linear subspace N. Projections are important because one of the ways of analysing linear operators is to express them, roughly, as linear combinations of projections. The problems below explore the basic relationships between projections and direct sum decompositions.

10S (Purely algebraic) Let P be a projection on X and I be the identity map $Ix = x$ $(x \in X)$.

(i) $Q = I - P$ is a projection.
(ii) $\operatorname{im} P = \{x : Px = x\} = \ker Q$; similarly $\ker P = \operatorname{im} Q$.
(iii) $X = \operatorname{im} P \oplus \ker P$.
(iv) Conversely given a direct sum decomposition $X = M \oplus N$ there is a unique projection R (called the **projection on** M **along** N) such that $\operatorname{im} R = M$ and $\ker R = N$.

10T Let X be a Banach space. Show that

(i) A projection P on X is bounded iff im P and ker P are closed.

(ii) A closed linear subspace M of X is direct $\Leftrightarrow$ there is a bounded projection whose range is $M \Leftrightarrow$ there is an element S of $\mathcal{B}(X/M, X)$ with $Q \circ S = I$, where Q is the quotient map and I is the identity map on X/M.

10U Let X be the space in Problem 10E and define closed subspaces

$$N = \{x \in X: x_1 = x_3 = x_5 = \ldots = 0\},$$

$$M = \{x \in X: x_1 = x_2, 2x_3 = x_4, 3x_5 = x_6, \text{etc.}\}.$$

Show that $X = M \oplus N$ but the projection on M along N is unbounded.

4

HILBERT SPACES AND RELATED TOPICS

11. Infinite sums in Banach spaces, second grade

In §4 we took over, verbatim, to Banach spaces the definition of an infinite series used in elementary analysis, whereby the sum $\sum_{n=1}^{\infty} x_n$ of a sequence of terms $x_1, x_2, \ldots$ is the limit, if it exists, of the sequence of partial sums $x_1, x_1 + x_2, x_1 + x_2 + x_3$, ... In many situations however one needs an alternative definition of infinite sum which is meaningful for a set of terms $\{x_i\}_{i \in I}$ (where I is an infinite index set, possible even uncountable) quite independently of any counting off $x_{i_1}, x_{i_2}, \ldots$ of the terms. Since this notion is an important tool of analysis we examine it fairly carefully, but its only use in this book is in the Hilbert space theory of the next chapter for which purpose 11.1, 2 and 3 will give the reader sufficient knowledge for a first reading.

11.1 Let $\{x_i\}_{i \in I}$ be an indexed family of vectors in a Banach space X. We start with the elementary fact that when J is a *finite* subset of I, the sum of the x_i's for i in J, denoted $\sum_{i \in J} x_i$ has an obvious, unambigous meaning, namely $x_{i_1} + \ldots + x_{i_n}$ where J is enumerated as $\{i_1, i_2, \ldots, i_n\}$ in any way we like. One extends this to infinite sums by the following

Definition The vectors $\{x_i\}_{i \in I}$ are **summable** to the vector $x \in X$ if for each $\epsilon > 0$ there is a finite subset $J_0 = J_0(\epsilon)$ of I such that whenever J is a finite subset of I containing J_0 one has

$$\left\| \sum_{i \in J} x_i - x \right\| < \epsilon.$$

As a helpful notation we shall denote $\sum_{i \in J} x_i$, where J is a finite subset of I, by s_J and call it one of the **partial sums** of the x_i. If we temporarily write $\mathcal{F}$ = {finite subsets of I} then the following shows the close relation between this definition and the one for ordinary series:

$\{x_i\}$ summable to $x \Leftrightarrow$ for all $\epsilon > 0$ there is $J_0 \in \mathcal{F}$ such that $J \in \mathcal{F}, J \supset J_0$ implies $\|s_J - x\| < \epsilon$.

$\sum_1^\infty x_n$ converges to x_n for all $\epsilon > 0$ there is $n_0 \in \mathbf{N}$ such that $n \in \mathbf{N}, n \geqslant n_0$ implies $\|s_n - x\| < \epsilon$,

where s_n is the usual nth partial sum $x_1 + \dots + x_n$.

Thus both definitions combine finite sums with an appropriate limiting process.

If x is a vector, and J_0 a finite subset of I, such that $\|s_J - x\| < \epsilon$ for each finite subset J containing J_0, we shall say that J_0 **sums the x_i to x within ϵ.** Thus the x_i are summable to x iff given ϵ they can be summed to x within ϵ by a suitable J_0.

The limit vector x when it exists is easily seen to be unique, and is denoted $\sum_{i \in I} x_i$. We shall use, synonymously, the phrases: $\{x_i\}_{i \in I}$ is (or 'the x_i are' – the index set I being understood) summable to x; $\sum_i x_i$ converges to x; $\sum_{i \in I} x_i = x$.

The distinction between $\sum_i$ and $\sum_{i \in I}$ need not be taken too seriously, but it is intended to suggest that $\sum_i$ is regarded as a mapping that transforms the family $\{x_i\}$ into the family of partial sums $\{s_J\}$, which converges, in some sense, to a vector called $\sum_{i \in I} x_i$.

It is elementary that *finite* sums obey the obvious computation rules

$$\sum_{i \in I} (x_i + y_i) = \sum_{i \in I} x_i + \sum_{i \in I} y_i$$
$$\sum_{i \in I} \lambda x_i = \lambda \sum_{i \in I} x_i$$

11.2 Proposition Let $\{x_i\}$, $\{y_i\}$ be vectors in X indexed over the same set I, and let λ be a scalar. Then the above computation rules hold whenever the sums on the right hand side are defined.

Proof Assume that $\{x_i\}$ is summable to x and $\{y_i\}$ to y. Given $\epsilon > 0$ we can choose finite $J_0 \subset I$ to sum the x_i to x within $\epsilon/2$, and finite $K_0 \subset I$ to sum the y_i to y within $\epsilon/2$. The computation rules for finite sums, and the triangle inequality, then easily show that $J_0 \cup K_0$ sums $\{x_i + y_i\}$ to $x + y$ within ϵ. This is true for any $\epsilon > 0$, and therefore $\sum_{i \in I} (x_i + y_i) = x + y = \sum_{i \in I} x_i + \sum_{i \in I} y_i$. The proof for $\Sigma \lambda x_i$ is similar.

The basic theorem on summability is about a sort of 'Cauchy property' of the partial sums.

11.3 Theorem $\sum_i x_i$ converges $\Leftrightarrow$ given $\epsilon > 0$ there is a finite $J_0 \subset I$ such that $\|s_K\| < \epsilon$ for every finite subset K of I disjoint from J_0.

Proof The sufficiency is proved by constructing a convergent sequence of partial sums and showing that its limit is the sum of the x_i. We use the obvious fact that, for *disjoint, finite* subsets J, K of I one has $s_{J \cup K} = s_J + s_K$.

For each $n = 1, 2, \ldots$ choose a finite subset, say J_n, of I such that $\|s_K\| < n^{-1}$ for any finite subset K of I disjoint from J_n. Replacing J_n by $J_0 \cup \ldots \cup J_n$ we may, and do, assume that $J_1 \subset J_2 \subset J_3 \subset \ldots$ Then if $m > n$,

$$\|s_{J_m} - s_{J_n}\| = \|s_{J_m \sim J_n}\| < n^{-1} \tag{1}$$

because $J_m \sim J_n$ is disjoint from J_n. Thus $\{s_{J_n}\}$ is a Cauchy sequence in X, and since we assume X to be a Banach space, $x = \lim_n s_{J_n}$ exists.

Holding n fixed in (1) and letting $m \to \infty$ gives

$$\|x - s_{J_n}\| \leqslant n^{-1} \quad (n = 1, 2, \ldots).$$

For any $\epsilon > 0$, choose n so that $2/n < \epsilon$; then whenever J is a finite subset of I, and $J \supset J_n$, we have

$$\|x - s_J\| = \|x - s_{J \sim J_n} - s_{J_n}\|$$

$$\leqslant \|x - s_{J_n}\| + \|s_{J \sim J_n}\| < \frac{1}{n} + \frac{1}{n} < \epsilon.$$

Thus J_n sums the x_i to x within ϵ, and this proves $\sum_I x_i$ converges to x.

Conversely suppose $\sum_I x_i$ converges to $x \in X$. Given $\epsilon > 0$ we can choose J_0 to sum the x_i to x within $\epsilon/2$. Then for any finite subset K of I disjoint from J_0 we have

$$\|x - s_{J_0} - s_K\| = \|x - s_{J_0 \cup K}\| < \epsilon/2 \quad \text{since } J_0 \cup K \supset J_0$$

and also

$$\|x - s_{J_0}\| < \epsilon/2$$

from which

$$\|s_K\| \leqslant \|x - s_{J_0}\| + \|x - s_{J_0} - s_K\| < \epsilon.$$

The next few results examine the connexion between different kinds of infinite sum in the special case where I is the set N of natural numbers, or where X is the scalar field **F**. In the case where $I = \mathbf{N}$ *and* $X = F$ we shall show that summability is the same as absolute convergence.

First, if $\{x_i\}$ is a *sequence*, that is if $I = \mathbf{N}$, one needs to know the relation between the phrases '$\{x_i\}$ is summable' and '$\sum_1^\infty x_i$ converges' (in the elementary sense). We emphasize again that the latter phrase, and others like it, refer to a property of the family $\{x_i\}$ and not to some hypothetical object called a 'series' or 'sum'.

11.4 Proposition Let $\{x_i\}$ be a sequence in a Banach space X. If $\{x_i\}$ is summable to x then $\sum_1^\infty x_i$ converges to x in the elementary sense that $\|\sum_{i=1}^n x_i - x\| \to 0$ as $n \to \infty$.

The proof follows at once upon taking $I = \mathbf{N}$ and J_n to be the subset $\{1, 2, \ldots, n\}$ of $\mathbf{N}$ in the next result, which brings out the basic idea more clearly.

11.5 Lemma If I is countable and $\{x_i\}_{i \in I}$ is summable to x then $x = \lim s_{J_n}$ for any increasing sequence of finite subsets J_n of I such that $\bigcup_n J_n = I$.

Proof Given $\epsilon > 0$ choose a finite subset K of I which sums the x_i to x within ϵ. Since the J_n are increasing, and $K \subset \bigcup_n J_n$, there is some n_0 such that $K \subset J_{n_0}$. Then also $K \subset J_n$ for $n \geqslant n_0$ and so $\|s_{J_n} - x\| < \epsilon$ for $n \geqslant n_0$. Hence $s_{J_n} \to x$.

This simple result makes it easy to write down a sequence of partial sums converging to the limit vector in many cases. For instance if I is the set $\mathbf{Z}$ of integers the family $\{x_i\} = \{\dots, x_{-1}, x_0, x_1, \dots\}$ is what we may call a 'two-sided sequence' and it is clear that $\sum_{i \in I} x_i$, when it exists, is the limit of the 'two-sided partial sums' $s_n = \sum_{i=-n}^{n} x_i$ as $n \to \infty$. It also follows that summable families of vectors share what the reader probably knows to be one of the most useful properties of absolutely convergent series in elementary analysis: roughly, that it is allowed to add up the terms in any order one likes. For if I is countable and $\{i_1, i_2, \dots\}$ is any enumeration of I, the Lemma implies that $\sum_{i \in I} x_i$, when it exists, equals $\lim_{n\to\infty} (x_{i_1} + \dots + x_{i_n})$.

The converse to Proposition 11.4 is false, as we shall see below.

As a preliminary to defining absolute summability we turn to the following case: here the index set I is arbitrary.

11.6 Proposition

(i) Let $\{x_i\}_{i\in I}$ be a family of non-negative real numbers. Then the x_i are summable iff the set of their partial sums $\{s_J : J$ a finite subset of $I\}$ is bounded above, and $\sum_{i \in I} x_i$ is then equal to the supremum of this set.

(ii) When $I = \mathbf{N}$, a family (sequence) of non-negative real numbers x_i is summable to x iff $\sum_1^\infty x_i$ converges to x in the elementary sense.

Proof (i) Suppose the set of partial sums is bounded above and let s be its supremum. Thus, given $\epsilon > 0$ we can choose a finite subset J_0 of I such that $s_{J_0} > s - \epsilon$. For each finite subset J of I containing J_0 it is clear, since the x_i are non-negative, that $s_J \geqslant s_{J_0}$. Since also $s_J \leqslant s$ this gives $|s_J - s| < \epsilon$ and consequently J_0 sums the x_i to s within ϵ. Since ϵ was arbitrary, $\sum_{i \in I} x_i = s$.

Conversely if $\{x_i\}$ is summable choose a J_0 to sum the x_i within (say) 1, and let $s_{J_0} = m$. It is then easy to verify that $s_J \leqslant m + 1$ for all finite $J \subset I$, so the partial sums are bounded above.

(ii) is a simple consequence of (i) and the last Lemma, and is left as an exercise.

Not surprisingly, in the case where the x_i are non-negative reals, we shall write $\sum_I x_i < \infty$ to mean that the x_i are summable, and $\sum_I x_i = \infty$ to mean the opposite. Now, turning to the case of an arbitrary indexed family of vectors in a Banach space we say that $\{x_i\}_{i \in I}$ is **absolutely summable** if $\sum_I \|x_i\| < \infty$. That this notion generalizes the elementary idea of an absolutely convergent series is shown by the next result.

11.7 Theorem

(i) If the family $\{x_i\}_{i \in I}$ of vectors in a Banach space is absolutely summable then it is summable.

(ii) A sequence of vectors $\{x_i\}_{i \in \mathbf{N}}$ is absolutely summable to x iff $\sum_1^\infty x_i$ converges absolutely to x in the elementary sense.

Proof (i) By Theorem 11.3 applied to the norms of the x_i, given $\epsilon > 0$ there exists a finite $J_0 \subset I$ such that $\sum_{i \in K} \|x_i\| < \epsilon$ whenever K is a finite subset of I disjoint from J_0. The triangle inequality gives $\|s_K\| = \|\sum_{i \in K} x_i\| \leqslant \sum_{i \in K} \|x_i\| < \epsilon$ for any such K, and applying the same Theorem in the reverse direction completes the proof.

(ii) Proposition 11.6 (ii) applied to the norms of the x_i shows that $\{x_i\}_{i \in \mathbf{N}}$ is absolutely summable iff $\sum_1^\infty x_i$ converges absolutely in the elementary sense; that the limit vector is the same in either case comes from Proposition 11.4.

The next result shows, in the case of a sequence of numbers, that summability and elementary series convergence are not equivalent. In case it makes the reader suspect, on the other hand, that summability and absolute summability are always equivalent, we point out this also is not so (Problem 11D).

11.8 Proposition A family of scalars $\{x_i\}_{i \in I}$ is summable iff it is absolutely summable.

Proof By the last theorem, it suffices to show that a summable family is absolutely summable. First suppose that the x_i are *real*: then for any finite $K \subset I$ one can partition K into $K_+ = \{i \in K: x_i \geqslant 0\}$, $K_- = \{i \in K: x_i < 0\}$. If we write s_J for the partial sum $\sum_{i \in K} x_i$ in the usual way, then it is clear that

$\sum_{i \in K} |x_i| = s_{K+} - s_{K-}$.

By Theorem 11.3, given $\epsilon > 0$ there exists a finite $J_0 \subset I$ such that $|s_K| < \epsilon/2$ whenever K is a finite subset of I not meeting J_0. For any such K, as K_+, K_- are also finite subsets not meeting J_0, one has

$$\sum_{i \in K} |x_i| = s_{K_+} - s_{K_-} \leqslant |s_{K_+}| + |s_{K_-}| < \epsilon$$

and Theorem 11.3 now shows that $\sum_I |x_i|$ converges. The extension to complex scalars is left as an exercise.

Thus any sequence of scalars $\{x_n\}$ such that $x_1 + x_2 + \ldots$ is convergent but not absolutely convergent such as $\{1, -\frac{1}{2}, \frac{1}{3}, -\frac{1}{4}, \ldots\}$ furnishes an example of a convergent series that is not summable.

11.9 It has probably not escaped the reader that there is a strong similarity between the notion of summability in the case where X is the scalar field, and the definition of the Lebesgue integral. In fact a set of scalars $\{x_i\}_{i \in I}$ is summable to x iff

$$\int_I x_i \, dn(i) = x$$

where the latter denotes integration with respect to counting measure n on the index set I. The verification of this fact is quite easy and is left as Problem 11 G.

11.10 We finish with a summary of the situation when a *sequence* of vectors $\{x_n\}$ in a Banach space X is under consideration:

$$\begin{array}{ccc} & (\Rightarrow \text{when } X = \mathbf{F}) & \\ \{x_n\} \text{ summable} & \Leftarrow & \{x_n\} \text{ abs. summable} \\ \Downarrow & & \Updownarrow \\ \Sigma x_n \text{ convergent} & & \Sigma x_n \text{ abs. convergent} \end{array}$$

Problems

Unless otherwise stated, vectors in a Banach space X are being considered.

11A Show that if $\{x_i\}_{i \in I}$ is summable then $\{x_i\}_{i \in J}$ is summable for each subset J of I.

11B Show that if $\{x_i\}$ is summable to x and if T is a bounded linear map from X to another normed space, then $\{Tx_i\}$ is summable to Tx.

11C Show that if $\dim X < \infty$ then $\{x_i\}$ is summable to x iff $\{x_i\}$ is absolutely summable to x.

[Show that $\{x_i\}$ summable implies $\Sigma|f(x_i)| < \infty$ for each f in X^*.]

11D Let $e_1, e_2, \ldots$ be the standard basis vectors in ℓ_p (see 5F) and let $\alpha_1, \alpha_2, \ldots$ be scalars. Show that $\{\alpha_n e_n\}$ is summable iff

$$\{\alpha_n\} \in \ell_p \quad (1 \leqslant p < \infty);$$

$$\{\alpha_n\} \in c_0 \quad (p = \infty)$$

Hence give an example of a summable but not absolutely summable family of vectors.

11E Show that if $\{x_i\}$ is summable then $(i : x_i \neq 0)$ is countable.

[If J_0 sums the x_i within ϵ then $J \supset \{i : \|x_i\| \geqslant \epsilon\}$.]

11F It is clear from the definitions that the closed linear span M of a family $\{x_i\}_{i \in I}$ of vectors in X contains all vectors expressible as convergent sums $\sum_{i \in I} \alpha_i x_i$ where the α_i are scalars. Is it generally true that every $x \in M$ is of this form? [An instructive case is the family $\{x_n\}_{n \in \mathbf{N}}$ in c_0 where the vector x_n has 1 in the first n places and 0 thereafter; see also 11H(a).]

11G Let f be a scalar function on a nonempty set I. Show that $\{f(i)\}_{i \in I}$ is summable to c iff the integral

$$\int_I f\, dn$$

with respect to counting measure, exists and equals c.

[Use 11.8 to reduce the problem to the case $f \geqslant 0$; then go back to the definitions.]

Further problems

11H In the space $\mathcal{C}[0,1]$, find necessary and sufficient conditions on the scalars $\{\alpha_n\}$ to ensure that $\{\alpha_n f_n\}$ is summable, in the following cases:

(a) $f_n(t) = t^n$ $(n = 1, 2, \ldots)$;
(b) f_n $(n = 1, 2, \ldots)$ is the 'spike-function' shown in Diagram 14.

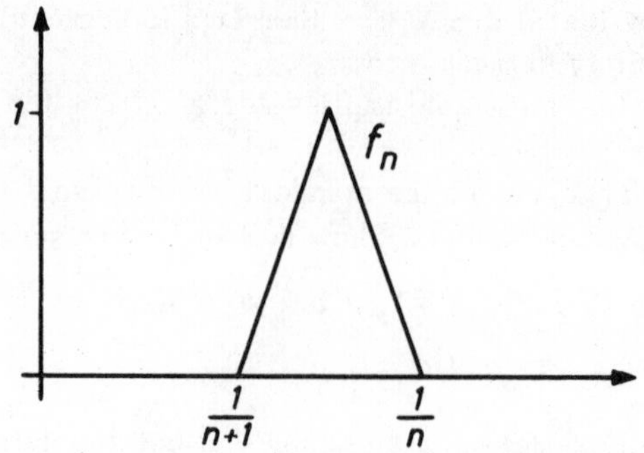

Diagram 14.

11I The Rearrangement Theorem Let $\{x_i\}_{i \in I}$ be summable to x, let the index set I be partitioned into disjoint subsets $\{I_j\}_{j \in J}$, and for each j define $y_j = \sum_{i \in I_j} x_i$ (see Problem 11A). Show that $\{y_j\}_{j \in J}$ is summable to x, or in other symbols,

$$\sum_{j \in J} \sum_{i \in I_j} x_i = \sum_{i \in I} x_i$$

[If $I_0 \subset I$ sums the x_i to x within ϵ, and $I_{j_1}, \ldots, I_{j_n}$ are chosen to cover I_0, show that $J_0 = \{j_1, \ldots, j_n\}$ sums the y_j within ϵ.]

11J Use the Rearrangement Theorem to deduce the following, which are its most important special cases:

(i) If $\{x_{mn}\}_{m \in N, n \in N}$ is a double sequence of elements in a Banach space X such that

$$\sum_{m=1}^{\infty} \sum_{n=1}^{\infty} \|x_{mn}\| < \infty,$$

then

$$\sum_{m=1}^{\infty} \sum_{n=1}^{\infty} x_{mn} = \sum_{n=1}^{\infty} \sum_{m=1}^{\infty} x_{mn} .$$

(ii) If x is an element of a Banach algebra A, and the scalar sequences $\{a_n\}, \{b_n\}$ are such that $\sum_{n=0}^{\infty} a_n x^n$ and $\sum_{n=0}^{\infty} b_n x^n$

converge absolutely, say to y and z, then

$$\sum_{n=0}^{\infty} c_n x^n, \text{ where } c_n = \sum_{r=0}^{n} a_r b_{n-r},$$

converges absolutely to yz. (In other words, power series in a Banach algebra can be multiplied just like the ordinary kind.)

[In either case, first form a sum indexed over $\mathbf{N}^2$ and show that it converges. For (i), partition $\mathbf{N}^2$ into rows or columns; for (ii) consider the diagonals $\{(r, s): r + s = \text{constant}\}$.]

11K Generalised ℓ_p spaces Let S be any nonempty set, X a normed space and $1 \leqslant p < \infty$. The space $\ell_p(S, X)$ is defined to consist of all X-valued functions f on S such that

$$\|f\|_p = \left(\sum_{s \in S} \|f(s)\|^p\right)^{1/p}$$

is finite.

(i) Prove that $\ell_p(S, X)$ is closed under the obvious linear operations and that $\| \;\|_p$ is a norm on it which makes it complete if X is complete.
(ii) What does $\ell_p(S, X)$ reduce to when (a) X is the scalar field and/or S is (b) a finite set or (c) the natural numbers?
(iii) Define $\ell_\infty(S, X)$, $c(S, X)$, $c_0(S, X)$ in the analogous way and examine their properties.

12. Hilbert spaces

The earliest Banach spaces to be studied in detail (well before Banach's famous book *Opérations Linéaires* appeared on the scene in 1932) were not the $\mathcal{C}(T)$ spaces or the general L_p spaces, but those in which there exists a natural idea of the angle between two vectors – or more importantly, the notion of *perpendicularity* of vectors. The reader probably knows that in elementary geometry it is customary to introduce the 'dot product' of two vectors $\mathbf{a} = \overline{OA}$, $\mathbf{b} = \overline{OB}$ defined by $\mathbf{a}.\mathbf{b} = OA.\,OB.\cos\theta$ (Diagram 15) and that this is a handy tool for all problems which have to do with lengths and angles. Clearly $\mathbf{a}$ and $\mathbf{b}$ are perpendicular iff $\mathbf{a}.\mathbf{b} = 0$; more unexpectedly the dot product is distributive, that is $\mathbf{a}.(\mathbf{b} + \mathbf{c}) = \mathbf{a}.\mathbf{b} + \mathbf{a}.\mathbf{c}$. Hilbert spaces have a structure which

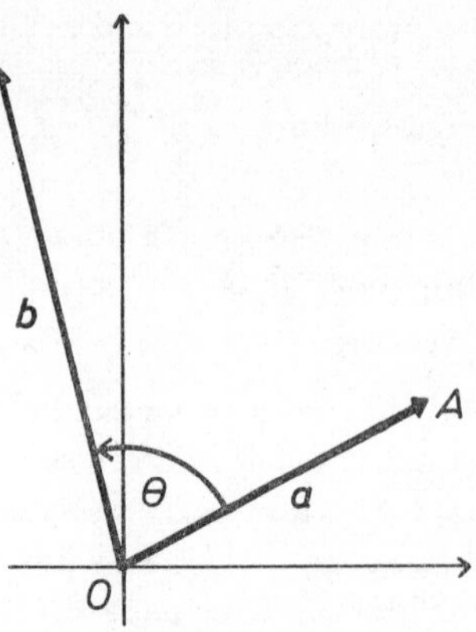

Diagram 15.

copies the algebraic properties of the dot product, and this gives them a rich geometric structure which is essentially Euclidean. They are ideally suited to certain problems of analysis, and even today it is possible to spend all one's mathematical life investigating the properties of operators on Hilbert spaces. This section treats the basic properties of the space itself, and we take up the rather distinct topic of operators in §18.

12.1 A **real inner-product space** is a real linear space X on which is defined a function $\phi: X \times X \to \mathbf{R}$, the **inner product**, whose value $\phi(x, y)$ is written as $\langle x, y\rangle$ (or $x.y$, especially in applied mathematics), obeying the rules

P1 $\langle \alpha x_1 + \beta x_2, y\rangle = \alpha \langle x_1, y\rangle + \beta \langle x_2, y\rangle$.

P2 $\langle x, y\rangle = \langle y, x\rangle$,

P3 $\langle x, y\rangle \geqslant 0$.

P4 $\langle x, x\rangle = 0$ if and only if $x = 0$.

We always assume that an inner-produce space carries the norm defined by

$$\|x\| = \langle x, x\rangle^{\frac{1}{2}}.$$

That it is really a norm is proved in Corollary 12.8.

From P1 and P2 we see that ϕ also satisfies

$$\langle x, \alpha y_1 + \beta y_2\rangle = \alpha\langle x, y_1\rangle + \beta\langle x, y_2\rangle .$$

Thus ϕ is linear in each argument separately and is called a **bilinear functional** (or **bilinear form**).

12.2 Example 1 The set of all vectors in ordinary space, with the dot product, is a 3-dimensional real inner-product space, the norm coinciding with the usual idea of length. (In fact all of Euclidean geometry can be derived from the preceding sentence.)

Example 2 R^n is a real inner-product space X with the inner product

$$\langle x, y\rangle =' \sum_{j=1}^{n} x_j y_j .$$

This gives $\|x\| = \langle x, x\rangle^{\frac{1}{2}} = (\Sigma x_j^2)^{\frac{1}{2}}$, so that as a normed space X is identical with ℓ_2^n.

Example 3 Let A be an $n \times n$ real matrix. Then R^n becomes a real inner product space under

$$\langle x, y\rangle = x^T A y$$

if, and only if, A is **symmetric** and **positive definite.** Example 2 is the case $A = I$. (Here y denotes y regarded as a column matrix, x^T denotes x regarded as a row matrix; if the reader doesn't know what a symmetric positive definite matrix is it doesn't matter.)

The deeper applications of inner-product spaces are mostly to two (related) areas: Fóurier series and the spectral analysis of linear operators. In order to get a satisfactory theory it is desirable in the first and essential in the second that the scalars be complex: this turns out to be closely connected with the fact that every nonconstant polynomial with complex coefficients has a complex zero, whereas there exist polynomials with real coefficients and no real zeros.

Let us see how we would generalize the inner product on R^n in Example 2 to the complex space $\mathbf{C}^n$. To define $\langle x, y\rangle = \sum_1^n x_j y_j$ is not much use, because in general $\langle x, x\rangle$ is complex; worse, it may be zero even when x is nonzero (let $x = (1, i) \in \mathbf{C}^2$), so that $\langle x, x\rangle^{\frac{1}{2}}$ doesn't look much like a norm. The obvious solution is to define

$$\langle x, y\rangle = \Sigma x_j \overline{y_j}$$

This has the property that $\langle x, x\rangle = \Sigma x_j \overline{x_j} = \Sigma |x_j|^2$, so that $\|x\|$ is the usual Euclidean norm. On the other hand property P2 is no longer true, but instead we have

P2′ $\langle x, y\rangle = \langle \overline{y, x}\rangle$.

12.3 With this in mind we define a **complex inner-product space** to be a complex linear space X with a function $\phi : X \times X \to \mathbf{C}$, denoted by $\langle x, y \rangle$ as before, such that:

$$\text{P1, P2}', \text{P3, P4 hold.}$$

From P1 and P2$'$ follows $\langle x, \alpha y_1 + \beta y_2 \rangle = \overline{\langle \alpha y_1 + \beta y_2, x \rangle} = \overline{(\alpha \langle y_1, x \rangle + \beta \langle y_2, x \rangle)} = \bar{\alpha} \overline{\langle y_1, x \rangle} + \bar{\beta} \overline{\langle y_2, x \rangle} = \bar{\alpha} \langle x, y_2 \rangle + \bar{\beta} \langle x, y_2 \rangle$; thus a complex inner-product is *linear* in the first argument and **conjugate-linear** in the second. The student is warned to take care: because of conjugate-linearity, many 'obvious' manipulations are not valid.

12.4 To show the difference between a real and a complex inner product we compare $\boldsymbol{R}^4$ with $\mathbf{C}^2$ – which as linear spaces over $\boldsymbol{R}$ can be regarded as identical via the correspondence $(x_1, x_2, x_3, x_4) \leftrightarrow (x_1 + ix_2, x_3 + ix_4)$. We have

$$\langle (x_1, x_2, x_3, x_4), (y_1, y_2, y_3, y_4) \rangle = x_1 y_1 + x_2 y_2 + x_3 y_3 + x_4 y_4$$

while

$$\langle (x_1 + ix_2, x_3 + ix_4), (y_1 + iy_2, y_3 + iy_4) \rangle$$
$$= x_1 y_1 + x_2 y_2 + x_3 y_3 + x_4 y_4 + i(x_2 y_1 - x_1 y_2 + x_4 y_3 - x_3 y_4)$$

so that one is the real part of the other.

12.5 Example 4 $\mathbf{C}^n$ is a complex inner-product space, as described above.

Example 5 The complex space ℓ_2 becomes an inner-product space under

$$\langle x, y \rangle = \sum_{j=1}^{\infty} x_j \bar{y}_j .$$

That the series converges follows from Hölder's inequality for $p = 2$, or, more directly, from the fact that if $a, b \geqslant 0$ then $ab \leqslant \max\{a^2, b^2\} \leqslant a^2 + b^2$ so that $\Sigma |x_j \bar{y}_j| \leqslant \Sigma(|x_j|^2 + |y_j|^2) < \infty$. Note that the inner product norm coincides with the usual ℓ_2 norm.

Example 6 Any complex L_2 space becomes an inner product space under

$$\langle f, g \rangle = \int f\bar{g}.$$

Arguing as in Example 5 we see that the integral exists, and the

$$\langle f, f\rangle^{\frac{1}{2}} = \left(\int f\bar{f}\right)^{\frac{1}{2}}$$

which is the usual L_2 norm.

12.6 A **Hilbert space** is a real or complex inner-product space which is complete in the inner-product norm. All the examples given so far are clearly Hilbert spaces, while

Example 7 (Complex) $\mathcal{C}[0,1]$ with the inner product $\int_0^1 f(t)\overline{g(t)}\,dt$ is an inner product space which, being a dense proper subspace of $L_2[0,1]$, must be incomplete.

We shall assume henceforth that the scalars are complex unless stated otherwise; this is the most important case and the reader can translate most of the results into the real case by ignoring 'complex conjugate' and 'real part' wherever they occur.

12.7 Theorem **(Cauchy–Schwarz inequality)** Let x, y be elements of an inner-product space X. Then

$$|\langle x, y\rangle| \leqslant \|x\|\ \|y\|.$$

Proof Choose $\alpha \in \mathbf{C}$ with $|\alpha| = 1$ such that $\alpha\langle x, y\rangle = |\langle x, y\rangle|$, and let $a = \|x\|^2$, $b = |\langle x, y\rangle|$, $c = \|y\|^2$. Then for all $t \in \boldsymbol{R}$, properties P1, P2$'$ and P3 give

$$\begin{aligned} 0 \leqslant \langle t\alpha x + y, t\alpha x + y\rangle &= t\alpha\langle x, t\alpha x + y\rangle + \langle y, t\alpha x + y\rangle \\ &= t\alpha . \overline{t\alpha}\langle x, x\rangle + t\alpha\langle x, y\rangle + \overline{t\alpha}\langle y, x\rangle + \langle y, y\rangle \\ &= at^2 + 2bt + c \end{aligned}$$

The usual 'discriminant test' for quadratic functions now gives $b^2 \leqslant ac$, and taking the square root of each side gives the result.

Variants of this quadratic function argument occur very frequently in proving inequalities in Hilbert spaces.

12.8 Corollary The inner-product norm is indeed a norm on X, and the inner product is jointly continuous with respect to this norm.

Proof The only norm property that is not trivial is the triangle inequality, $\|x + y\| \leqslant \|x\| + \|y\|$. If $\|x + y\| = 0$ this is obvious, and otherwise it follows from

$$\begin{aligned} \|x + y\|^2 = |\langle x + y, x + y\rangle| &\leqslant |\langle x + y, x\rangle| + |\langle x + y, y\rangle| \\ &\leqslant \|x + y\|\ \|x\| + \|x + y\|\ \|y\|. \end{aligned}$$

To say that $\langle\ ,\ \rangle$ is jointly continuous amounts to saying that $x_n \to x$, $y_n \to y$ imply $\langle x_n, y_n\rangle \to \langle x, y\rangle$. This follows from the identity $\langle x_n, y_n\rangle - \langle x, y\rangle = \langle x_n - x, y_n\rangle + \langle x, y_n - y\rangle$

and the consequent inequality

$$|\langle x_n, y_n\rangle - \langle x, y\rangle| \leqslant \|x_n - x\| \, \|y_n\| + \|x\| \, \|y_n - y\|.$$

The next few simple facts are analogues of theorems in elementary Euclidean geometry. We say two vectors x, y in an inner-product space are *orthogonal*, and write $x \perp y$, (pronounced 'x perp y') if $\langle x, y\rangle = 0$. Clearly $x \perp y \Leftrightarrow y \perp x$, and 0 is the only vector orthogonal to itself, as $x \perp x \Rightarrow 0 = \langle x, x\rangle = \|x\|^2$.

12.9 Proposition

(i) (**Pythagoras' Theorem**) If $x \perp y$ then $\|x + y\|^2 = \|x\|^2 + \|y\|^2$.

(ii) (**Generalized Pythagorean Theorem**) If $x_1, \ldots, x_n$ are pairwise orthogonal, that is if $x_i \perp x_j$ for $i \neq j$, then

$$\| \sum_1^n x_j \|^2 = \sum_1^n \|x_j\|^2$$

(iii) (**Parallelogram Identity**) For any x, y

$$\|x + y\|^2 + \|x - y\|^2 = 2(\|x\|^2 + \|y\|^2)$$

Proof The first of these follows by writing the left hand side as an inner product: $\|x + y\|^2 = \langle x + y, x + y\rangle = \langle x, x + y\rangle + \langle y, x + y\rangle = \langle x, x\rangle + \langle x, y\rangle + \langle y, x\rangle + \langle y, y\rangle = \|x\|^2 + \|y\|^2$ since the middle terms vanish. The others are similar and we leave them as an exercise.

More surprising is the fact that the inner product can be expressed in terms of the norm. For real spaces we have

$$4\langle x, y\rangle = \|x + y\|^2 - \|x - y\|^2$$

whereas for complex spaces it is easy to verify that this formula – which after all must be real – only gives $4\,\mathrm{Re}\langle x, y\rangle$, and instead we have

$$4\langle x, y\rangle = \|x + y\|^2 - \|x - y\|^2 + i\|x + iy\|^2 - i\|x - iy\|^2.$$

The reader is recommended to prove these forthwith, as useful practice in the manipulation of inner products.

Orthogonality and functionals in Hilbert space

Throughout this section H will denote a Hilbert space. The notation assumes complex scalars, but all the results and proofs are valid, with the appropriate changes of notation, for real Hilbert spaces. We begin a sequence of results which are of considerable interest and importance in their own right, but which for our purposes are mainly directed to showing that the dual space of H is (more-or-less) H itself. In fact we shall prove the following result (due independently to Fréchet and Riesz, 1907).

12.10 Theorem For each $a \in H$ the complex function on H defined by

$$f_a(x) = \langle x, a \rangle$$

is a bounded linear functional. Moreover the map

$$T: a \mapsto f_a$$

is a conjugate-linear congruence of H onto H^*. (By this we mean that $T(a + b) = Ta + Tb$, $T(\alpha a) = \bar{\alpha}\, Ta$, and $\|Ta\| = \|a\|$.)

Proof (Easy part) That f_a is linear comes immediately from property P1 of the inner product. That f_a is bounded with $\|f_a\| \leqslant \|a\|$ comes from the Cauchy–Schwarz inequality. That equality holds here is clear if $\|a\| = 0$ and otherwise it follows from

$$\|f_a\|\;\|a\| \geqslant |f_a(a)| = |\langle a, a \rangle| = \|a\|^2.$$

Turning now to the map $T: a \mapsto f_a$ we see that we have just proved that T is norm-preserving. That T is conjugate-linear follows similarly from the inner-product rules; it amounts to the identities

$$f_{a+b}(x) = \langle x, a + b \rangle = \langle x, a \rangle + \langle x, b \rangle = f_a(x) + f_b(x),$$

and

$$f_{\alpha a}(x) = \langle x, \alpha a \rangle = \bar{\alpha} \langle x, a \rangle = \bar{\alpha} f_a(x).$$

The only fact left unproved is that T is onto – in other words, that every $f \in H^*$ is an f_a for suitable a. This lies deeper, and we postpone it till 12.13.

Note that when the scalars are real, T is just a congruence in the ordinary sense.

The next theorem, a totally geometric one, nevertheless is the crucial fact needed to complete the proof of the last theorem, and has profound consequences also for the study of linear operators on a Hilbert space.

12.11 Theorem Let C be a closed convex subset of H. Then there is a unique point u of smallest norm in C. This point has the property that $\mathrm{Re}(u, x - u) \geqslant 0$ $(x \in C)$.

Remark The corresponding inequality when the scalars are real is $\langle u, x - u\rangle \geqslant 0$ for $x \in C$, which corresponds to the fact that the angle θ between the line joining 0 to u and the line joining u to x is acute ($\cos\theta \geqslant 0$) for each $x \in C$ (Diagram 16).

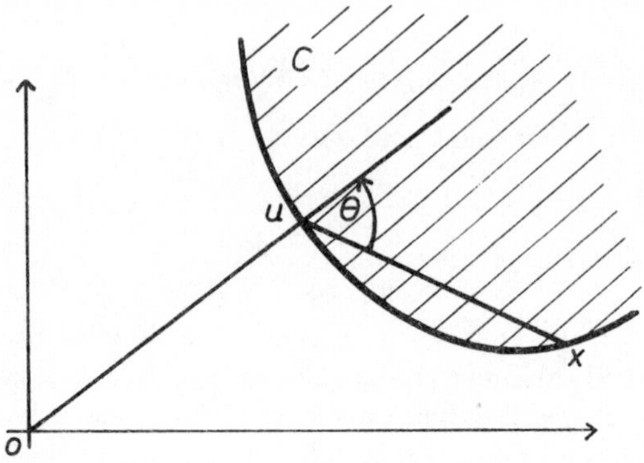

Diagram 16.

Proof of theorem Let $\alpha = \inf\{\|x\| : x \in C\}$ and consider the subset of C defined by $C_n = \{x \in C : \|x\| \leqslant \alpha + n^{-1}\}$. For any $x, y \in C$ we have $\frac{1}{2}(x + y) \in C$ since C is convex, so that $\|\frac{1}{2}(x + y)\|^2 \geqslant \alpha^2$ and $\|x + y\|^2 \geqslant 4\alpha^2$. By the Parallelogram identity,

$$\begin{aligned}\|x - y\|^2 &= 2\|x\|^2 + 2\|y\|^2 - \|x + y\|^2 \\ &\leqslant 2(\alpha^2 + \frac{1}{n}) + 2(\alpha^2 + \frac{1}{n}) - 4\alpha^2 = \frac{4}{n}.\end{aligned}$$

This shows that that diameter of C_n is at most $2n^{-\frac{1}{2}}$ and so the C_n form a sequence of (clearly closed and nonempty) decreasing subsets of C with diam $C_n \to 0$. Since H is complete, Cantor's Intersection Theorem 1.9 shows there is a unique point u lying in C_n for all n, which amounts to saying there is a unique point u of C with $\|u\| = \alpha$, as required.

For any $x \in C$, $0 < t \leqslant 1$ we have $u + t(x - u) = tx + (1 - t)u \in C$, so that $\|u\|^2 = \alpha^2 \leqslant \|u + t(x - u)\|^2 = \langle u + t(x - u), u + t(x - u)\rangle$, which, on expanding, becomes $\|u\|^2 + 2t\,\mathrm{Re}\langle u, x - u\rangle + t^2\|x - u\|^2$. Subtracting $\|u\|^2$ and dividing by $2t$ gives

$$0 \leqslant \mathrm{Re}\langle u, x - u\rangle + \tfrac{1}{2}t\|x - u\|^2 \quad (0 < t \leqslant 1)$$

which implies $\mathrm{Re}\langle u, x - u\rangle \geqslant 0$ $(x \in C)$ on letting $t \to 0$.

12.12 If A is an arbitrary nonempty subset of H, and $x \in H$, we write $x \perp A$, and say x is orthogonal to A, if $x \perp a$ for all $a \in A$, and we define the **orthogonal complement** $A^\perp$ of A by

$$A^\perp = \{x \in H : x \perp A\}.$$

The proof of the next theorem is closely connected with the fact that in ordinary geometry the shortest distance from a point to a line or plane is obtained by 'dropping a perpendicular'.

12.13 Theorem Let M be a closed linear subspace of H and x any point of H. Then there is a unique point v of M such that $x - v \in M^\perp$. (Of course $x \in M$ iff x and v coincide.)

Following geometrical language we call v the **(orthogonal) projection** of x onto M. The map $x \mapsto u$ is denoted P_M; it transpires (Problem 12F) that P_M is indeed a projection in the sense of Problems 10S and 10T.

Proof of Theorem In the last theorem take C to be the closed convex set $x + M$. Let u be the point of $x + M$ with smallest norm, and let $v = x - u$, so that $v \in M$. If $x - v = u$ were not in $M^\perp$ there would exist $m \in M$ with $\langle u, m\rangle \neq 0$, and multiplying m by a suitable scalar we may assume $\langle u, m\rangle = -1$. Now $u + m = x + (m - v) \in x + M$ so by the second part of the last theorem, $0 \leqslant \mathrm{Re}\langle u, (u + m) - u\rangle = \mathrm{Re}\langle u, m\rangle = -1$, a contradiction which shows $u \in M^\perp$, (see Diagram 17).

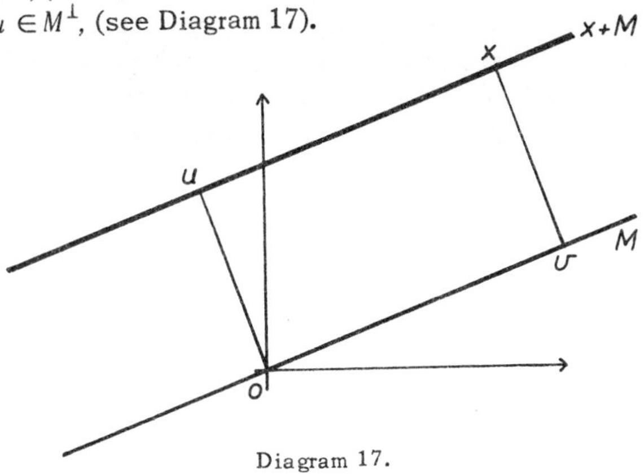

Diagram 17.

Uniqueness is shown by the fact that if v, w are two points of M with $x - v, x - w \perp M$ then since $v - w \in M$,

$$0 = \langle x - w, v - w \rangle - \langle x - v, v - w \rangle = \langle v - w, v - w \rangle$$

so $v = w$.

Completion of proof of theorem 12.10 We only have to show that every $f \in H^*$ is, in the notation of the theorem, an f_a for some $a \in H$. If $f = 0$ this is trivial. Otherwise, since $\ker f = N$ is then a proper closed subspace of H, the construction of Theorem 12.13 provides a *nonzero* $u \in N^{\perp}$ (choose $x \notin N$, and then $u = x - v \neq 0$). For any scalar α it is clear that $f(x)$ and $f_{\alpha u}(x)$ agree (both being zero) for all points x in N. For suitable choice of α we can make them agree at $x = u$; in fact $f(u) = f_{\alpha u}(u) = \langle u, \alpha u \rangle = \bar{\alpha} \|u\|^2$ shows that we must choose $\alpha = \overline{f(u)/\|u\|^2}$. Since $u \notin N$ and N is a hyperplane, $\text{lin}(u, N) = H$; thus f and $f_{\alpha u}$ agree on a spanning set and so are equal by Lemma 3.18. We have thus shown that if

$$a = \overline{f(u)}\, u/\|u\|^2$$

then $f_a = f$, and the theorem is proved.

Galois dualities

As we shall soon see, the orthogonal complement $A^{\perp}$ introduced in 12.12 is an important computational tool in Hilbert-space theory. We prove some preliminary results in this section which are nothing but very elementary set-theory, but which occur, in various versions, time and again in functional analysis – and many other areas of mathematics – as we see later in the book.

12.14 Let X and Y be any nonempty sets, and let $\sim$ be an arbitrary relation between members of X and members of Y. (Formally, $\sim$ is an arbitrary subset of $X \times Y$, and $x \sim y$ is short for $(x, y) \in \sim$.) Given $a \in X$, $B \subset Y$ we agree that $a \sim B$ is to mean that $a \sim y$ for all $y \in B$. We make the convention that $a \sim \emptyset$ for every a – which accords with the obvious fact that the smaller B is, the easier it is for $a \sim B$ to be true. Similarly if $A \in X$, $b \in Y$ we say that $A \sim b$ if $x \sim b$ for all $x \in A$, with the same convention when $A = \emptyset$. For our immediate purposes, $\sim$ is the relation 'perp' between vectors of H, in which case of course $X = Y$, and $A \sim b$ is the same as $b \sim A$, but in general this need not be so.

Given a subset A of X, we define its **dual set** to be the subset of Y defined by

$$A^{\sim} = \{y \in Y : A \sim y\}$$

and similarly, given $B \subset Y$ we define its dual set in X by

$$B_{\sim} = \{x \in X : x \sim B\}.$$

These operations, which assign to each set in X a set in Y and vice versa, are the **Galois duality operations** associated with the relation $\sim$. Clearly $A^{\sim}$ is the intersection, as x runs over A, of all the sets $\{x\}^{\sim} = \{y \in Y : x \sim y\}$ and from this follow some simple properties:

12.15 Lemma If A, B are subsets of X,

(i) $A \subset B \Rightarrow A^{\sim} \supset B^{\sim}$ and $(A \cup B)^{\sim} = A^{\sim} \cap B^{\sim}$.
(ii) $\emptyset^{\sim} = Y$.
(iii) $A \subset A^{\sim}{}_{\sim}$ (the latter is short for $(A^{\sim})_{\sim}$), and $A^{\sim}{}_{\sim}{}^{\sim} = A^{\sim}$.

Proof (i) is trivial from the remark before the Lemma, and (ii) follows from the convention that $\emptyset \sim y$ for all $y \in Y$. For (iii), if $x \in A$ then $x \sim y$ for all $y \in A^{\sim}$ by the definition of $A^{\sim}$, so that $x \sim A^{\sim}$, $x \in (A^{\sim})_{\sim}$, showing $A \subset A^{\sim}{}_{\sim}$. Replacing A by $A^{\sim}$ in the analogous formula for subsets of Y gives $A^{\sim} \subset (A^{\sim})_{\sim}{}^{\sim}$; replacing B by $A^{\sim}{}_{\sim}$ in the first formula of (i) gives the reverse inequality and therefore $A^{\sim} = A^{\sim}{}_{\sim}{}^{\sim}$. Note that we have implicitly used the fact that $A^{\sim}{}_{\sim}{}^{\sim}$, $(A^{\sim})_{\sim}{}^{\sim}$ *and* $(A^{\sim}{}_{\sim})^{\sim}$ are the same; since they are all short for $((A^{\sim})_{\sim})^{\sim}$, this is clear.

Where confusion is unlikely one usually uses the same symbol $A^{\sim}$ for both duality operations, whether A is a set in X or in Y, and we shall proceed to do so.

Especially significant are the **$\sim$-sets**, that is the sets M in either X or Y for which $M^{\sim\sim} = M$. We shall denote by $\mathfrak{X}, \mathfrak{Y}$ the classes of $\sim$-sets in X and Y respectively.

12.16 Proposition

(i) Let $M \subset X$. Then $M^{\sim\sim} = M \Leftrightarrow M = B^{\sim}$ for some $B \subset Y$.
(ii) For any $A \subset X$, $A^{\sim\sim}$ is the smallest $\sim$-set containing A.
Similar statements hold for subsets of Y.

Proof (i) If $M^{\sim\sim} = M$ then $M = B^{\sim}$ where $B = M^{\sim}$. Conversely if $M = B^{\sim}$ then $M^{\sim\sim} = B^{\sim\sim\sim} = B^{\sim} = M$ by part (iii) of the last lemma. (ii) $A^{\sim\sim}$ is certainly a $\sim$-set containing A. If M is any other such set then $M \supset A$ implies $M^{\sim} \subset A^{\sim}$, $M^{\sim\sim} \supset A^{\sim\sim}$, so that $A^{\sim\sim}$ is the smallest such set.

12.17 Corollary The duality operation $M \mapsto M^{\sim}$ is a one-to-one mapping of $\mathcal{X}$ onto $\mathcal{Y}$, and vice versa.

Proof It follows easily from part (i) of the Proposition and the definition of a $\sim$-set that

$$M \in \mathcal{X} \text{ and } N = M^{\sim} \Leftrightarrow N \in \mathcal{Y} \text{ and } M = N^{\sim}$$

from which the result is clear.

The last proposition and corollary show that in any specific case it is interesting and useful to find a simple explicit description of the $\sim$-sets. A more useful exercise still, when one has a set X and a particularly interesting class $\mathcal{X}$ of subsets of X, is to construct, as simply as possible, a relation $\sim$ between X and another set Y such that $\mathcal{X}$ is exactly the class of $\sim$-sets in X. The relation of this idea to the Hahn–Banach theorem is taken up in §14 and the problems at the end of that section.

Orthogonal complements in H

We now apply the results above to the case of the relation $x \perp y$ for vectors in a Hilbert space H.

12.18 Lemma

(i) $A \subset B \Rightarrow A^{\perp} \supset B^{\perp}$; $(A \cup B)^{\perp} = A^{\perp} \cap B^{\perp}$; $A \subset A^{\perp\perp}$; $A^{\perp\perp\perp} = A^{\perp}$.

(ii) If $x \in A \cap A^{\perp}$ then $x = 0$.

(iii) $\{0\}^{\perp} = H, H^{\perp} = \{0\}$.

(iv) $A^{\perp}$ is a closed linear subspace of H.

Proof

(i) follows at once from the previous section.

(ii) If $x \in A \cap A^{\perp}$ then $x \perp x$, so $x = 0$.

(iii) $\{0\}^{\perp} = H$ is clear, and $H^{\perp} = \{0\}$ follows from (ii).

(iv) The remark before Lemma 12.15 implies that $A^{\perp}$ is the intersection, as varies over A, of the sets $\{a\}^{\perp} = \{x \in H: \langle x, a \rangle = 0\}$, and the linearity and continuity (12.8) of the inner product show that each such set is a closed linear subspace, so that the same is true of $A^{\perp}$.

The next result is nothing but a restatement of Theorem 12.13 but nevertheless it is of prime importance. If L, M, N are subspaces of a linear space one says L is the **direct sum** of M and N, and writes $L = M \oplus N$, if each element of L is *uniquely* the sum of an element of M and an element of N. It is easily seen that this occurs iff $M + N = L$ and $M \cap N = \{0\}$.

12.19 Theorem If M is a closed linear subspace of H then $H = M \oplus M^{\perp}$.

Proof Clear from Theorem 12.13.

12.20 Corollary The sets such that $M^{\perp\perp} = M$ – that is, the $\perp$-sets of H – are precisely the closed linear subspaces.

Proof We have seen that every set M such that $M^{\perp\perp} = M$ – being the form $N^{\perp}$ where $N = M^{\perp}$ – is a closed linear subspace. Conversely if M is a closed linear subspace, our previous results give us

$$\text{(a) } H = M + M^{\perp} \quad \text{(b) } \{0\} = M^{\perp} \cap M^{\perp\perp} \quad \text{(c) } M \subset M^{\perp\perp}.$$

For any $x \in M^{\perp\perp}$, by (a) we can write $x = u + v$ with $u \in M$, $v \in M^{\perp}$. By (c), $u \in M^{\perp\perp}$, so that $v = x - u$ must be in both $M^{\perp\perp}$ and $M^{\perp}$ and so is zero by (b). Hence $x = u \in M$, proving that $M^{\perp\perp} = M$.

The following important result is now a triviality:

12.21 Theorem For any $A \subset H$, $A^{\perp\perp} = \overline{\text{lin}}\, A$. That is, a vector $x \in H$ can be approximated by linear combinations of vectors in A if and only if x is orthogonal to every vector which is orthogonal to every vector in A.

Proof Corollary 12.20 and Proposition 12.16 (ii).

It is worth pointing out that most of the above results on orthogonality are by no means trivial even in finite-dimensional spaces (and are equally useful tools there), but that many of the proofs are much easier – indeed, many of them can be established by purely algebraic arguments. In Problem 12L we ask the reader to fill in the details of these simpler proofs.

Orthonormal sets

In ordinary Euclidean space one introduces a rectangular coordinate system with coordinates (x_1, x_2, x_3) by expressing vectors in the form $x = \Sigma x_j e_j$ relative to three mutually orthogonal unit vectors – that is vectors of norm equal to unity – e_1, e_2, e_3. In a general Hilbert space, the generalization of this notion of rectangular coordinates plays an important role. The theory of this section contains the (non-trivial) case of finite-dimensional inner-product spaces as a special case, but it is infinite-dimensional spaces that concern us most, and the essential difference is topological; the expansion $x = \Sigma x_j e_j$ is then an *infinite* sum.

Most of the theorems, as well as some of the terminology, originated in the theory of L_2 *convergence of Fourier series*, and several bear the names of mathematicians who did the pioneering work on L_2 spaces around the turn of the century.

12.22 An **orthonormal set** in an inner-product space X is a subset E of X consisting of unit vectors such that any two distinct elements of E are orthogonal: in other words such that for $e, f \in E$.

$$\langle e, f \rangle = \begin{cases} 0, & e \neq f \\ 1, & e = f. \end{cases}$$

For convenience of notation we shall usually think of E as an indexed family $\{e_j\}_{j \in J}$ over some suitable index set J.

12.23 Example 1 In ℓ_2 or ℓ_2^n the standard basis vectors $e_1 = (1, 0, 0, \dots)$, $e_2 = (0, 1, 0, \dots)$, ... form an orthonormal set, with respect to the usual inner product $\langle x, y \rangle = \Sigma x_j \bar{y}_j$.

Example 2 In complex ℓ_2^n the vectors defined by

$$f_k = n^{-\frac{1}{2}} (1, \omega^k, \omega^{2k}, \dots \omega^{(n-1)k}) \; (k = 1, 2, \dots, n)$$

where $\omega = e^{2\pi i/n}$ is an nth root of unity, form a more interesting example of an orthonormal set in finite dimensions. Though this has no analogue in ℓ_2 it clearly has some similarity to

Example 3 In the complex space $L_2[-\pi, \pi]$ the functions e_n $(n \in \mathbf{Z})$ defined by $e_n(t) = (2\pi)^{-\frac{1}{2}} e^{int}$ $(t \in [-\pi, \pi])$ are easily verified to be an orthonormal set with respect to the usual inner product

$$\langle f, g \rangle = \int_{-\pi}^{\pi} f(t) \overline{g(t)} \, dt.$$

First we prove some simple results:

12.24 Proposition Let $\{e_1, e_2, \dots, e_n\}$ be a finite orthonormal set in an inner product space X, and M be its linear span. Then for any x in X,

(i) The vector $u = x - \sum_1^n \langle x, e_j \rangle e_j$ is in $M^\perp$;

(ii) $\sum_1^n |\langle x, e_j \rangle|^2 \leqslant \|x\|^2$ (**Bessel's Inequality**).

Proof (i) For each k we have

$$\begin{aligned}\langle u, e_k\rangle &= \langle x, e_k\rangle - \sum_{j=1}^{n} \langle x, e_j\rangle \langle e_j, e_k\rangle \\ &= \langle x, e_k\rangle - \langle x, e_k\rangle = 0\end{aligned}$$

since the only nonvanishing term in the sum is the one for $j = k$. Thus u is orthogonal to each e_k and hence to their linear span M.

(ii) By (i) the vectors $\langle x, e_j\rangle e_j$, $(1 \leqslant j \leqslant n)$, and u are pairwise orthogonal, so the generalized Pythagorean theorem gives

$$\begin{aligned}\|x\|^2 &= \sum_1^n \|\langle x, e_j\rangle e_j\|^2 + \|u\|^2 \\ &= \sum_1^n |\langle x, e_j\rangle|^2 + \|u\|^2 \geqslant \sum_1^n |\langle x, e_j\rangle|^2.\end{aligned}$$

12.25 Proposition An orthonormal set is linearly independent.

Proof If $e_1, \dots, e_n$ are vectors from some orthonormal set, and if $\lambda_1 e_1 + \dots + \lambda_n e_n = 0$, the generalized Pythagorean theorem applied to the mutually orthogonal vectors $\lambda_1 e_1, \dots, \lambda_n e_n$ gives $0 = \|0\|^2 = \sum_{j=1}^{n} \|\lambda_j e_j\|^2 = \Sigma |\lambda_j|^2$ so that all the λ_j are zero.

The reader's knowledge of ordinary space should tell him that, already in R^2, orthonormal sets exist in large numbers. The next Proposition describes the standard way of constructing an orthonormal set from a given (finite or infinite) countable set of vectors enumerated in some definite order. We describe it for an infinite set, the modification for a finite set being obvious.

12.26 Proposition (The **Gram–Schmidt Algorithm**) Let $x_1, x_2, \dots$ be a linearly independent sequence of vectors in an inner-product space X. Then the inductive rules

$$e_1 = \frac{x_1}{\|x_1\|},$$

$$e_n = \frac{y_n}{\|y_n\|} \text{ where } y_n = x_n - \sum_{j=1}^{n-1} \langle x_n, e_j\rangle e_j \qquad \text{for } n \geqslant 2,$$

define an orthonormal sequence $e_1, e_2, \dots$ such that for each n,

$$\operatorname{lin}\{x_1, \dots, x_n\} = \operatorname{lin}\{e_1, \dots, e_n\}.$$

Proof By hypothesis x_1 must be nonzero, so e_1 is well-defined. Assume, inductively, that orthonormal vectors $e_1, \ldots, e_{n-1}$ have been defined and that they span the same subspace, say M_{n-1}, as $x_1, \ldots, x_{n-1}$. By hypothesis $x_n \notin M_{n-1}$, so that y_n, as defined above, must be nonzero. By Proposition 12.24, y_n is orthogonal to $e_1, \ldots, e_{n-1}$ and hence $e_n = y_n/\|y_n\|$ is a well-defined unit vector such that $\{e_1, \ldots, e_n\}$ is orthonormal. It is clear that e_n is a linear combination of x_n and a vector in M_{n-1}, and so e_n, as well as $e_1, \ldots, e_{n-1}$, lie in $M_n = \text{lin}\{x_1, \ldots, x_n\}$. Since the latter space has dimension n and the $e_1, \ldots, e_n$ are a linearly independent set in it by Proposition 12.25, they must span it, which completes the induction step and the proof.

12.27 Corollary Every finite-dimensional inner-product space X has an orthonormal basis – that is, an orthonormal set that is a basis for X.

Proof Take $x_1, \ldots, x_n$ to be a basis for X in the finite version of the Gram–Schmidt Algorithm. The resulting $e_1, \ldots, e_n$ are also a basis.

For applied mathematicians the Gram–Schmidt process is important both as a practical computational tool and for its connexion with the study of *orthogonal polynomials* and other special functions. For instance if X is the real polynomials on $[-1,1]$ with the inner product $\langle f, g \rangle = \int_{-1}^{1} f(t)\, g(t)\, dt$, application of the Gram–Schmidt process to the functions $1, t^2, t^2, \ldots$ yields the **normalized Legendre polynomials.** Our study of orthonormal sets will centre however on the two-way relation between a vector x and its **coordinates**, or as they are termed for historical reasons its **Fourier coefficients**, c_j – that is the coefficients, if such exist, for which $x = \sum_{j \in J} c_j e_j$.

12.28 Recall from 11.2 that the identity for infinite sums of vectors

$$\sum_{j \in J} (\alpha x_j + \beta y_j) = \alpha \sum_{j \in J} x_j + \beta \sum_{j \in J} y_j$$

holds whenever the sums on the right converge. In H we can add to this the identity

$$\langle \sum_{j\in J} x_j , y\rangle = \sum_{j\in J} \langle x_j , y\rangle \quad \text{if } \sum_J x_j \text{ converges} \qquad (1)$$

(note that on the left we have a sum of vectors, on the right a sum of scalars). This follows easily from the fact that for any *finite* subset K of J,

$$\left|\langle x, y\rangle - \sum_{j\in K} \langle x_j , y\rangle\right| = \left|\langle x - \sum_{j\in K} x_j, y\rangle\right|$$
$$\leq \|x - \sum_{j\in K} x_j\| \, \|y\|,$$

and we leave the reader to give the details. Similarly it is true that

$$\langle y, \sum_{j\in J} x_j \rangle = \sum_{j\in J} \langle y, x_j\rangle \quad \text{if } \sum_J x_j \text{ converges}$$

We assume in the next few results that $\{e_j\}_{j\in J}$ is a given orthonormal set in a Hilbert space H.

12.29 Theorem Let $\{c_j\}$ be a family of scalars indexed over J. Then $\Sigma\, c_j\, e_j$ converges to some x in H iff $\sum_J |c_j|^2 < \infty$. The c_j can then be recovered from x by the formula $c_j = \langle x, e_j\rangle$.

Conversely for any $x \in H$, if $c_j = \langle x, e_j\rangle$ $(j \in J)$ then $\Sigma\, c_j\, e_j$ converges, but not necessarily to x.

Proof Let us use the notation s_K for $\sum_{j\in K} c_j\, e_j$ whenever K is a finite subset of J. Since the vectors $c_j\, e_j$ $(j \in K)$ are orthogonal, the generalized Pythagorean theorem gives

$$\|s_K\|^2 = \sum_{j\in K} \|c_j\, e_j\|^2 = \sum_{j\in K} |c_j|^2 \qquad (2)$$

for any such K.

First suppose that $\sum_J |c_j|^2 < \infty$. By the 'necessity' part of Theorem 11.3, there exists for each $\epsilon > 0$ a finite subset J_0 of J such that $\sum_{j\in K} |c_j|^2 < \epsilon^2$ for any finite subset K of J disjoint from J_0; by (2) this implies $\|s_K\| < \epsilon$ for any such K, and the 'sufficiency' part of Theorem 11.3 now shows that $\sum_J c_j\, e_j$ converges. Exactly the same argument in reverse shows that conversely, if $\sum_J c_j\, e_j$ converges then $\sum_J |c_j|^2 < \infty$. Suppose that $\sum_J c_j\, e_j$ converges to x. Then using the fact that $\langle e_j, e_k\rangle = \delta_{jk}$ we have by (1)

$$\langle x, e_k \rangle = \sum_{j \in J} \langle c_j e_j, e_k \rangle = \sum_{j \in J} c_j \langle e_j, e_k \rangle = c_k.$$

Finally let x be any vector in H, and define $c_j = \langle x, e_j \rangle$ $(j \in J)$. Then Proposition 12.24 (ii) implies that the partial sums of $\sum_J |c_j|^2$ are bounded above, in fact $\sum_{j \in K} |c_j|^2 \leqslant \|x\|^2$ for any finite subset K of J. Consequently $\sum_J |c_j|^2 < \infty$ and, by the first part, $\sum c_j e_j$ converges to some $y \in H$. That y need not equal x is easy to see: for instance x might be orthogonal to all the e_j, and then $c_j = 0$ $(j \in J)$, so $y = 0$.

It is clear in fact that $\sum_{j \in J} |c_j|^2 \leqslant \|x\|^2$ in the last part of this proof, so that Bessel's inequality is valid for infinite orthonormal sets as well as finite ones.

Note that one cannot use the theorem that an absolutely convergent sum is convergent (11.7) in the first part of the proof because it is not necessarily true that $\sum |c_j|^2 < \infty \Rightarrow \sum \|c_j e_j\| = \sum |c_j| < \infty$. For a simple example let $e_1, e_2, \ldots$ be the standard basis vectors (12.23) in ℓ_2 and let $c_n = 1/n$. Then $\sum |c_n|$ diverges, $\sum |c_n|^2$ converges, and $\sum c_n e_n$ is easily seen to converge to the vector $x = (1, \frac{1}{2}, \frac{1}{3}, \ldots)$ in ℓ_2.

12.30 Theorem 12.29 reduces the problem of representing a given vector x in terms of the orthonormal set $\{e_j\}$ to that of studying the series.

$$\sum_J \langle x, e_j \rangle \, e_j$$

which is called the **Fourier series** of x relative to the e_j. The scalars $c_j = \langle x, e_j \rangle$ are called the **coordinates** of x; geometrically they can be thought of as the components of x in the directions determined by the e_j, and Bessel's inequality can be interpreted as saying that the sum of the squares of the components of a vector in various mutually perpendicular directions cannot exceed the square of its length. Such interpretations are helpful in the abstract theory, but can become rather less so when applied to concrete Hilbert spaces such as $L_2[-\pi, \pi]$ (Example 3 above). In that example the Fourier series reduces to

$$\sum_{-\infty}^{\infty} c_n e_n(t) = \frac{1}{\sqrt{2\pi}} \sum_{-\infty}^{\infty} c_n e^{int}$$

where

$$c_n = \langle f, e_n \rangle = \int_{-\pi}^{\pi} f \, \bar{e}_n = \frac{1}{\sqrt{2\pi}} \int_{-\pi}^{\pi} f(t) \, e^{-int} \, dt,$$

f being a given function in $L_2[-\pi, \pi]$. This is called the **classical Fourier series** of f (strictly, the 'complex form' of this) and the coordinates c_n are the normalized classical **Fourier coefficients** of f – 'normalized' referring to the factor $(2\pi)^{-\frac{1}{2}}$ which is introduced to transform the e^{int} functions to unit vectors. By the last theorem we know that the above Fourier series is summable, in the L_2 norm, to *some* function in L_2, but for the purposes of analysis this is of meagre usefulness: what we want to know, in this and other concrete cases, is that the Fourier series of f converges to f, for every f. The relevant notion here is that of a *complete* orthonormal set. This important concept can be defined in several equivalent ways, as the next theorem shows: we shall proceed as follows. The class of all orthonormal subsets of H is clearly partially ordered by the relation $\subset$. A **complete orthonormal set** is defined to be one that is maximal in this partial ordering, in other words is not properly contained in any other orthonormal set.

12.31 Theorem Let $E = \{e_j\}_{j \in J}$ be an orthonormal set in H. Then the following are equivalent:

(i) E is complete.
(ii) The only vector orthogonal to e_j for all j is the zero vector, or in other words $E^\perp = \{0\}$.
(iii) The closed linear span of the e_j is H, in other words E is fundamental.
(iv) $x = \sum_{j \in J} \langle x, e_j \rangle \, e_j$ for all x in H.
(v) $\|x\|^2 = \sum_{j \in J} |\langle x, e_j \rangle|^2$ for all x in H (**Parseval's Equation**).

Proof (i) $\Rightarrow$ (ii): Assume E is complete. If there were some nonzero x such that $x \perp e_j$ for all e_j then $E \cup \{e\}$, where $e = x/\|x\|$, would form a strictly larger orthonormal set, a contradiction.

(ii) $\Rightarrow$ (iii): If $E^\perp = \{0\}$ then by Theorem 12.21, $\overline{\text{lin}}\, E = E^{\perp\perp} = \{0\}^\perp = H$, while conversely if $\overline{\text{lin}}\, E = H$ then $E^\perp = (E^{\perp\perp})^\perp = (\overline{\text{lin}}\, E)^\perp = H^\perp = \{0\}$.

(ii) $\Rightarrow$ (iv): Assume (ii) and let x be any element of H, and let $c_j = \langle x, e_j \rangle$ for each j. By Theorem 12.21, $\sum c_j e_j$ converges to an element v of H, such that for each j,

$$\langle v, e_j \rangle = c_j = \langle x, e_j \rangle .$$

Thus $\langle x-v, e \rangle = 0$ for each j, so $x - v = 0$, $x = v$ by (ii).

(iv) ⇒ (v): Assume (iv), let $x \in H$ and write $c_j = \langle x, e_j \rangle$. By the inner product identity (1) of 12.28.

$$\|x\|^2 = \langle x,x \rangle = \langle x, \sum_{j \in J} c_j e_j \rangle = \sum_{j \in J} \langle x, c_j e_j \rangle = \sum_{j \in J} c_j \langle x, e_j \rangle$$
$$= \sum_{j \in J} |c_j|^2.$$

(v) ⇒ (i): Assume (v). If E fails to be complete there is a unit vector e such that $E \cup \{e\}$ forms a larger orthonormal set. But then $e \perp e_j$ for all j, so (v) gives the contradiction $1 = \|e\|^2 = \sum_{j \in J} |\langle e, e_j \rangle|^2 = 0.$

Thus E must be complete.

It should be pointed out that though the equivalence of condition (iii) with the remaining ones was proved with the help of the quite deep theorem 12.21, it is not hard to prove directly (Problem 12N).

12.32 It is of interest from the point of view of the abstract theory to know that complete orthonormal sets in H always *exist*. This is a consequence of Corollary 12.27 when H has finite dimension, and of Theorem 12.34 when H is separable and infinite-dimensional. In the general case it follows from an easy Zorn's Lemma argument, which the reader is invited to provide. A more important question in practice is: when faced by an orthonormal set, how to determine whether it is complete? Criteria (ii) and (iii) above are the most useful for the purpose. Some orthonormal sets are obviously complete: for instance the standard basis vectors $\{e_1, \dots, e_n\}$ in ℓ_2^n, or the corresponding $\{e_n\}$ in ℓ_2, for which criterion (ii) clearly holds. On the other hand the completeness of the orthonormal set of functions $e_n : t \mapsto (2\pi)^{-\frac{1}{2}} e^{int}$ in $L_2[-\pi, \pi]$ (Theorem 13.4) is far from obvious, and is a deep and powerful result with many important applications.

Separable Hilbert spaces

We conclude this chapter by proving that all separable infinite dimensional Hilbert spaces with the same scalars are, from the point of view simply of their inner-product structure, identical: by which we mean that for any two such spaces H and K there is an inner-product preserving linear isomorphism of H onto K. To bring home the startling nature of this result let us note some of the spaces in this class of 'abstractly identical' Hilbert spaces: the sequence space ℓ_2; the spaces $L_2[-\pi, \pi]$, $L_2(\boldsymbol{R})$ and $L_2(\boldsymbol{R}^2)$ (with respect to one- or two-dimensional Lebesgue measure); any closed infinite-dimensional subspace of any of these. For the reader with some knowledge of complex analysis we mention an important example of such a subspace:

12.33 Example The set H_2 of all functions f analytic in the open unit disc Δ and such that $\int_\Delta |f|^2 \, d\alpha < \infty$, where α denotes two-dimensional Lebesgue measure (area measure), is a complex separable infinite-dimensional Hilbert space with the inner product $\langle f, g\rangle = \int_\Delta f\bar{g} \, d\alpha$. It is clear that H_2 is a linear subspace of the Hilbert space $L_2(\Delta)$ of all α-measurable complex functions f on Δ with $\int_\Delta |f|^2 \, d\alpha < \infty$; what is interesting is that H_2 is *closed* in $L_2(\Delta)$, and therefore complete.

12.34 Proposition A Hilbert space is separable and infinite-dimensional $\Leftrightarrow$ it has a countably infinite complete orthonormal set.

Proof If H has a countably infinite complete orthonormal set $E = \{e_1, e_2, \dots\}$ then it is infinite-dimensional by Proposition 12.25 and separable by Theorem 12.31 (iii) and Proposition 9.15. Conversely, suppose H is separable and infinite-dimensional, and let $C = \{x_1, x_2, \dots\}$ be a countable dense set in H. By discarding any x_n which is a linear combination of $x_1, \dots, x_{n-1}$ we produce a subset $\{y_1, y_2, \dots\}$ of $\{x_n\}$ (which must be infinite since H is infinite-dimensional) with the property that every x_n is a linear combination of the y_m's, and the additional property of linear independence. Applying the Gram–Schmidt process 12.26 to $\{y_n\}$ produces an infinite orthonormal sequence $E = \{e_1, e_2, \dots\}$ such that each y_n, and hence each x_n, is a linear combination of the e_n's. Thus $\operatorname{lin} E$ contains C and so is dense. By Theorem 12.31 (iii), E is a complete orthonormal set.

12.35 Theorem Let H be a separable, infinite dimensional (real or complex) Hilbert space. Then there is a linear isomorphism T of H onto (real or complex) ℓ_2 with the property that

$$\langle Tx, Ty\rangle = \langle x, y\rangle \quad (x, y \in H).$$

We call such a T an **inner-product isomorphism.** In particular, taking $x = y$, we see that T is a congruence.

Proof By the previous proposition H has a countably infinite complete orthonormal set $\{f_1, f_2, \ldots\}$. If we associate to each $x \in H$ its Fourier coefficients

$$c_n = \langle x, f_n \rangle \quad (n = 1,2, \ldots)$$

then by Theorem 12.31 (v)

$$\|x\|^2 = \sum_1^\infty |c_n|^2.$$

This proves that the sequence $c = (c_1, c_2, \ldots)$ is a member of ℓ_2 and that $\|c\|_2 = \|x\|$, so that the map T: $x \mapsto c$, which is clearly linear, is a congruence (in particular a linear isomorphism) of H into ℓ_2. Now for any $a = (a_1, a_2, \ldots) \in \ell_2$, Theorem 12.29 shows that $\sum_1^\infty a_n f_n$ converges to some $x \in H$ and that $\langle x, f_n \rangle = a_n$ – in other words $Tx = a$, so that T is onto.

That $\langle Tx, Ty \rangle = \langle x, y \rangle$ amounts to saying that if x, y have Fourier coefficients a_n, b_n respectively then $\langle x, y \rangle = \sum_{n=1}^\infty a_n \bar{b}_n$. We leave this as an exercise for the reader, using the inner-product identities 12.28. An alternative proof, which the reader may find instructive, uses the identities after 12.9 which express the inner product in terms of the norm.

It is clear that the relation of (inner-product) isomorphism between two Hilbert spaces is an equivalence relation, and thus if two Hilbert spaces H, K are both isomorphic to ℓ_2 in the above sense then they are isomorphic to each other, so that we have proved the

12.36 Corollary Any two separable infinite-dimensional Hilbert spaces are inner-product isomorphic.

It is easy to see from the proof of the Theorem that each vector f_n maps to the standard nth basis vector e_n in ℓ_2, and that once the complete orthonormal set $\{f_n\}$ has been specified, the operator T of the Theorem is the unique $T \in \mathcal{B}(H, \ell_2)$ with the property $Tf_n = e_n \, (n = 1, 2, \ldots)$. It is clear that there is a very large number of different isomorphisms T of one separable infinite-dimensional Hilbert space onto another – roughly as many as there are complete orthonormal sets. In concrete cases it will usually turn out that one is more convenient to work with than another: anyone who has solved problems in three-dimensional geometry will agree that half the battle is in choosing the right isomorphism of Euclidean space onto R^3.

† *Application: The Radon–Nikodym Theorem*

Our study of Hilbert spaces has centered on the Riesz–Fréchet theorem on the representation of functionals; the orthogonal complement operation; and expansions in terms of orthonormal sets. We shall see applications of the latter two topics shortly, and we show in this section how the Riesz–Fréchet theorem can be used in a neat proof of an important result of measure theory, the Radon–Nikodym Theorem.

12.37 Let $(S, \mathcal{S})$ be a measurable space and let μ and ν be two measures on the σ-algebra $\mathcal{S}$. We say ν is **absolutely continuous** with respect to μ, and write $\nu \ll \mu$, if for any $A \in \mathcal{S}$,

$$\mu(A) = 0 \Rightarrow \nu(A) = 0,$$

that is, the μ-null sets are ν-null. The name originated in the classical theory of functions of a real variable (see Rudin **11**, ch 8. for details). A fact one should know about absolute continuity, though we have no need for it here, is that if $\nu \ll \mu$ and ν is finite then given $\epsilon > 0$ there exists $\delta > 0$ such that $\nu(A) < \epsilon$ whenever $\mu(A) < \delta$; showing that a kind of continuity is indeed involved.

If μ is given it is easy to construct large numbers of ν such that $\nu \ll \mu$:

12.38 Lemma Let f be any measurable function with $0 \leqslant f < \infty$ μ-almost everywhere, and define

$$\nu(A) = \int_A f \, d\mu \quad (A \in \mathcal{S})$$

Then ν is a measure and $\nu \ll \mu$.

Proof If $\{A_n\}$ is a sequence of pairwise disjoint measurable sets whose union is A then $\chi_A f$ is the monotone limit of the functions $\chi_{A_1} f + \ldots + \chi_{A_n} f$ as $n \to \infty$ and it follows by the Monotone Convergence Theorem that

$$\nu(A) = \int \chi_A f \, d\mu = \lim_n \int (\chi_{A_1} f + \ldots + \chi_{A_n} f) \, d\mu$$

$$= \lim_n \sum_1^n \int \chi_{A_n} f \, d\mu$$

$$= \sum_1^\infty \int_{A_n} f \, d\mu = \sum_1^\infty \nu(A_n),$$

with the usual conventions about extended real numbers, so that ν is a measure. It is obvious that $\nu \ll \mu$.

Note that ν will be finite iff f is μ-integrable.

It is convenient (especially when μ is Lebesgue measure, the commonest case) to use a 'differential' notation. If ν is defined by the above construction one writes

$$d\nu = f \, d\mu;$$

a logical notation, since $\int_A d\nu = \nu(A) = \int_A f \, d\mu$, and more generally because of the

12.39 Proposition

(i) For any non-negative measurable g,

$$\int g \, d\nu = \int g f \, d\mu$$

(ii) A finite real or complex measurable function g is in $L_1(\nu)$ iff gf is in $L_1(\mu)$ and then

$$\int g \, d\nu = \int g f \, d\mu.$$

We leave the details of the proof as an exercise: (i) is immediate for characteristic functions, hence for non-negative simple functions, hence by the Monotone Convergence Theorem for the general case; and (ii) follows from (i).

The main theorem is a representation theorem, showing that under suitable conditions the converse of the Lemma is true. It must be admitted that the Hilbert space proof given below does not really 'show what is going on', and the student is strongly advised to compare it with a direct proof such as that in Bartle (1, p. 85).

In the course of the proof several different measures are under consideration, so that care has to be taken with phrases like 'almost everywhere' to specify the measure concerned.

12.40 Radon–Nikodym Theorem Let $(S, \mathfrak{S})$ be a measurable space on which are defined finite measures μ, ν with $\nu \ll \mu$. Then there is a finite non-negative measurable function h such that $d\nu = h\,d\mu$.

Remark The result is valid also when μ and ν are σ-finite, which is important because Lebesgue measure in σ-finite. In the problems we suggest methods of deriving the σ-finite from the finite case. On the other hand the Theorem fails in general for non-σ-finite measures.

Proof of theorem Let $\lambda = \mu + \nu$; then λ is also a finite measure. Define a linear functional on the Hilbert space $L_2(\lambda)$ – where it will be convenient to assume the scalars are *real* – by

$$\phi(f) = \int f\,d\nu, \quad f \in L_2(\lambda).$$

The constant function 1 is in $L_2(\lambda)$ and by Hölder's inequality

$$|\phi(f)| \leqslant \int |f|\,d\nu \leqslant \int |f|\,d\lambda \leqslant \|f\|_2\,\|1\|_2$$

which shows ϕ is bounded. By the Riesz–Fréchet theorem 12.10 there is a function $g \in L_2(\lambda)$ such that

$$\phi(f) = \int f\,g\,d\lambda, \quad f \in L_2(\lambda).$$

In other words we have

$$\int f\,d\nu = \int f\,g\,d\mu + \int f\,g\,d\nu, \quad f \in L_2(\lambda). \tag{1}$$

Taking f to be $-g_-$, where g_- is the negative part of g, we have $fg = g_-^2$ and so

$$0 \geqslant \int -g_-\,d\nu = \int g_-^2\,d\mu + \int g_-^2\,d\nu \geqslant 0$$

which shows that g_-^2 is a μ-null and ν-null function and hence $g \geqslant 0$ μ-a.e. and ν-a.e. By re-defining g on a suitable null set we may assume $g \geqslant 0$ everywhere.

Since λ is finite, every non-negative simple function ϕ is in $L_2(\lambda)$. Since every non-negative measurable function f is a monotone limit of simple functions, the Monotone Convergence Theorem, and the fact that $g \geqslant 0$, extend (1) to the case where f is any non-negative measurable function. Now by repeated substitution in (1) we have for any such f

$$\int f\,d\nu = \int f\,g\,d\mu + \int f\,g\,d\nu \tag{2}$$
$$= \int f\,g\,d\mu + \int f\,g.g\,d\mu + \int f\,g.g\,d\nu$$
$$= \dots \quad \dots$$
$$= \int f(g + g^2 + \dots + g^n)\,d\mu + \int f\,g^n\,d\nu$$

Taking $f = 1$ and noting that all quantities in sight are $\geqslant 0$, we see that $\int (g + g^2 + \dots + g^n)\,d\mu$ is bounded above as $n \to \infty$. This is only possible if

$$g < 1 \quad \mu\text{-a.e.}$$

and therefore

$$g < 1 \quad \nu\text{-a.e.}$$

(It is only at this innocent-looking point that the proof uses the absolute continuity of ν with respect to μ.) If we define

$$h(s) = \begin{cases} \sum_1^\infty g(s)^n & \text{where } g(s) < 1 \\ 0 & \text{otherwise} \end{cases}$$

then $g + g^2 + \dots + g^n \to h$ monotonely μ-a.e., while $g^n \to 0$ monotonely ν-a.e., so that taking $f = \chi_A$ in (2) and applying the Monotone Convergence Theorem to the integrals on the right gives

$$\int \chi_A\,d\nu = \int \chi_A h\,d\mu + \int \chi_A 0\,d\nu$$

or

$$\nu(A) = \int_A h\,d\mu \qquad (A \in S)$$

completing the proof.

Problems

Inner products and geometrical properties

12A

(i) Show that the following defines an inner product on $\mathcal{C}[0,1]$:

$$\langle f, g \rangle = \sum_{n=1}^{\infty} n^{-2} f(t_n)\,\overline{g(t_n)},$$

where $\{t_1, t_2, \ldots\}$ is a fixed enumeration of the rational points in $[0,1]$.

(ii) Show that $\mathcal{C}[0,1]$ is not complete in the associated norm, and that this norm is not equivalent to that derived from the inner product $\langle f, g \rangle = \int_0^1 f(t)\overline{g(t)}\,dt$.

12B Complete the proof of Proposition 12.9.

12C Define a (non-linear) transformation V on the set of non-zero vectors in an inner-product space by $Vx = x/\|x\|^2$. Show that

$$\|Vx - Vy\| = \frac{\|x - y\|}{\|x\|\,\|y\|},$$

and deduce that if a, b, c, d are four points of the space then

$$\|a-c\|\,\|b-d\| \leqslant \|a-b\|\,\|c-d\| + \|a-d\|\,\|b-c\|.$$

[First reduce to the case $a = 0$] The map V, an **inversion**, has many applications in Euclidean geometry.

12D Show that vectors $u_1, \ldots, u_n$ in an inner product space are linearly independent iff their **Gram matrix** $A = (a_{ij})$ where $a_{ij} = \langle u_i, u_j \rangle$, $(i, j = 1, \ldots, n)$ is nonsingular. (See Problem 12R for the significance of the Gram matrix.)

Orthogonality and functionals

12E Deduce from Theorem 12.11 that every closed convex set C in a Hilbert space has the **unique nearest point property,** namely that for any x in H there is a unique point of C closest to x. Verify also that the orthogonal projection of a point x on a closed subspace M is the point of M nearest to x (Theorem 12.13.)

12F Let P denote the orthogonal projection map P_M onto a closed nonzero linear subspace of a Hilbert space H (Theorem 12.13).

Verify that:

(a) P is a linear operator satisfying $P^2 = P$ (i.e. $P(Px) = Px$) – see Problems 10S, 10T.

(b) $\langle Px, y\rangle = \langle x, Py\rangle$ for all x, y.

(c) $\|P\| = 1$.

Show conversely that any operator P satisfying (a), (b) is the (orthogonal) projection on some closed subspace.

12G Let $\mathcal{P}_n$ denote the set of real polynomial functions of a real variable, with degree $\leqslant n$. Show that there is a unique $p \in \mathcal{P}_n$ such that

$$\int_{-1}^{0} f(t)\,dt = \int_{0}^{1} f(t)\,p(t)\,dt \qquad (f \in \mathcal{P}_n).$$

[Introduce the inner product $\langle f, g \rangle = \int_0^1 f(t)\,g(t)\,dt$.]
Find p in the case $n = 2$.

Galois dualities

12H Identify $A^{\sim}{}_{\sim}$, for any subset A of X, in the following examples of Galois dualities:

(i) $X = Y = \mathbf{R}^2$, $(x_1, x_2) \sim (y_1, y_2)$ means $x_1 \leqslant x_2$ and $y_1 \leqslant y_2$.

(ii) X is a topological space, Y is the class of all closed subsets of X, and $x \sim F$ means $x \in F$.

12I Given a Galois duality between sets X and Y, and a pair of $\sim$-sets A, B in X, show that $(A \cap B)^{\sim} = (A^{\sim} \cup B^{\sim})^{\sim\sim}$. Deduce that, for closed linear subspaces M, N of a Hilbert space,

$$(M \cap N)^{\perp} = (M^{\perp} + N^{\perp})^{-}.$$

Orthonormal sets

12J Verify that the vectors f_k in 12.23 Example 2 are an orthonormal basis for $\mathbf{C}^n$.

12K In the space $\mathcal{P}_3$ (Problem 12G) carry out the Gram-Schmidt process on the functions $1, t, t^2, t^3$ in that order, with respect to the inner product $\langle f, g \rangle = \int_{-1}^{1} f(t)\,\overline{g(t)}\,dt$. The result is the first four **normalized Legendre polynomials.**

12L (See the remarks after Theorem 12.21.) Suppose H is a finite dimensional inner product space. Prove the following by purely algebraic means:

(i) For any linear subspace M and any x in H there is a unique $u \in M$ such that $x - u \perp M$. Consequently $M^{\perp} \oplus M = H$. [Choose an orthonormal basis for M and use 12.24 (ii).] (See 12.13, 12.19.)

(ii) The $\perp$-sets in H are just the linear subspaces (12.20).

(iii) $A^{\perp\perp} = \operatorname{lin} A$ for any subset A of H. (12.21).

12M (To Theorem 12.29) Show that for any orthonormal set $\{e_j\}_{j \in J}$ in a Hilbert space H, and any $x \in H$, the Fourier series

$$\sum_J \langle x, e_j \rangle\, e_j$$

converges to the projection of x on the closed linear span of the e_j.

12N In theorem 12.31 prove the equivalence of (iii) with the remaining conditions without using Theorem 12.13 or its consequences.

[Use the last problem.]

12O Let $\{e_1, \ldots, e_n\}$ be a finite orthonormal set in an inner-product space X and let $x \in X$. If $\alpha_1, \ldots, \alpha_n$ are arbitrary scalars, show by direct computation that $\|x - \sum_{j=1}^{n} \alpha_j e_j\|$ attains its minimum value iff $\alpha_j = \langle x, e_j \rangle$ for each j.

[Expand $\|x - \Sigma \alpha_j e_j\|^2$, add and subtract $\Sigma |\langle x, e_j \rangle|^2$, and obtain an expression of the form $\Sigma |\langle x, e_j \rangle - \alpha_j|^2$ in the result.]

Generalize to infinite orthonormal sets, assuming X complete.

12P Agree that an **orthogonalization** of a linearly independent sequence of vectors $x_1, x_2, \ldots$ in an inner-product space means any orthonormal sequence $\{f_1, f_2, \ldots\}$ such that lin $\{f_1, \ldots, f_n\}$ = lin $\{x_1, \ldots, x_n\}$ for each n. Show that the Gram–Schmidt orthogonalization $\{e_n\}$ of 12.26 is 'almost unique' in the sense that if $\{f_n\}$ is another orthogonalization of $\{x_n\}$ then there exist scalars α_n of modulus 1 such that $f_n = \alpha_n e_n$ for each n.

Further problems

12Q Let the vectors x, y in a Hilbert space H have coordinates $\{c_j\}$, $\{d_j\}$ with respect to a complete orthonormal set $\{e_j\}_{j \in J}$. Show that

$$\langle x, y \rangle = \sum_{j \in J} c_j \bar{d}_j .$$

12R Let M be a finite-dimensional subspace of an inner-product space X and w a vector in X; and suppose a (not necessarily orthonormal) spanning set $u_1, \ldots, u_n$ of M is known. An oft-recurring problem in numerical analysis is that of computing scalars $\alpha_1, \ldots, \alpha_n$ such that $y = \sum_{j=1}^{n} \alpha_j u_j$ is the point of M nearest to w. (This is the mathematical model of what statisticians call linear multiple regression.)

(i) Let (a_{ij}) be the Gram matrix (Problem 12D) of the u_j and define $b_i = \langle w, u_i \rangle$ $(i = 1, \ldots, n)$. Show that, whether or not the u_j are linearly independent, the set of equations

$$\sum_{j=1}^{n} a_{ij} \alpha_j = b_i \ (i = 1, \ldots, n)$$

has solutions and that any solution $(\alpha_1, \ldots, \alpha_n)$ solves the problem above.

(ii) Another approach is clearly to apply the Gram–Schmidt process to obtain an orthonormal basis $\{e_1, \ldots, e_k\}$, say, of M, and then to set $y = \sum_1^k \langle w, e_j \rangle \, e_j$ (see Problem 12M) and to disentangle the α_j from this formula. Discuss conditions which might render either of these methods preferable to the other one.

12S Let M be a finite-dimensional linear subspace of $\mathcal{C}[a, b]$, and let M carry the inner product $\int_a^b f(t)\,\overline{g(t)}\,dt$. Show that the function K on the square $[a, b] \times [a, b]$ (the **Bergman kernel** of M). defined by

$$K(s, t) = \sum_{r=1}^{n} \overline{e_r(s)}\, e_r(t) \qquad (s, t \in [a, b])$$

is always the same for any orthonormal basis $\{e_1, \ldots, e_n\}$ of M.

Show also that the Bergman kernel has the property

$$f(s) = \int_a^b K(s, t)\, f(t)\, dt$$

for all f in M and s in $[a, b]$.

Find the kernel of $\mathcal{P}_2$ (Problem 12G) regarded as a subspace of $\mathcal{C}[-1, 1]$.

12T Show that a norm on a linear space, for which the parallelogram identity 12.9(iii) holds, can be derived from some inner product on the space.

Radon–Nikodym Theorem

12U Two ways of extending Theorem 12.40 to the case where μ and ν are σ-finite measures:

(i) Show that there is a partition of S into a sequence of measurable sets S_n on each of which both μ and ν are finite. Apply the Theorem to get a function f_n on S_n with

$$\nu(A) = \int_{S_n} f_n \chi_A \, d\mu \qquad (A \subset S_n)$$

and piece the f_n together to a function f on S.

(ii) Show there is a strictly positive measurable function p on S which is both μ- and ν-integrable. Define μ' and ν' by $d\mu' = p\, d\mu$, $d\nu' = p\, d\nu$; then μ' and ν' are both finite measures and $\nu' \ll \mu'$. Hence there is f such that $d\nu' = f\,d\mu'$. It follows easily that $d\nu = f\,d\mu$.

13. Two topics in Fourier series

13.1 In 12.30 we introduced the Fourier series of a real or complex function on the interval $[-\pi, \pi]$ relative to the set of functions $e_n(t) = (2\pi)^{-\frac{1}{2}} e^{int}$. It will be more convenient in this section to drop the normalizing factor $(2\pi)^{-\frac{1}{2}}$, so that the **classical Fourier series** of f is

$$\sum_{n=-\infty}^{\infty} c_n e^{int} \tag{1}$$

where the c_n are the (non-normalized) **classical Fourier coefficients** given by

$$c_n = \frac{1}{2\pi}\int_{-\pi}^{\pi} f(t)\, e^{-int}\, dt \qquad (n \in Z). \tag{2}$$

Arguing as in 6.26 one obtains the equivalent **real form** of the Fourier series

$$\tfrac{1}{2} a_0 + \sum_{n=1}^{\infty} (a_n \cos nt + b_n \sin nt) \tag{3}$$

which is often more convenient, especially in applied mathematics, and was the form used in the example in 10.7. We leave it as an easy exercise for the reader to verify that the a_n, b_n are determined by the integral formulae given in 10.7. In the last chapter it was natural to think of f as belonging to the Hilbert space $L_2[-\pi, \pi]$ of square-integrable functions, but the definition of the c_n makes sense whenever f belongs to the (larger) class $L_1[-\pi, \pi]$ of integrable functions. Purely as formal notation we shall write

$$f \sim \sum c_n e^{int}, \quad \text{or } f(t) \sim \sum c_n e^{int}, \quad \text{or } f \sim \{c_n\}$$

to mean that $f \in L_1[-\pi, \pi]$ and the c_n are the Fourier coefficients of f.

Convergence of the series

13.2 The student should lay it to heart that *in general* there is no guarantee that the series (1) converges pointwise or in any other sense, whether to f or to any other function. The basic problem of Fourier series is to find when $\sim$ can be replaced by $=$ in some appropriate kind of convergence. For instance we can ask:

If $f \in L_p$ does its Fourier series converge in the L_p norm?
(Yes, when $p = 2$; no, in general.)
If f is continuous does its Fourier series converge uniformly?
(No, in general.)

Some quite simple questions in this area are still unanswered, and may even be unanswerable: for instance the problem of finding necessary and sufficient conditions for the Fourier series of an L_1 function to converge pointwise. On the other hand there are simple tests for pointwise and uniform convergence which dispose of these problems for most functions that the applied mathematician is likely to come across (Problem 13L).

13.3 The reason for choosing the c_n in the series (1) according to formula (2) was explained in 12.30, and derives from the *orthogonality* of the functions e^{int}; but to bring this down to earth we remind the reader that the latter fact amounts to the formula

$$\int_{-\pi}^{\pi} e^{int}\, e^{-imt}\, dt = \left\{ dt = \begin{array}{ll} 0, & m \neq n \\ 2\pi & m = n \end{array} \right.$$

and thus if

$$f(t) = \sum_{n=-\infty}^{n} c_n\, e^{int}$$

is to be true in a reasonable enough sense for term-by-term integration to be valid, then on multiplying by e^{-imt} and integrating over $[-\pi, \pi]$ we see that all the terms of the series vanish except the mth, which equals $2\pi c_m$, and (2) follows.

It is often convenient to deal with functions defined on the whole of $\boldsymbol{R}$ and periodic (2π). If f is defined on $[-\pi, \pi]$ we shall denote by $\tilde{f}$ the (unique) periodic (2π) function on $\boldsymbol{R}$ which agrees with f on $[-\pi, \pi)$. Note that f and $\tilde{f}$ generally differ at $t = \pi$; thus $\tilde{f}$ will be continuous if and only if f is continuous and $f(-\pi) = f(\pi)$; that is (as noted in Corollary 6.28) iff f belongs to the space $\mathcal{C}_p = \mathcal{C}_p[-\pi, \pi] = \{f \in \mathcal{C}[-\pi, \pi]\colon f(-\pi) = f(\pi)\}$.

A useful result of the $\tilde{f}$ notation, and of the periodicity of the functions e^{int}, is that the Fourier coefficients of f can be computed by integration over *any* interval of length 2π:

$$c_n = \frac{1}{2\pi}\int_{a-\pi}^{a+\pi} \tilde{f}(t)\, e^{-int}\, dt \tag{4}$$

as the reader can easily verify.

The **standard** k**th partial sum** s_k of the Fourier series is the function $s_k(t) = \sum_{n=-k}^{k} c_n e^{int}$ and it is easy to see that this is the same as the kth partial sum $\frac{1}{2}a_0 + \sum_{n=1}^{k} (a_n \cos nt + b_n \sin nt)$ of the real form (3) of the Fourier series. Though one could consider more general partial sums (such as $\sum_{-50}^{100}$) it is the s_k, and certain modifications of them that we shall see, which have proved most fruitful to study.

The two theorems we shall prove, the $\boldsymbol{L_2}$ **Convergence Theorem** and **Féjer's Theorem**, are important in their own right and also furnish excellent applications of the methods of general functional analysis which we have developed in the preceding chapters. It is for the latter reason that this section includes three separate proofs of Féjer's Theorem.

13.4 Theorem The functions $e_n (n \in \mathbf{Z})$, where $e_n(t) = (2\pi)^{-\frac{1}{2}} e^{int}$, form a complete orthonormal set in $L_2[-\pi, \pi]$.

Proof Since the linear span of the e_n is nothing but the set $\mathcal{T}[-\pi, \pi]$ of trigonometric polynomials on $[-\pi, \pi]$, Theorem 12.31(iii) shows that it is sufficient to prove that $\mathcal{T}[-\pi, \pi]$ is dense in $L_2[-\pi, \pi]$ with respect to the L_2 norm. Now by Theorem 9.9, given any $f \in L_2[-\pi, \pi]$ there exists a sequence $\{f_n\}$ of continuous functions on $[-\pi, \pi]$ such that $\|f_n - f\|_2 \to 0$, and by the remark after that Theorem we may, and do, assume that $f_n(-\pi) = f_n(\pi) = 0$. Then, by Weierstrass' Approximation Theorem for trigonometric polynomials 6.28, we can choose a sequence $\{p_n\}$ in $\mathcal{T}[-\pi, \pi]$ with $\|p_n - f_n\|_\infty < 1/n$ for each n. In view of the inequality

$$\int_{-\pi}^{\pi} |p_n - f_n|^2 \leqslant \int_{-\pi}^{\pi} (\|p_n - f_n\|_\infty)^2 \leqslant 2\pi (\|p_n - f_n\|_\infty)^2$$

this gives $\|p_n - f_n\|_2 \to 0$, and therefore $\|p_n - f\|_2 \to 0$, which completes the proof.

The significance of this result becomes apparent when we combine it with the general theorems about Hilbert spaces in the last chapter, to obtain the following facts – which tell one all that one needs to know about the L_2 convergence of Fourier series.

13.5 Theorem

(i) A necessary and sufficient condition for a set of complex numbers $\{\ldots c_{-1}, c_0, c_1, c_2, \ldots\}$ to be the Fourier coefficients of some $f \in L_2[-\pi, \pi]$ is that $\sum_{-\infty}^{\infty} |c_n|^2 < \infty$.

(ii) Let $f \in L_2$ and let $\{s_k\}$ be the sequence of standard partial sums defined above. Then $\|s_k - f\|_2 \to 0$.

(iii) Let $f \in L_2$ and $f \sim \{c_n\}$. Then

$$\frac{1}{2\pi}\int_{-\pi}^{\pi} |f(t)|^2\, dt = \sum_{-\infty}^{\infty} |c_n|^2 \quad \text{(Parseval's Equation)}$$

(iv) Let $f, g \in L_2$ and $f \sim \{c_n\}$, $g \sim \{d_n\}$. Then

$$\frac{1}{2\pi}\int_{-\pi}^{\pi} f(t)\,\overline{g(t)}\, dt = \sum_{-\infty}^{\infty} c_n \bar{d}_n,$$

the latter series being absolutely convergent.

The proofs follow almost immediately (note the adjusting factor $(2\pi)^{-\frac{1}{2}}$) from Theorem 12.29, the remarks after Lemma 11.5, Theorem 12.31 (v) and Problem 12Q, for parts (i, ii, iii, iv) respectively.

It follows from part (ii) and Corollary 5.11 that the partial sums s_k of any L_2 function f contain a subsequence convergent to f pointwise a.e. One of the remarkable recent results in Fourier series, due to Carleson (1966), is that for any L_2 function the sequence $\{s_k\}$ itself converges pointwise a.e. The proof is hard.

13.6 Parseval's equation (iii) can be used to prove certain classical identities which seem to resist proof by elementary means. As an amusing and typical example we prove the formula

$$\sum_{n=1}^{\infty} \frac{1}{n^2} = \frac{\pi^2}{6}.$$

The only proofs known to the author rely on Fourier series techniques or on Cauchy's Residue Theorem in complex analysis: if the student can give a direct proof he ought to be writing, not reading, this book.

Consider the Fourier series of the function $u(t) = t$ on $[-\pi, \pi]$. It is easy to verify that $\|u\|_2^2 = 2\pi^3/3$ while

$$c_n = \frac{1}{2\pi}\int_{-\pi}^{\pi} t\, e^{-int}\, dt = \begin{cases} 0\ (n = 0) \\ \frac{i}{n}(n = \pm 1, \pm 2, \dots), \end{cases}$$

so that Parseval's equation gives

$$\frac{1}{2\pi}\cdot\frac{2\pi^3}{3} = 0 + \sum_{n \in Z \sim \{0\}} \frac{1}{n^2} = 2\sum_{1}^{\infty}\frac{1}{n^2}$$

from which the result follows.

13.7 We now move on to the second of our topics. One of the pieces of heavy machinery used in the proof of Theorem 13.4 is the Stone–Weierstrass theorem, via Weierstrass' theorem 6.28, which amounts to the fact that $\mathcal{C}_p$ is just the set of uniform limits of functions in $\mathcal{T}[-\pi,\pi]$. The importance of Féjer's theorem below is that it bypasses the Stone–Weierstrass theorem by giving an explicit construction for a sequence of trigonometric polynomials converging uniformly to a given $f \in \mathcal{C}_p$. The first proof below uses good old-fashioned analysis. We then show for good measure how Weierstrass' Theorem furnishes a short proof of Féjer's Theorem by an argument very similar to those of Theorem 9.11 and Problems 9B, C, and finally extract from the ideas behind this second proof a remarkable result called Korovkin's Theorem, which reduces Féjer's Theorem to a three line corollary.

We first derive an *integral formula* for the partial sums s_n of the Fourier series of any integrable function f on $[-\pi,\pi]$. Using the notation of 13.3 whereby $\tilde{f}$ denotes the periodic extension of f to $\boldsymbol{R}$, we have for any t (not necessarily in $[-\pi,\pi]$)

$$\begin{aligned} s_n(t) &= \sum_{r=-n}^{n} c_r\, e^{irt} \\ &= \frac{1}{2\pi}\sum_{r=-n}^{n} e^{irt}\int_{t-\pi}^{t+\pi} \tilde{f}(v)\, e^{-irv}\, dv \qquad \text{by formula (4) of 13.3} \\ &= \frac{1}{2\pi}\int_{t-\pi}^{t+\pi} \tilde{f}(v) \sum_{r=-n}^{n} e^{ir(t-v)}\, dv \end{aligned}$$

which if we set $u = v - t$, reduces to

$$s_n(t) = \int_{-\pi}^{\pi} \tilde{f}(t+u)\, D_n(u)\, du \tag{5}$$

where $D_n(u) = (2\pi)^{-1} \sum_{r=-n}^{n} e^{-iru}$ is **Dirichlet's kernel.** By summing the geometric series one obtains the more convenient form

$$D_n(u) = \frac{1}{2\pi}\left\{\frac{e^{-(n+\frac{1}{2})u} - e^{-i(n+\frac{1}{2})u}}{e^{i\frac{1}{2}u} - e^{-i\frac{1}{2}u}}\right\} = \frac{\sin(n+\frac{1}{2})u}{2\pi \sin\frac{1}{2}u} \qquad (6)$$

This is not the place for a fuller study of formula (5) so we merely note that it is fundamental to the study of the *pointwise convergence* of the s_n and move on to

13.8 Féjer's Theorem (1904) Let $f \in \mathcal{C}_p$, and with the above notation define $\sigma_n = n^{-1} \sum_{0}^{n-1} s_r$. Then the σ_n are a sequence of trigonometric polynomials converging uniformly on $\boldsymbol{R}$ to $\tilde{f}$, and in particular uniformly on $[-\pi, \pi]$ to f.

Remark We again make the passing note that the trick of 'smoothing' divergent series by taking arithmetic (as well as more complicated) means has a well-developed theory. The functions σ_n are called the **(C, 1) sums** of the Fourier series after Césaro, who first studied them in 1890.

Proof of theorem It is clear that each σ_n is a trigonometric polynomial and from (5) we get the integral formula

$$\sigma_n(t) = \int_{-\pi}^{\pi} \tilde{f}(t+u) \cdot \frac{1}{n} \sum_{r=0}^{n-1} D_r(u)\, du = \int_{-\pi}^{\pi} \tilde{f}(t+u)\, C_n(u)\, du \qquad (7)$$

where the function $C_n(u) = n^{-1} \sum_{0}^{n-1} D_r(u)$ is **Féjer's kernel.** From formula (6) – by induction, or by considering the imaginary part of the geometric series

$$\sum_{r=0}^{n-1} e^{i(r+\frac{1}{2})u} = \frac{e^{inu} - 1}{e^{i\frac{1}{2}u} - e^{-i\frac{1}{2}u}} = \frac{\cos nu - 1 + i \sin nu}{2i \sin\frac{1}{2}u}$$

– we easily obtain

$$C_n(u) = \frac{\sin^2 \frac{1}{2}nu}{2n\pi \sin^2 \frac{1}{2}u} \qquad (8)$$

The important properties of C_n are:

C1 $C_n \geqslant 0$, which is obvious.

C2 $\int_{-\pi}^{\pi} C_n(u)\, du = 1$ for each n. This follows from (7) by taking $f = 1$; then clearly $s_n = 1$, and so $\sigma_n = 1$, for all n.

C3 For each $\delta > 0$, $C_n(u) \to 0$ as $n \to \infty$, uniformly on $[-\pi, \pi] \sim (-\delta, \delta)$. For $\sin \frac{1}{2}u$ has its minimum on this set at $u = \delta$, so that $|C_n(u)| \leqslant (2n\pi \sin^2 \frac{1}{2}\delta)^{-1}$.

Let $\epsilon > 0$ be given. Let $M = \|f\|_\infty$. Since $\tilde{f}$ is continuous and periodic it is easily seen to be uniformly continuous on $\boldsymbol{R}$, so that we can choose δ such that $0 < \delta \leqslant \pi$ and $|f(t) - f(t+u)| < \epsilon$ for any $t \in \boldsymbol{R}$ and $u \in (-\delta, \delta)$. Split $[-\pi, \pi]$ into two parts, $I = (-\delta, \delta)$ and $J = [-\pi, \pi] \sim (-\delta, \delta)$, and using C3 choose n_0 such that $|C_n(u)| < \epsilon/4\pi M$ whenever $n \geqslant n_0$ and $u \in J$.

Then for $n \geqslant n_0$ and for any t we have, using C2,

$$\begin{aligned} |\tilde{f}(t) - \sigma_n(t)| &= |\tilde{f}(t) \int_{-\pi}^{\pi} C_n(u)\, du - \int_{-\pi}^{\pi} \tilde{f}(t+u)\, C_n(u)\, du| \\ &= |\int_{-\pi}^{\pi} [f(t) - f(t+u)]\, C_n(u)\, du| \\ &\leqslant \int_{-\pi}^{\pi} |f(t) - f(t+u)|\, C_n(u)\, du \end{aligned}$$

using C1 at the last step. Dividing the integral into $\int_I + \int_J$ we have $\int_I \leqslant \int_I \epsilon\, C_n(u)\, du \leqslant \int_{-\pi}^{\pi} \epsilon\, C_n(u)\, du = \epsilon$

$$\int_J \leqslant \int_J 2M.\frac{\epsilon}{4\pi M}\, du \leqslant \int_{-\pi}^{\pi} \frac{\epsilon}{2\pi}\, du = \epsilon$$

so, adding, we obtain $|\tilde{f}(t) - \sigma_n(t)| \leqslant 2\epsilon$ $(n \geqslant n_0, t \in \boldsymbol{R})$ from which the result follows

13.9 Corollary (Weierstrass' Theorem 6.27) Any periodic (2π) continuous function on $\boldsymbol{R}$ is the uniform limit of trigonometric polynomials.

If the reader is wondering why the same sort of proof will not work for the s_n just as well as the σ_n, a look at formula (6) shows that the Dirichlet kernel D_n possesses neither of the properties C1, C3 of Féjer's kernel C_n and it is just this that marks the essential difference between the two kinds of partial sum. A graph of the two kernels makes this plain (in view of the symmetry only the right-hand half is shown in Diagram 18).

The importance of the positivity of Féjer's kernel, property C1, comes out again in the second proof:

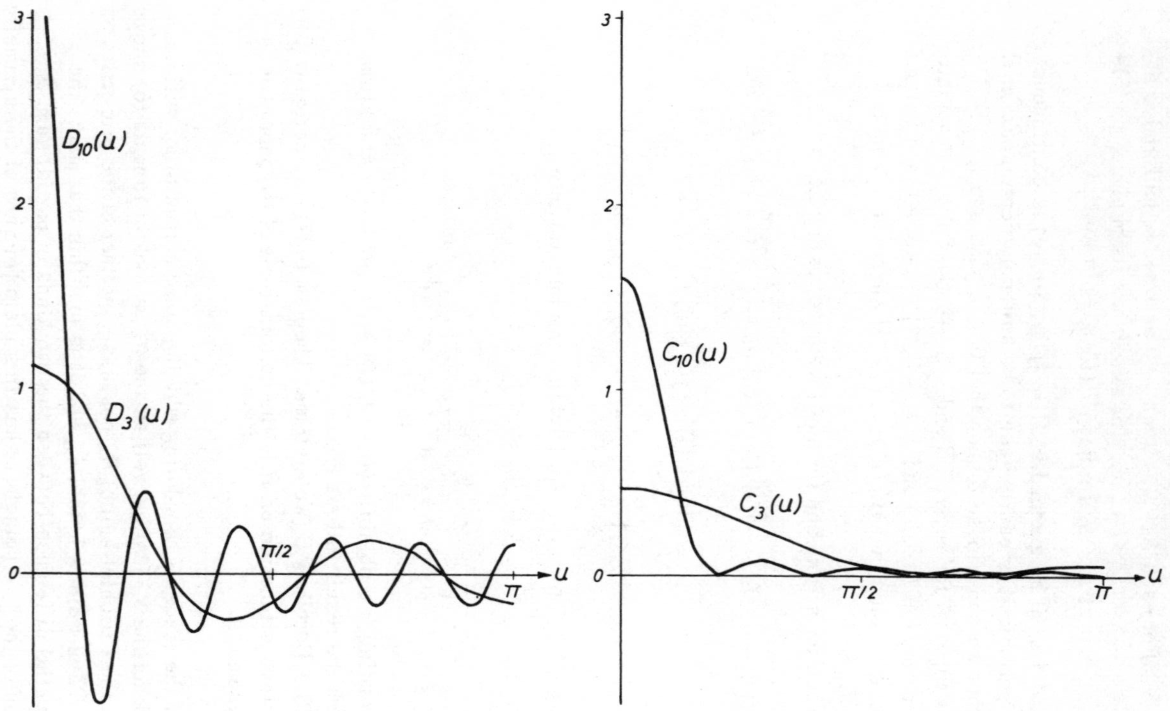

Diagram 18.

13.10 Féjer's theorem, proof no. 2 Consider the normed space $\mathcal{C}_p[-\pi,\pi]$ with the sup norm, and for fixed n let T_n be the (obviously linear) map of $\mathcal{C}_p$ into itself taking f to its corresponding σ_n, that is

$$(T_n f)(t) = \int_{-\pi}^{\pi} \tilde{f}(t+u)\, C_n(u)\, du, \qquad t \in [-\pi,\pi]$$

Since

$$|(T_n f)(t)| \leqslant \int_{-\pi}^{\pi} |\tilde{f}(t+u)|\, C_n(u)\, du \leqslant \|f\|_\infty \int_{-\pi}^{\pi} C_n(u)\, du$$

$$= \|f\|_\infty \text{ (by properties C1, 2)}$$

we have $\|T_n f\| \leqslant \|f\|$ for each $f \in \mathcal{C}_p$, so that the T_n form a uniformly bounded sequence in $\mathcal{B}(\mathcal{C}_p)$, with $\|T_n\| \leqslant 1$ for all n. Féjer's theorem amounts to asserting that $T_n f \to f$, in the $\mathcal{C}_p$ norm, for all $f \in \mathcal{C}_p$. By Theorem 9.3 it suffices to show that this occurs for all f in some fundamental subset of $\mathcal{C}_p$. A simple computation shows that for $f(t) = e^{ikt}$ (k any integer) one has $s_n = 0$ for $n < |k|$, $s_n = f$ for $n \geqslant |k|$, and therefore $\sigma_n = (1 - |k|/n) f$ for $n \geqslant |k|$. Thus clearly $T_n f \to f$ whenever f is one of the functions e^{ikt}. By Weierstrass' Theorem, these functions are indeed fundamental in $\mathcal{C}_p$, and the result follows.

† *Positive operators on $\mathcal{C}(T)$*

We now give a third proof of Féjer's theorem. Because this section cannot really be called basic functional analysis the student may skip the details on a first reading, but he ought at least to grasp the definitions and the statements of the results in order to see their significance.

13.11 Let T be a compact topological space. A linear mapping P of the *real* Banach space $\mathcal{C}(T)$ into itself is called **positive** if $Pf \geqslant 0$ whenever $f \geqslant 0$. Since $f \geqslant g \Leftrightarrow f - g \geqslant 0$, an equivalent condition is: $Pf \geqslant Pg$ whenever $f \geqslant g$. (For this reason the name **monotone operator** is often used.) A number of the interesting and quite elementary properties of positive operators are given in the problems – for instance, that a positive operator is automatically bounded – but we shall not need them. A **minimizing** subspace of $\mathcal{C}(T)$ is a linear subspace M such that $1 \in M$ and for any $t \in T$ there is a function in M having a strict minimum at t.

13.12 Korovkin's Theorem Let $\{P_n\}$ be a sequence of positive operators on $\mathcal{C}(T)$ and suppose $P_n f \to f$ (in norm) for every f in some minimizing subspace M. Then $P_n f \to f$ for all $f \in \mathcal{C}(T)$.

Remark The subspace M may be quite tiny – for example the 3-dimensional subspace of quadratic functions $a + bt + ct^2$ is a minimizing subspace of $\mathcal{C}[0, 1]$. The reader, bearing this in mind, should compare this theorem with Theorem 9.3 to see the strength of the assumption of positivity.

Proof The first part of the proof is to show that given $f \in \mathcal{C}(T)$, $t_0 \in T$ and $\epsilon > 0$ we can find $g \in M$ such that $g(t_0) \leqslant f(t_0) + \epsilon$ but $g > f$ everywhere. In fact, choose $h \in M$ having a strict minimum at t_0; since $1 \in M$ we can suppose by subtracting a constant function that $h(t_0) = 0$, so that $h(t) > 0$ everywhere else.

Let $\alpha = f(t_0)$; for any $t \in T$ we have

$$\lim_n (n\, h(t) + \alpha + \epsilon) = \begin{cases} \infty > f(t) & (t \neq t_0) \\ \alpha + \epsilon > f(t) & (t = t_0) \end{cases}$$

so that the expanding sequence of open sets

$$U_n = \{t: n\, h(t) + \alpha + \epsilon > f(t)\}$$

covers T. By compactness there is n such that $U_n = T$, in other words the function g in M defined by $g = nh + (\alpha + \epsilon)1$ is greater than f on the whole of T. Clearly $g(t_0) = f(t_0) + \epsilon$ so that g is the required function.

Using this fact we now pick, for each $t \in T$, a $g_t \in M$ with $g_t > f$, $g_t(t) \leqslant f(t) + \epsilon/2$. By continuity the set of points where $g_t(s) < f(s) + \epsilon$ is an open neighbourhood V_t of t. Thus the V_t cover T, and by compactness there is a cover by finitely many $V_{t_1}, \ldots, V_{t_k}$. If we label the corresponding g_{t_i} as $g_1, \ldots, g_k$ then for each $s \in T$ there is a g_i such that $g_i(s) \leqslant f(s) + \epsilon$ because s is in some V_{t_i}. Hence $g_1 \wedge \ldots \wedge g_k \leqslant f + \epsilon 1$. On the other hand $f \leqslant g_i$ for each i. (This bit is exactly parallel to part of the proof of the Stone–Weierstrass theorem.)

By hypothesis we can find n_0 such that for each of the g_i and for $n \geqslant n_0$, $\|P_n g_i - g_i\| \leqslant \epsilon$, so that in particular

$$P_n g_i \leqslant g_i + \epsilon 1$$

For any $n \geqslant n_0$, we have by positivity

$$P_n f \leqslant P_n g_i \leqslant g_i + \epsilon 1 \quad (i = 1, \ldots, k)$$

so

$$P_n f \leqslant (g_1 + \epsilon 1) \wedge \ldots \wedge (g_k + \epsilon 1) = (g_1 \wedge \ldots \wedge g_k) + \epsilon 1$$
$$\leqslant (f + \epsilon 1) + \epsilon 1$$

Thus

$$P_n f \leqslant f + 2\,\epsilon 1 \quad (n \geqslant n_0)$$

By an analogous argument or by replacing f by $-f$ we see that there exists n_0' such that

$$P_n f \geqslant f - 2\,\epsilon 1 \quad (n \geqslant n_0')$$

It follows that $\|P_n f - f\| \leqslant 2\epsilon$ $(n \geqslant \max\{n_0, n_0'\})$ and since f, ϵ were arbitrary the theorem follows.

If $P_n f \to f$ for all f in a set A, it is clear that $P_n f \to f$ for all f in the linear span of A, so that one has the immediate corollaries.

13.13 Theorem In order that $P_n f \to f$ for all $f \in \mathcal{C}[0,1]$, where the P_n are positive operators, it is sufficient that $P_n f \to f$ for the functions $1, u, u^2$ where $u(t) = t$.

13.14 Theorem In order that $P_n f \to f$ for all $f \in \mathcal{C}_p[0, 2\pi]$, where the P_n are positive operators, it is sufficient that $P_n f \to f$ for the functions $f = 1$, sin, cos.

Proofs $1, u, u^2$ span the minimizing subspace of functions $a + bt + ct^2$ in $\mathcal{C}[0,1]$. To get the second theorem we have to regard $\mathcal{C}_p[-\pi, \pi]$ as $\mathcal{C}(\Gamma)$, Γ the unit circle, by 'wrapping $[-\pi, \pi]$ round Γ' and identifying $-\pi$ with π. It then follows that functions of the form $a + b \sin t + c \cos t$ form a minimizing subspace (see the remark after Theorem 6.27).

13.15 Corollary (Féjer's theorem, proof 3) Define the operators T_n on $\mathcal{C}_p[-\pi, \pi]$ as in proof 2. It is clear from the positivity of the Féjer kernel C_n that the T_n are positive operators. Simple computations, as in proof 2, show that $T_n f \to f$ when $f = 1$, sin, cos, and the result follows. The Korovkin theorems assume that the functions involved are real, but the complex case follows trivially by taking real and imaginary parts.

Problems

13A Verify equation (4) in 13.3.

The L_2 Theorem

13B To avoid messy square roots in computational work it is often convenient to use a **complete orthogonal set** (COS) rather than an orthonormal one; this means a set $E \subset H \sim \{0\}$ such that $\{e/\|e\|: e \in E\}$ is a complete orthornormal set.

(i) Show that for a COS $\{e_j\}_{j \in J}$ one has the expansion

$$x = \sum_{j \in J} \frac{\langle x, e_j \rangle}{\langle e_j, e_j \rangle} e_j \quad (x \in H).$$

Using Theorem 13.4 verify that

(ii) The functions $\{\cos nt \ (n \geqslant 0), \sin nt \ (n \geqslant 1)\}$ form a COS in $L_2[-\pi, \pi]$.

(iii) In $L_2[0, \pi]$ each of the families $C = \{\cos nt \ (n \geqslant 0)\}$ and $S = \{\sin nt \ (n \geqslant 1)\}$ is a COS. [To show C is complete, consider the **even extension** f_e of a function $f \in L_2[0, \pi]$ to $[-\pi, \pi]$, defined to be $f(t)$ for $t \geqslant 0$ and $f(-t)$ for $t < 0$, and examine the form of the Fourier coefficients of f_e. Similarly with S. Expansions in terms of C, S are called the **Fourier cosine** and **sine series** respectively of a function on $[0, \pi]$.

13C Apply Parseval's equation to the complex Fourier series of the function $f(t) = e^{st}$ on $[-\pi, \pi]$ to deduce that

$$\frac{\pi}{s} \coth \pi s = \sum_{n=-\infty}^{\infty} \frac{1}{s^2 + n^2}.$$

Féjer's Theorem

13D Suppose the partial sums s_n of a series $\sum_1^\infty a_n$ converge to s. Show that the $(C,1)$ sums $\sigma_n = (s_1 + \dots + s_n)/n$ also converge to s.

13E Give an induction proof of formula (8) for the Féjer kernel C_n.

13F Let S_n denote the map sending a function f to the nth partial sum s_n of its Fourier series as in 13.3. Show that, considered as operators on $L_2[-\pi,\pi]$, the S_n are orthogonal projections and that $S_n \circ S_m = S_m \circ S_n = S_m$ whenever $m \leqslant n$.

Further Problems

13G Fill in the details of the proof, sketched in 10.7, that the Fourier series of some $f \in \mathcal{C}_{2\pi}(\mathbf{R})$ diverges at $t = 0$. [The functional ϕ_n that associates to f the value at $t = 0$ of the nth partial sum s_n can, by formula 5 of 13.7, be written

$$\phi_n(f) = \int_{-\pi}^{\pi} f(u)\, D_n(u)\, du.$$

By formula 6 and Problem 8B one has

$$\|\phi_n\| = \int_{-\pi}^{\pi} \left| \frac{\sin(n+\frac{1}{2})t}{2\pi \sin\frac{1}{2}t} \right| dt.$$

Using the inequality $|\sin\frac{1}{2}t| \leqslant \frac{1}{2}|t|$ on $[-\pi,\pi]$ and the substitution $v = (n+\frac{1}{2})t$, show the integral is $\geqslant \frac{4}{\pi}(1 + \frac{1}{2} + \dots + \frac{1}{n})$.]

The last problem shows the importance of finding bounds on the norms of a sequence of operators or functionals. The next few problems circle round this idea.

13H Let $T_1, T_2, \dots$ be operators from a Banach space X to a normed space Y. Prove the following strengthening of Theorem 9.3: Suppose $\lim_n T_n x$ exists for all x in a fundamental subset of X. Then $\lim_n T_n x$ exists for *all* x in X iff $\sup_n \|T_n\| < \infty$.

13I Let X be a Banach space and let $T_n \in \mathcal{B}(X)$, $n = 1, 2, \dots$. Show that the subspace $M = \{x \in X\colon \lim_n T_n x = x\}$ becomes a Banach space under the norm $|x| = \sup_n \|T_n x\|$.

13J Show that, for $1 \leqslant p < \infty$, the Fourier series of a function f in $L_p[-\pi,\pi]$ is $(C,1)$ summable to f in the L_p norm.

[Let q be the conjugate index, and T_n be as in 13.10 (but acting on L_p). By writing the integral as a double integral and reversing the order show that for any g in L_q

$$\left| \int_{-\pi}^{\pi} T_n f(t)\, g(t)\, dt \right| \leqslant \|f\|_p \, \|g\|_q.$$

Deduce that $T_n \in \mathcal{B}(L_p)$ with $\|T_n\| \leqslant 1$. Now use Theorem 9.3.]

A standard test for uniform convergence of Fourier series

13K Let the function $f \in \mathcal{C}_p[-\pi, \pi]$ be piecewise smooth: that is there are points $-\pi = t_0 < t_1 < \ldots < t_n = \pi$ such that f is continuously differentiable on $[t_{r-1}, t_r]$, $1 \leqslant r \leqslant n$, in other words f is smooth except for a finite number of 'corners'. Show that the Fourier series of f converges absolutely and uniformly to f.

[Hard till you see how: use integration by parts and Cauchy–Schwarz.]

Korovkin's Theorem

13L Let $f \in \mathcal{C}[0,1]$ and define the nth **Bernstein polynomial** $B_n f$ associated with f by

$$(B_n f)(t) = \sum_{k=0}^{n} \binom{n}{k} t^k (1-t)^{n-k} f(k/n).$$

Show that $B_n f \to f$ uniformly on $[0,1]$ as $n \to \infty$, thus giving a 'constructive' proof of Weierstrass' Theorem. The construction is due to Bernstein (1912).

[Clearly the maps $f \to B_n f$ are positive linear operators on $\mathcal{C}[0,1]$. By 13.13 we only need show $B_n f \to f$ for $f = 1, u, u^2$. By the binomial theorem,

$$\sum_{k=0}^{n} \binom{n}{k} x^k (1-y)^{n-k} = (1 + x - y)^n. \tag{1}$$

Differentiate with respect to x and multiply by x/n to get

$$\sum_{k=0}^{n} \binom{n}{k} x^k (1-y)^{n-k} (k/n) = x(1 - x + y)^{n-1}. \tag{2}$$

Repeat to get

$$\sum_{k=0}^{n} \binom{n}{k} x^k (1-y)^{n-k} (k^2/n^2)$$

$$= \frac{(n-1)}{n} x^2 (1 + x - y)^{n-2} + \frac{x}{n}(1 + x - y)^{n-1}. \tag{3}$$

Set $x = y$ to deduce $B_n(1)$, $B_n(u)$ and $B_n(u^2)$, and let $n \to \infty$].

5

DUAL SPACES

14. Dual spaces revisited

A number of questions in analysis are basically about the structure of the dual of some normed linear space X. This chapter describes some results and constructions which form the basic computational methods when one works with dual spaces: the **canonical embedding** of X into its second dual; the **annihilator operation** (and various equally important variants of it that are described in the problems) which generalizes the orthogonal complement operation in Hilbert space; and the identification of the dual of a quotient with a subspace of the dual and vice versa. The Hahn–Banach Theorem is very much in evidence – in other words these are 'geometrical theorems'. They are by no means trivial even when X is finite-dimensional, and indeed the descriptions of the duals of a subspace or a quotient can be regarded as versions of the *duality theorem* of linear programming which is a fundamental result in finite-dimensional convexity theory.

The *sequence spaces*, as usual, furnish enlightening examples of the general theory and we finish the chapter with explicit descriptions of the duals of c_0 and ℓ_p, leaving the reader to compute some other duals in the problems.

The space X^{**}

14.1 If X is a normed space, so also is X^*. Thus X^* in turn has a dual space, which is itself a Banach space, denoted by X^{**}. It is difficult to visualize what X^{**} 'looks like' when considered purely as a set: for X^* is a space of scalar functions on X, and

X^{**} in turn consists of scalar functions defined on a space of scalar functions. This need not worry us, for we can say a good deal about the structure of X^{**} when considered as a normed linear space. The main result is that X^{**} contains a subspace which is congruent, in a natural way, to X: this is the simplest example of a construction of fundamental importance in many areas of analysis, whose properties are studied more closely in the next chapter.

We shall use Greek letters $\phi, \psi, \ldots$ to denote functionals in X^{**}. The norm in X^{**} is given by

$$\|\phi\| = \sup\{|\phi(f)|: f \in X^*, \|f\| \leqslant 1\}$$

and the linear operations are defined, of course, pointwise so that

$$(\phi + \psi)(f) = \phi(f) + \psi(f), \quad (\alpha\phi)(f) = \alpha\psi(f).$$

14.2 Theorem For each x in X the scalar-valued function ϕ_x on X^* defined by

$$\phi_x(f) = f(x) \quad (f \in X^*)$$

is an element of X^{**}. The mapping $x \mapsto \phi_x$ is a congruence of X onto a linear subspace $\hat{X}$ of X^{**}. The subspace $\hat{X}$ is closed iff X is a Banach space.

Proof Using only the fact that the linear operations in X^* are defined pointwise we have

$$\phi_x(\alpha f + \beta g) = (\alpha f + \beta g)(x) = \alpha f(x) + \beta g(x) = \alpha\phi_x(f) + \beta\phi_x(g)$$

showing that ϕ_x is a *linear* function on X^*. Moreover

$$|\phi_x(f)| = |f(x)| \leqslant \|x\| \, \|f\| \quad (f \in X^*)$$

and so ϕ_x is bounded with $\|\phi_x\| \leqslant \|x\|$, showing that ϕ_x is an element of X^{**}.

Now using the fact that the linear operations in X^{**} are defined pointwise, and that elements of X^* are linear functions on X we have for each $f \in X^*$

$$\phi_{\alpha x + \beta y}(f) = f(\alpha x + \beta y) = \alpha f(x) + \beta f(y) = \alpha\phi_x(f) + \beta\phi_y(f)$$

which shows that $\phi_{\alpha x + \beta y} = \alpha\phi_x + \beta\phi_y$. In other words $x \mapsto \phi_x$ is a linear mapping of X into X^{**}.

We have seen that $\|\phi_x\| \leqslant \|x\|$, and to prove that $x \mapsto \phi_x$ is a congruence we have only to show that $\|\phi_x\| \geqslant \|x\|$. This is trivial when $x = 0$, and when $x \neq 0$, Theorem 8.10 shows that there exists $f \in X^*$ with $\|f\| = 1$ and $f(x) = \|x\|$. Thus

$$\|x\| = |f(x)| = |\phi_x(f)| \leqslant \|\phi_x\| \, \|f\| = \|\phi_x\|,$$

as required.

The range $\hat{X} = \{\phi_x : x \in X\}$ of the mapping $x \mapsto \phi_x$ is a subspace of X^{**} congruent to X. Since X^{**} is always complete, $\hat{X}$ is closed in X^{**} iff $\hat{X}$ is complete (Proposition 1.10). In virtue of the congruence, $\hat{X}$ is complete iff X is complete. The last assertion of the theorem follows immediately.

14.3 The element ϕ_x associated with x is called the **evaluation functional** of x (because it evaluates each f in X^* at the point x of X) and the congruence $x \mapsto \phi_x$ will be denoted by e and called the **evaluation map** or the **natural** (or **canonical**) **embedding** of X into X^{**}. Since there is a sense in which congruent spaces can be regarded as identical one can say that X 'is' a subspace of X^{**}. The student should treat the word 'is' here with extreme caution, for it is often nonsense to ignore the map e which identifies X with $\hat{X}$.

14.4 If it should happen that $\hat{X} = X^{**}$, that is the *only* bounded linear functionals on X^* are the evaluations, then X is called **reflexive.** This happens in many important cases: any finite-dimensional space is reflexive (using Theorem 8.4 one has $\dim \hat{X} = \dim X = \dim X^* = \dim X^{**}$); so is any Hilbert space (Problem 14B: an easy but not trivial consequence of the Riesz–Fréchet Theorem 12.10). The sequence spaces ℓ_p, and more generally any L_p space, are reflexive for $1 < p < \infty$; while ℓ_1, ℓ_∞, c and c_0 are not reflexive, nor is $\mathcal{C}(T)$ if T is a compact Hausdorff space with infinitely many points. Since X^{**} is necessarily complete it is obvious that no incomplete normed space can be reflexive. For a long time it was thought that if X was a non-reflexive Banach space then $\hat{X}$ must be 'very much smaller' than X^{**}, until James in 1951 produced an 'almost reflexive' Banach space with the property that $\hat{X}$ is a hyperplane in X^{**}, so that $X^{**}/\hat{X}$ is one-dimensional.

A reflexive space and its dual show an enjoyable symmetry: X^* is the dual of X, and X is congruent to the dual of X^*. A notation which highlights this symmetry, and which we shall use henceforth (whether or not X is reflexive) is to write $\hat{x}$ instead of ϕ_x: that is $\hat{x}$ is defined by

$$\hat{x}(f) = f(x) \quad (x \in X, f \in X^*)$$

As one would expect we have the

14.5 Proposition Let X be a Banach space. Then X is reflexive $\Leftrightarrow$ X^* is reflexive.

The proof, whose main difficulty lies in the intricacies of notation, is left as Problem 14C.

Annihilators

14.6 Let X be a normed space. Given $x \in X$ and $f \in X^*$ we write $x \perp f$ if $f(x) = 0$. Although this has not the same geometric meaning of perpendicularity as the corresponding relation between vectors in a Hilbert space, it has many similar properties as a computational tool. We say that x **annihilates** f (or that f annihilates x). Adopting the same notation as in the Hilbert space chapter we say that a vector x in X annihilates a subset B of X^*, and write $x \perp B$, if $x \perp f$ for all $f \in B$; conversely if $f \in X^*$, $A \subset X$ we write $A \perp f$ if $x \perp f$ for all $x \in A$.

The **annihilator** of $B \subset X^*$ is the subset of X defined by

$$B^\perp = \{x \in X : x \perp B\}$$

and similarly the annihilator of $A \subset X$ is the subset of X^* defined by

$$A^\perp = \{f \in X^* : A \perp f\}$$

This is an example of the Galois duality defined in §12, and from that section we have at once the relations: $A \subset B \Rightarrow A^\perp \supset B^\perp$; $(A \cup B)^\perp = A^\perp \cap B^\perp$; $A \subset A^{\perp\perp}$; $A^\perp = A^{\perp\perp\perp}$; where A, B here represent either two sets in X or two sets in X^*.

The most significant subsets for this duality are the **$\perp$-sets**, that is the sets A (in either X or X^*) such that $A^{\perp\perp} = A$: according to Proposition 12.16 they are precisely the sets which are of the form $B^\perp$ for some B (in X^* or X). The next theorem, which is an important application of the Hahn–Banach theorem, describes the $\perp$-sets in X completely.

14.7 Theorem Let $A \subset X$. Then

(i) $A = A^{\perp\perp} \Leftrightarrow A$ is a closed linear subspace of X.

(ii) Consequently, for any $A \subset X$, $A^{\perp\perp}$ is the closed linear span of A.

Proof (i) Suppose $A = A^{\perp\perp}$, that is $A = \{x \in X: \ x \perp f$ for all $f \in A^{\perp}\}$. Thus A is the intersection, as f runs over $A^{\perp}$, of the sets

$$\begin{aligned} &\{x \in X: x \perp f\} \\ &= \{x \in X: f(x) = 0\} \\ &= \ker f \end{aligned}$$

and so, being an intersection of closed linear subspaces ker f, is itself of the same sort. Conversely suppose A is a closed linear subspace of X. Since $A \subset A^{\perp\perp}$ is always true, it suffices to show that if $x_0 \notin A$ then $x_0 \notin A^{\perp\perp}$. By Theorem 8.13, if $x_0 \notin A$ there exists $f \in X^*$ such that $f = 0$ on A (in other words $f \in A^{\perp}$) but $f(x_0) \neq 0$ (in other words for which $x_0 \perp f$ is false). Thus it is false that $x_0 \perp A^{\perp}$, so that $x_0 \notin A^{\perp\perp}$ and the result follows.

(ii) By Proposition 12.16, $A^{\perp\perp}$ is the smallest $\perp$-set containing A, and in view of (i) this is the same as the closed linear span of A.

Unfortunately the complete symmetry between X and X^* breaks down here: the obvious dual version of this theorem is false when X is non-reflexive. In fact the $\perp$-sets in X^* are precisely those linear subspaces of X^* that are closed in the *weak*-topology* to be introduced in the next chapter, but a proof of this important theorem is beyond the scope of this book.

The dual of a subspace and a quotient

Let X be a normed space and M a linear subspace of X. The next two theorems are somewhat technical but the basic idea is simple and elegant: we identify

$$M^* \text{ with } X^*/M^{\perp}$$

and

$$(X/M)^* \text{ with } M^{\perp}$$

provided that in the latter case M is closed.

14.8 Theorem For each $g \in M^*$ let C_g denote the set of all extensions of g to an element f of X^*. Then C_g is a coset of $M^{\perp}$ in X^* and the mapping $g \mapsto C_g$ is a congruence of M^* with $X^*/M^{\perp}$.

Proof By the Hahn–Banach theorem 8.9, C_g is nonempty: indeed we can choose an $f_0 \in C_g$ with

$$\|f_0\| = \|g\| \tag{1}$$

For any $f \in X^*$ we now have

$$f \in C_g \Leftrightarrow f - g = 0 \text{ on } M \Leftrightarrow f - f_0 = 0 \text{ on } M \Leftrightarrow f - f_0 \in M^\perp,$$

so that C_g is precisely the coset $f_0 + M^\perp$. We leave to the reader the simple task of showing that $g \mapsto C_g$ is a linear map.

Any $f \in C_g$, being an extension of g, must have $\|f\| \geqslant \|g\|$. By the definition of quotient norm this gives

$$\|C_g\| = \inf_{f \in C_g} \|f\| \geqslant \|g\|$$

On the other hand by (1), $\|C_g\| \leqslant \|f_0\| = \|g\|$ so that $\|C_g\| = \|g\|$. This proves that $g \mapsto C_g$ is a congruence of M^* *into* $X^*/M^\perp$. Finally if $C = f + M^\perp$ is an arbitrary element of $X^*/M^\perp$ it is clear that $C = C_g$ where g is the restriction of f to M. This shows T is onto and completes the proof.

The closedness of M is never used in the next proof: it is only needed to ensure that X/M is a bona fide normed space according to Theorem 4.24.

14.9 Theorem Let M be closed. The map $T: g \mapsto g \circ Q$, where Q is the quotient map of X onto X/M, is a congruence of $(X/M)^*$ with the subspace $M^\perp$ of X^*.

Proof Since Q is defined by $Qx = x + M$, the definition of T says that for each $g \in (X/M)^*$, Tg is the functional f on X defined by $f(x) = g(x + M)$. Clearly f is linear, and we have

$$|f(x)| = |g(x + M)| \leqslant \|g\| \, \|x + M\| \leqslant \|g\| \, \|x\|$$

from the definition of quotient norm, which shows that $f \in X^*$ and $\|f\| \leqslant \|g\|$. If $x \in M$ then Qx is the zero element in X/M, so that $f(x) = g(Qx) = 0$; thus $f \in M^\perp$. We have thus shown that T maps $(X/M)^*$ into $M^\perp$, and again we leave it to the reader to show that T is linear.

Now, given any f in $M^\perp$ we note that whenever x and y belong to the same coset C of M we have

$$x - y \in M \Rightarrow f(x - y) = 0 \Rightarrow f(x) = f(y)$$

so that we can unambiguously define a functional g on X/M by

$$g(C) = f(x), \text{ where } x \in C.$$

This gives $|g(C)| \leqslant \|f\|\,\|x\|$ $(x \in C)$ and hence $|g(C)| \leqslant \|f\| \inf\{\|x\| : x \in C\} = \|f\|\,\|C\|$, so that, since g is clearly linear, g is in $(X/M)^*$ with $\|g\| \leqslant \|f\|$. It is obvious that $Tg = f$ and g is the unique element of $(X/M)^*$ with this property. Thus T maps $(X/M)^*$ one-to-one onto $M^\perp$. In the course of the above working we have proved that $\|Tg\| \leqslant \|g\|$ and $\|g\| \leqslant \|Tg\|$, so that T is a congruence.

The duals of ℓ_p and c_0

14.10 The sequence spaces give useful illustrations of reflexive spaces, and of the natural embedding of a normed space into its second dual. Elements of these spaces are scalar sequences $x = (x_1, x_2, \ldots)$: we shall prove theorems that represent bounded linear functionals on them in the general form

$$f(x) = \sum_{j=1}^{\infty} a_j x_j \tag{2}$$

where $a = (a_1, a_2, \ldots)$ is a sequence of a specified kind. In fact the series (2) converges absolutely, so that as noted in 2.23 and 11.9 it can be regarded as an integration with respect to counting measure. It is convenient to choose a notation that makes the connexion with integration obvious, so we shall write Σx for the sum $\sum_{j=1}^{\infty} x_j$ of an absolutely convergent series; following the notation for functions we write xy for the sequence $(x_1y_1, x_2y_2, \ldots)$; similarly $|x|$, $\overline{\text{sgn}}\, x$ and so on.

Let e_n denote the standard basis vector with 1 in the nth place, 0 elsewhere. The e_n span the subspace ℓ_0 of sequences with only *finitely many* nonzero terms. It is easy to see that ℓ_0 is dense in ℓ_p for $1 \leqslant p < \infty$, and also in c_0, but not in ℓ_∞: it is for this reason that the proof below breaks down for the case $p = \infty$.

14.11 Theorem Let $1 \leqslant p < \infty$ and let q be the conjugate index. Then every $y \in \ell_q$ defines a bounded linear functional f_y on ℓ_p by the formula

$$f_y(x) = \Sigma yx \quad \left(\text{that is, } \sum_{j=1}^{\infty} y_j x_j\right) \tag{3}$$

and the map $T\colon y \mapsto f_y$ is a congruence of ℓ_q onto $(\ell_p)^*$.

Proof It follows from Hölder's inequality 5.4 that Σyx exists – in other words the series converges absolutely – and that $|f_y(x)| \leqslant \|y\|_q \|x\|_p$ so that f_y, which is obviously linear, is bounded and

$$\|f_y\| \leqslant \|y\|_q \tag{4}$$

The obvious identity $\Sigma(\alpha y + \beta z)x = \alpha\Sigma yx + \beta\Sigma zx$ for all y, z in ℓ_q, all x in ℓ_p and all scalars α, β shows that

$$f_{\alpha y+\beta z} = \alpha f_y + \beta f_z$$

so that T is linear. Proposition 5.5 is precisely what is needed to establish equality in (4), so that T is a congruence of ℓ_q *into* $(\ell_p)^*$.

Now let f be any element of $(\ell_p)^*$ and define a sequence y by

$$y_n = f(e_n), \quad n = 1, 2, \ldots \tag{5}$$

(the motivation for this being the easily verified fact that if $z \in \ell_q$ then $f_z(e_n) = z_n$). We show that $f = f_y$.

Any element $x = (x_1, x_2, \ldots, x_k, 0, 0 \ldots)$ of ℓ_0 can be written as $\sum_1^k x_j e_j$ so that

$$f(x) = \sum_1^k x_j f(e_j) = \sum_1^k y_j x_j = \Sigma yx \quad (x \in \ell_0) \tag{6}$$

the latter series having only finitely many nonzero terms. Replacing x by $x\,\overline{\text{sgn}}\,(yx)$, which is in ℓ_0, gives

$$\Sigma|yx| = \Sigma yx\,\overline{\text{sgn}}\,(yx) = f(x\,\overline{\text{sgn}}\,yx) \leqslant \|f\|\,\|x\|_p, \quad (x \in \ell_0)$$

because x and $x\,\overline{\text{sgn}}\,(yx)$ have the same norm. Theorem 5.7 now gives

$$\|y\|_q = \sup\{\Sigma|yx| : x \in \ell_0, \|x\|_p \leqslant 1\} \leqslant \|f\|$$

so that $y \in \ell_q$. Equation (6) amounts to saying that f and f_y agree on the dense subspace ℓ_0 of ℓ_p. By continuity, $f = f_y$, so that T is onto and the proof is complete.

The proof can be considerably simplified in the case $p = 1$; for instance the definition of the sequence y gives immediately

$$|y_n| \leqslant \|f\|\,\|e_n\|_1 = \|f\|, \text{ so that } \|y\|_\infty \leqslant \|f\|.$$

14.12 Theorem Every $y \in \ell_1$ defines a bounded linear functional on c_0 by the formula

$$f_y(x) = \Sigma yx$$

and the map $T: y \mapsto f_y$ is a congruence of ℓ_1 onto $(c_0)^*$.

Proof The proof is almost identical with the previous one up to equation (6). Then given n, let x be the element of ℓ_0 such that

$$x_j = \begin{cases} \overline{\operatorname{sgn}}\, y_j, & 1 \leqslant j \leqslant n \\ 0, & j > n. \end{cases}$$

From (6),

$$\sum_{j=1}^{n} |y_j| = |\Sigma yx| = |f(x)| \leqslant \|f\|\ \|x\|_\infty = \|f\|.$$

Letting $n \to \infty$ gives $y \in \ell_1$ with $\|y\|_1 = \Sigma|y| \leqslant \|f\|$, and the rest of the proof follows as in the last theorem.

14.13 We can describe the content of the last two theorems by saying that $(\ell_p)^*$ 'is' ℓ_q for $1 \leqslant p < \infty$, and $(c_0)^*$ 'is' ℓ_1, but it should be borne in mind that this is true modulo the canonical correspondence $y \leftrightarrow f_y$. More generally it is true that the dual of any L_p space ($1 \leqslant p < \infty$; in case $p = 1$ one has also to demand that the measure be σ-finite) can be identified with L_q. The first part of the proof of Theorem 14.12 goes through with only notational changes: in fact every function $y \in L_q$ defines an element f_y of $(L_p)^*$ by the formula

$$f_y(x) = \int yx\, d\mu$$

and the mapping $y \mapsto f_y$ is a congruence of L_q into $(L_p)^*$. The main snag in proving that every $f \in (L_p)^*$ is an f_y is that there is no longer any explicit formula like equation (5) for the function y. Instead, the crucial tool at this point is the Radon–Nikodym Theorem which recovers y from a certain absolutely continuous measure. We refer the reader to Bartle (**1**, pp. 89–92) for a detailed proof.

14.14 Theorems 14.11 and 14.12 show that c_0^{**} 'is' ℓ_1^* 'is' ℓ_∞, whereas Theorem 14.2 implies that c_0 'is' a subspace of c_0^{**}. Since c_0 is a bona fide subspace of ℓ_∞ it is natural to guess that the embedding of c_0 in c_0^{**} is 'the same' as its natural embedding (the identity map) into ℓ_∞. This is in fact true, but the student should not take such things for granted and is strongly advised to embark on a 'notation chase' to prove it. Let

T denote the congruence $\ell_\infty \to \ell_1^*$
S denote the congruence $\ell_1 \to c_0^*$
S^* denote the resulting congruence $c_0^{**} \to \ell_1^*$ (Problem 8R)
e the embedding $c_0 \to c_0^{**}$
I the identity map $c_0 \to \ell_\infty$.

To prove: $T^{-1} \circ S^* \circ e = I$ – see Diagram 19.

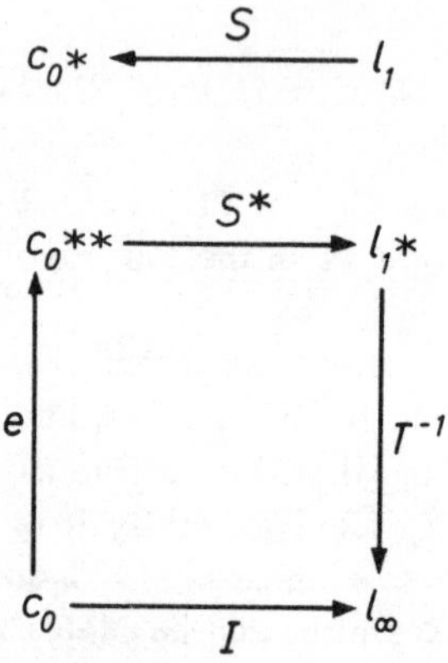

Diagram 19.

In the same vein it seems obvious that $(\ell_p)^{**}$ 'is' $(\ell_q)^*$ 'is' ℓ_p for $1 < p < \infty$, and therefore

14.15 Theorem ℓ_p is reflexive for $1 < p < \infty$.

This also the student should make rigorous by considering the maps involved.

Problems

*The space X***

14A Show that a normed space has infinite dimension iff its dual does so.

14B Show that every Hilbert space is reflexive. [In view of the correspondence $x \mapsto f_x$ of H with H^* (12.10), H^* becomes a

Hilbert space if one defines $\langle f_x, f_y \rangle = \langle y, x \rangle$. This gives rise to a similar correspondence $f \mapsto \phi_f$ of H^* with H^{**}. Show $\phi_{f_x} = \hat{x}$.]

14C Prove Proposition 14.5. [This is a notation-chase rather on the lines of 14.14.]

Annihilators

14D Let $a_1, a_2, \dots, a_n, b$ be vectors in a normed space X. Use Theorem 14.9 to show that the distance of b from $\text{lin}\{a_1, \dots, a_n\}$ is equal to

$$\sup\{|f(b)| : f \in X^*, \|f\| \leqslant 1, f(a_j) = 0 \text{ for each } j\}.$$

14E

(i) Show that every $\perp$-set in X^* is a closed linear subspace.

(ii) Suppose X is not reflexive, and let $\phi \in X^{**} \sim X$. Show that $\ker\phi$ is a closed linear subspace of X^* but not a $\perp$-set. [The only linear functionals ψ on X^*, such that $\ker\psi = \ker\phi$, are multiples of ϕ, cf. 8.6.]

14F Let M be a closed linear subspace of a reflexive space X. Show that for each $x \in X$ there is a (not generally unique) point of M closest to x. [Exploit the natural congruence of X/M with $(M^{\perp})^*$.]

Computing dual spaces (see also 15F)

14G Show that the dual of the space $\mathcal{c}$ of convergent sequences is congruent to ℓ_1 (so $\mathcal{c}^*$, $\mathcal{c}_0^*$ are congruent). [Show that every $f \in \mathcal{c}^*$ has the form

$$f(x) = \sum_{n=1}^{\infty} y_n x_n + y_0 \lim_n x_n$$

for suitable scalars y_n.]

14H (to 14.13) Let (S, S, μ) be a σ-finite measure space and let $1 \leqslant p < \infty$. Let ϕ be a positive ($\phi(f) \geqslant 0$ whenever $f \geqslant 0$) bounded linear functional on $L_p = L_p(S, \mathsf{S}, \mu)$. Adapt 14.11 using the Radon–Nikodym Theorem to prove ϕ is of the form $f \mapsto \int fg$ for some $g \in L_q$.

14I In $\mathcal{c}_0$, let $e_1, e_2, \dots$ be the standard unit vectors and let $x_1, x_2, \dots$ be vectors such that $\sum_{n=1}^{\infty} \|x_n - e_n\| < 1$. Show that the x_n form a linearly independent fundamental set in $\mathcal{c}_0$.

Further problems

Completions

14J A **completion** of a metric space X means a complete metric space $\bar{X}$ together with an isometric map i of X onto a dense subset of $\bar{X}$. One speaks of 'the completion $i: X \to \bar{X}$' or (loosely) of 'the completion $\bar{X}$'. Prove that

(i) Any two completions $i: X \to Y$ and $j: X \to Z$ of X are essentially the same in the sense that there exists an isometry ϕ of Y onto Z such that $\phi \circ i = j$. For this reason one usually speaks of 'the' completion of X, and identifying X with its image $i(X)$, one regards X as being a subset of its completion.

(ii) Every completion of a normed linear space carries a natural linear structure under which it becomes a Banach space.

14K Clearly any isometry i of X into a complete metric space Z gives rise to a completion: just let $\bar{X} = i(X)^{-}$. Verify that the constructions below yield completions in this way.

(i) X is any normed space, Z is X^{**}, i is the canonical map of X into X^{**}.

(ii) X is any metric space, a_0 a fixed point of X, Z is the Banach space of bounded continuous real functions on X with the sup-norm, and i is the map $a \mapsto f_a$ $(a \in X)$ where

$$f_a(x) = d(x, a) - d(x, a_0) \quad (x \in X).$$

Duality theory

X denotes a real Banach space; x, f denote typical elements of X and X^* respectively. Consider the relations $\circ, \omega, \pi$ between X and X^* defined below.

14L Let $x \circ f$ mean $|f(x)| \leqslant 1$. Show that the $\circ$-sets of X in the resulting Galois duality are precisely the nonempty closed convex symmetric subsets of X. (A°, $B_\circ$ are called the **symmetric polars** of $A \subset X$, $B \subset X^*$.)

14M A **wedge** in a linear space is a subset W such that αx, $x + y \in W$ whenever $x, y \in W$ and $\alpha \geqslant 0$. Prove the following:

(i) A wedge is convex; a convex set C is a wedge iff $\alpha x \in C$ whenever $x \in C$ and $\alpha \geqslant 0$.

(ii) Let W be a wedge in X and let $f \in X^*$. Then $f \geqslant 0$ on W iff f is bounded below on W.

(iii) Let $x\ \omega\ f$ mean $f(x) \geqslant 0$. Show that the ω-sets of X in the resulting duality are precisely the closed wedges. (A^ω, B_ω are called the **dual wedges** of A, B.)

(iv) Deduce **Farkas' Theorem:** Let $\mathbf{R}^n$ have the usual inner product and let $a_1, \ldots, a_n, b \in \mathbf{R}^n$. Then b can be written as $\beta_1 a_1 + \ldots + \beta_n a_n$ with all the $\beta_j \geqslant 0$ iff one has $\langle x, b\rangle \geqslant 0$ for every x with the property that $\langle x, a_j\rangle \geqslant 0$ for every j.

14N Let $x\ \pi\ f$ mean $f(x) \geqslant -1$. What are the π-sets in X?

15. Functionals on $\mathcal{C}(T)$

We now turn to the task of finding an explicit representation for bounded linear functionals ϕ on the real or complex space $\mathcal{C}(T)$, where to avoid undue complications we take T to be a compact metric space. The problem is attacked by starting with the important special case when ϕ is **positive** – that is, $\phi(f) \geqslant 0$ whenever f is real and $f \geqslant 0$ – and the main result, Theorem 15.4, is that there exists a measure μ on the Borel subsets of T (we call such a measure a **Borel measure** on T) such that ϕ consists precisely of integration with respect to μ; moreover μ is unique. The way in which this representation is extended to a general $\phi \in \mathcal{C}(T)^*$ is briefly discussed in 15.20.

Theorem 15.4 and its many generalizations are usually called the **Riesz Representation Theorem.** Of all the representation theorems for functionals this is the most useful – more so than the description of the dual of L_p or of a Hilbert space – and it accounts for much of the importance of measure theory in modern analysis. In 15.21–34 we describe, as an important and typical application, some properties of **barycentres** (centres of gravity) of measures.

15.1 A fixed positive linear functional ϕ on $\mathcal{C}(T)$ will be under consideration. Note that by Problem 8M, ϕ is necessarily bounded; in fact $\|\phi\| = \phi(1)$.

Let $\mathbf{S}$ denote the class of *all* subsets of T, $\mathbf{F}$ the class of closed sets and $\mathbf{B}$ the class of Borel sets; thus $\mathbf{F} \subset \mathbf{B} \subset \mathbf{S}$, and $\mathbf{B}$ can be described either as the smallest σ-algebra containing the open sets or as the smallest σ-algebra containing the closed sets. We begin the construction of μ by defining a set function on **S** associated with the functional ϕ: for any $A \subset T$ define $\mu^*(A)$, the **outer measure** of A, to be the infimum of $\sum_n \phi(p_n)$ taken over all (finite or countably infinite) families of non-negative functions $p_1, p_2, \ldots$ in $\mathcal{C}(T)$ such that $\sum_n p_n > \chi_A$ – that is, $\sum_n p_n(t) > 1$ for all $t \in A$ and > 0 for all $t \in T \sim A$.

We now define a distance function ρ on **S**, whereby the distance between A and B is the outer measure of their **symmetric difference** $A \,\Delta\, B = (A \sim B) \cup (B \sim A)$:

15.2 Definition $\rho(B, A) = \mu^*(A \,\Delta\, B)$ for any $A, B \subset T$.

The function ρ turns out to be more-or-less a metric on **S** (it satisfies $\rho(A, B) = \rho(B, A)$ and the triangle inequality, but it is possible to have distinct sets that are zero distance apart) and we define the class of measurable sets to be the *closure* of the class **F** of closed sets in this metric:

15.3 Definition A is **measurable** if for each $\epsilon > 0$ there is a closed F with $\rho(A, F) < \epsilon$.

15.4 Theorem The class **M** of measurable sets is a σ-algebra containing the Borel sets **B**, and the restriction of μ^* to **M** is a measure μ such that

$$\phi(f) = \int_T f d\mu \quad (f \in \mathcal{C}(T)).$$

The proof is contained in 15.5–17. The question 'is μ unique?' needs clarifying and is left until 15.18. Referring back to the definition of μ^* and forward to the next proposition we note that there is no special reason why a countable family of functions p need come ready indexed $p_1, p_2, \ldots$; it may, as in the proof below, be a double sequence $\{p_{nk}\}$. Since we are dealing with sums of non-negative numbers, we are justified in rearranging the sum in any way we like. (Note that the statement $\sum p_n > \chi_A$ is also about numbers since it merely means $\sum p_n(t) > \chi_A(t)$ for each t).

15.5 Proposition

(i) $\mu^*(\emptyset) = 0$ and $0 \leqslant \mu^*(A) \leqslant \phi(1)$ for all A.

(ii) $\mu^*(A) \leqslant \mu^*(B)$ if $A \subset B$.

(iii) For any sequence of sets A_n, $\mu^*(\bigcup A_n) \leqslant \sum \mu^*(A_n)$.

Proof (i) Obviously $\mu^*(\emptyset) = 0$ and $\mu^*(A) \geqslant 0$. For any A and any $\epsilon > 0$ the family of functions consisting of the single function $p_1 = (1 + \epsilon)1$ has the property $\sum p_n > \chi_A$, so $\mu^*(A) \leqslant \phi((1+\epsilon)1)$. Letting $\epsilon \to 0$ gives $\mu^*(A) \leqslant \phi(1)$.

(ii) Clearly if $A \subset B$, any $\{p_n\}$ such that $\Sigma p_n > \chi_B$ also has the property $\Sigma p_n > \chi_A$, so $\Sigma\phi(p_n) \geqslant \mu^*(A)$. Taking the infimum over all such $\{p_n\}$ gives the result.

(iii) Given $\epsilon > 0$ we can, for each n, find functions which we shall label

$$p_{n1}, p_{n2}, p_{n3}, \ldots$$

such that $\sum_k p_{nk} > \chi_{A_n}$ and $\Sigma\phi(p_{nk}) < \mu^*(A_n) + \epsilon/2^n$. Then the double sequence of functions $\{p_{nk}\}_{n,k\in\mathbf{N}}$ clearly has the property that $\sum_{n,k} p_{nk} > \chi_{\bigcup A_n}$ so that

$$\begin{aligned}\mu^*(\bigcup A_n) &\leqslant \sum_{n,k} \phi(p_{nk}) = \sum_{n=1}^{\infty} (\sum_k \phi(p_{nk}))\\ &< \sum_{n=1}^{\infty} (\mu^*(A_n) + \epsilon/2^n)\\ &= \sum_{n=1}^{\infty} (\mu^*(A_n) + \epsilon.\end{aligned}$$

Since ϵ is arbitrary the result follows.

15.6 This is the point to prove some properties of the 'metric' ρ, which depend only on the above properties of μ^* and on some set-theoretic relations. The latter are:

(i) $A \,\Delta\, B = B \,\Delta\, A$, and $A \,\Delta\, C \subset (A \,\Delta\, B) \cup (B \,\Delta\, C)$;
(ii) $A^c \,\Delta\, B^c = A \,\Delta\, B$;
(iii) For any sequences $\{A_n\}, \{B_n\}$ one has $\bigcap_n A_n \,\Delta\, \bigcap_n B_n \subset \bigcup_n (A_n \,\Delta\, B_n)$.

The first two are elementary; (iii) can be proved as follows. If $t \in \bigcap_n A_n \sim \bigcap_n B_n$ then $t \in A_n$ for all n, while there exists k such that $t \notin B_k$. For this k one has $t \in A_k \sim B_k \subset A_k \,\Delta\, B_k$, so that $t \in \bigcup_n (A_n \,\Delta\, B_n)$. Thus $\bigcap_n A_n \sim \bigcap_n B_n \subset \bigcup_n (A_n \,\Delta\, B_n)$. Similarly with $\bigcap_n B_n \sim \bigcap_n A_n$ and hence with $\bigcap_n A_n \,\Delta\, \bigcap_n B_n$.

From these and Proposition 15.5 one has at once

15.7 Lemma The distance function $\rho(A, B) = \mu^*(A \,\Delta\, B)$ satisfies

(i) $\rho(A, B) = \rho(B, A)$ and $\rho(A, C) \leqslant \rho(A, B) + \rho(B, C)$;
(ii) $\rho(A^c, B^c) = \rho(A, B)$;
(iii) For any sequences $\{A_n\}, \{B_n\}$ one has $\rho(\bigcap_n A_n, \bigcap_n B_n) \leqslant \sum_n \rho(A_n, B_n)$.

15.8 Corollary $\mu^*(A) \leqslant \rho(A, B) + \mu^*(B)$.
Proof Let $C = \emptyset$ in (i).

Parts (ii), (iii) of the Lemma are need in Proposition 15.16 to prove that the measurable sets form a σ-algebra.

The topology of T now makes its entry. It is a fundamental fact about this theory that a *compactness argument* has to be used somewhere: in our development it comes into (ii) below.

15.9 Proposition

(i) $\mu^*(A) = \inf\{\mu^*(U): U \text{ open}, U \supset A\}$.
(ii) If A is closed, only one p is needed in the formula defining μ^*, that is

$$\mu^*(A) = \inf\{\phi(p): p > \chi_A\}.$$

Proof (i) Given $\epsilon > 0$ choose $\{p_n\}$ with $\Sigma p_n > \chi_A, \Sigma\phi(p_n) < \mu^*(A) + \epsilon$. For each $t \in A$, since $\sum_1^\infty p_n(t) > 1$, there must be some k such that $\sum_1^k p_n(t) > 1$, which is to say that A is covered by the (clearly increasing) sequence of open sets

$$U_k = \{t: \sum_1^k p_n(t) > 1\} \quad (k = 1,2, \ldots).$$

Thus the open set $U = \bigcup U_k$ contains A and clearly $\Sigma p_n(t) > 1$ on U, so that $\mu^*(A) \leqslant \mu^*(U) \leqslant \Sigma\phi(p_n) < \mu^*(A) + \epsilon$. Since ϵ is arbitrary the result follows.

(ii) If A is closed it is compact, so, with the increasing sequence $\{U_k\}$ as in part (i), there must be some k such that $A \subset U_k$.

Define $q = \sum_1^k p_n$, then $q > \chi_A$ and

$$\mu^*(A) \leqslant \phi(q) = \sum_1^k \phi(p_n) \leqslant \sum_1^\infty \phi(p_n) < \mu^*(A) + \epsilon.$$

Again since ϵ is arbitrary the result follows.

15.10 Corollary $\mu^*(T) = \phi(1)$.
The proof is obvious.

15.11 Corollary If $A \in \mathsf{M}$ then $\mu^*(A) = \sup\{\mu^*(E)$: E closed, $E \subset A\}$.

Proof Let $A \in \mathsf{M}$ and $\epsilon > 0$. First we may choose closed F with $\rho(A, F) = \mu^*(A \,\Delta\, F) < \epsilon/3$. By part (i) above we may choose open $U \supset A \,\Delta\, F$ with $\mu^*(U) < \mu^*(A \,\Delta\, F) + \epsilon/3 < 2\epsilon/3$. Now let $E = F \sim U$, which is clearly a closed set. It is left to the reader to verify that $E \subset A$ and $E \,\Delta\, F \subset U$, so that $\rho(E, F) \leqslant \mu^*(U) < 2\epsilon/3$, and thus $\mu^*(A) - \mu^*(E) \leqslant \rho(A, E) \leqslant \rho(A, F) + \rho E, F) < \epsilon$, by Lemma 15.7 (i) and Corollary 15.8.

15.12 Proposition $\mu^*(F_1 \cup F_2) = \mu^*(F_1) + \mu^*(F_2)$ for disjoint closed F_1, F_2.

Proof By 15.5 (iii) we only need to prove $\mu^*(F_1 \cup F_2) \geqslant \mu^*(F_1) + \mu^*(F_2)$. The technique needed here is a way of decomposing any $p > \chi_{F_1 \cup F_2}$ into $p_1 + p_2$ with $p_1 > \chi_{F_1}$, $p_2 > \chi_{F_2}$. We can construct a function $g \in \mathcal{C}(T)$ with $0 \leqslant g \leqslant 1$, $g = 1$ on F_1 and $g = 0$ on F_2; for instance by setting $g(t) = d(t, F_1)/(d(t, F_1) + d(t, F_2))$, as the reader can verify. Writing $h = 1 - g$ we then have

$$g \geqslant \chi_{F_1},\ h \geqslant \chi_{F_2},\ g + h = 1$$

so if $p > \chi_{F_1 \cup F_2}$ and we set $p_1 = gp$, $p_2 = hp$ it is clear that p_1, p_2 have the required property.

Thus $\qquad \phi(p) = \phi(p_1) + \phi(p_2) \geqslant \mu^*(F_1) + \mu^*(F_2)$

and taking the infimum over all p gives the result.

15.13 Proposition If $\{F_n\}$ is an increasing sequence of closed sets then

$$\mu^*(\bigcup_n F_n) = \sup_n \mu^*(F_n)$$

Proof Choose any $\epsilon > 0$. We shall show there is an increasing sequence $p_1 \leqslant p_2 \leqslant p_3 \leqslant \ldots$ of continuous functions with

$$p_n > \chi_{F_n},\ \phi(p_n) < \mu^*(F_n) + \epsilon \text{ for all } n.$$

Clearly we can do this for $n = 1$. Inductively assume p_n has been chosen satisfying these conditions. Then $\mu^*(F_n) + \epsilon - \phi(p_n)$ is a positive number, say δ. Choose $q > \chi_{F_{n+1}}$ such that

$$\phi(q) < \mu^*(F_{n+1}) + \delta$$

and define $p_{n+1} = p_n \vee q$ ($\vee, \wedge$ are the max, min operations on real functions). Since $p_n > \chi_{F_n}$, $q > \chi_{F_{n+1}} \geqslant \chi_{F_n}$, we have $p_n \wedge q > \chi_{F_n}$ and in view of the equation $p_n + q = (p_n \vee q) + (p_n \wedge q)$ we have

$$\begin{aligned} \mu^*(F_{n+1}) + \delta + \mu^*(F_n) + \epsilon > \phi(q) &+ (\phi(p_n) + \delta) \\ &= \phi(p_n \vee q) + \phi(p_n \wedge q) + \delta \\ &\geqslant \phi(p_{n+1}) + \mu^*(F_n) + \delta \end{aligned}$$

whence $\phi(p_{n+1}) < \mu^*(F_{n+1}) + \epsilon$. Obviously $p_{n+1} \geqslant q > \chi_{F_{n+1}}$, so the induction step is complete.

With the p_n defined in this way, define functions $r_1 = p_1$, $r_2 = p_2 - p_1$, $r_3 = p_3 - p_2, \ldots$, so that the r_n are non-negative continuous functions such that for each n, $\sum_1^n r_j = p_n > \chi_{F_n}$. This gives $\sum_1^\infty r_j > \chi_{F_n}$ for every n, and hence $\sum_1^\infty r_j > \chi_{\cup F_n}$. Thus

$$\mu^*(\cup F_n) \leqslant \sum_1^\infty \phi(r_j) = \sup_n \sum_1^n \phi(r_n) = \sup_n \phi(p_n)$$

$$\leqslant \sup_n \mu^*(F_n) + \epsilon.$$

Since ϵ is arbitrary, $\mu^*(\cup F_n) \leqslant \sup_n \mu^*(F_n)$. The reverse inequality is clear, and this completes the proof.

15.14 Corollary If U is open then $\mu^*(U) = \sup\{\mu^*(F): F$ closed, $F \subset U\}$.

Proof U is the union of the increasing sequence of closed sets

$$F_n = \{t \in U: d(t, U^c) \geqslant \frac{1}{n}\}$$

and $\mu^*(U) = \sup \mu^*(F_n)$ by the last result.

As a further corollary we obtain a very important step in the construction:

15.15 Proposition Open sets are measurable.

Proof Let U be open, let $\epsilon > 0$, and using the last result choose closed $F \subset U$ with $\mu^*(F) > \mu^*(U) - \epsilon$. Clearly $U \Delta F =$

$U \sim F$, which is an open set V say, and by the definition of measurability it is enough to prove $\mu^*(V) \leqslant \epsilon$. If $\mu^*(V) > \epsilon$ then applying the last result again one would choose closed $E \subset V$ with $\mu^*(E) \geqslant \epsilon$. But then E, F would be closed, disjoint, and subsets of U, so by Proposition 15.12 we should have

$$\mu^*(U) \geqslant \mu^*(E \cup F) = \mu^*(E) + \mu^*(F) > \mu^*(U) - \epsilon + \epsilon = \mu^*(U),$$

a contradiction which proves the result.

15.16 Proposition $\mathbf{M}$ is a σ-algebra and the restriction μ of μ^* to $\mathbf{M}$ is a measure. Moreover $\mathbf{B} \subset \mathbf{M}$.

Proof To show $\mathbf{M}$ is a σ-algebra it is sufficient to show it is closed under taking countable intersections and complements. Therefore suppose first that $A_1, A_2, \ldots$ are measurable, and let $\epsilon > 0$. For each n choose a closed set F_n with $\rho(A_n, F_n) < \epsilon/2^n$. Then $\bigcap F_n$ is closed, and by Lemma 15.7 (iii),

$$\rho(\bigcap A_n, \bigcap F_n) \leqslant \Sigma\rho(A_n, F_n) < \Sigma\epsilon/2^n = \epsilon,$$

showing $\bigcap A_n$ is measurable.

Secondly, let A be measurable, let $\epsilon > 0$ and choose a closed set F with $\rho(A, F) < \epsilon/2$. Then $U = F^c$ is open and $\rho(A^c, U) = \rho(A, F)$ by Lemma 15.7 (ii). Now U is measurable, so choose a closed set F_1 with $\rho(U, F_1) < \epsilon/2$.

Then

$$\rho(A^c, F_1) \leqslant \rho(A^c, U) + \rho(U, F_1) < \epsilon/2 + \epsilon/2 = \epsilon,$$

showing A^c is measurable.

Next we show μ^* restricted to $\mathbf{M}$ is a measure. In view of the property $\mu^*(\bigcup A_n) \leqslant \Sigma\mu^*(A_n)$, it is sufficient to prove that $\mu^*(\bigcup A_n) \geqslant \Sigma\mu^*(A_n)$ whenever $A_1, A_2, \ldots$ are disjoint measurable sets. Let $\epsilon > 0$, then by Corollary 15.11 we can choose closed $F_n \subset A_n$ with $\mu^*(F_n) > \mu^*(A_n) - \epsilon/2^n$. The F_n are clearly disjoint, and, using the obvious extension of Proposition 15.12 by induction to any *finite* number of disjoint closed sets, one has for each k

$$\begin{aligned}\mu^*(\bigcup A_n) &\geqslant \mu^*(F_1 \cup \ldots \cup F_n) \\ &= \sum_{n=1}^{k} \mu^*(F_n) \\ &> \sum_{n=1}^{k} (\mu^*(A_n) - \epsilon/2^n)\end{aligned}$$

$$> \sum_{n=1}^{k} \mu^*(A_n) - \epsilon.$$

Since k and ϵ are arbitrary we obtain

$$\mu^*(\cup A_n) \geqslant \Sigma \mu^*(A_n)$$

as required.

Finally $\mathbf{M}$ is obviously a σ-algebra containing the closed sets $\mathbf{F}$ and hence contains the Borel sets $\mathbf{B}$ by definition.

There remains one necessary but tedious task, which will complete the proof of theorem 15.4: namely to show that

$$\int f\, d\mu = \phi(f) \quad (f \in \mathcal{C}(T))$$

Clearly it is sufficient to verify this for functions f such that $0 \leqslant f \leqslant 1$, since every f is a linear combination of (in the real case at most two, in the complex case at most four) such functions. Hence the problem is reduced to proving:

15.17 Lemma If $f \in \mathcal{C}(T)$ and $0 \leqslant f \leqslant 1$ then $\phi(f) = \int f\, d\mu$.

Proof The argument can be visualized as a cutting up of the area under the graph of f into horizontal slices (Diagram 20).

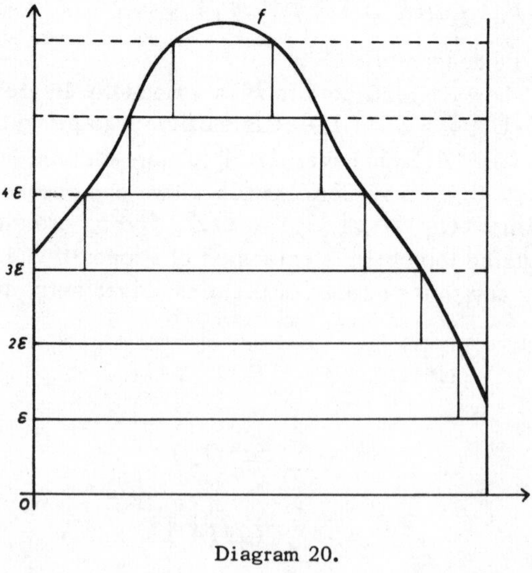

Diagram 20.

Choose any positive integer N and let $\epsilon = 1/N$. For $n = 1,2,\ldots,N$ define closed sets

$$F_n = \{t \in T: f(t) \geqslant n\epsilon\}$$

and continuous functions

$$g_1 = f \wedge \epsilon 1,\ g_2 = (f \wedge 2\epsilon 1) - (f \wedge \epsilon 1), \ldots,$$
$$g_N = (f \wedge N\epsilon 1) - (f \wedge (N-1)\epsilon 1)$$
$$= f - (f \wedge (N-1)\epsilon 1).$$

We leave as an exercise the following facts that should be clear from the diagram:

$$\sum_1^N g_n = f,\ g_n \geqslant \epsilon\chi_{F_n},\ \sum_1^N \epsilon\chi_{F_n} \geqslant f - \epsilon 1.$$

Since the definition of μ implies $\phi(g_n) \geqslant \epsilon\mu(F_n)$ we now have

$$\phi(f) = \sum_1^N \phi(g_n) \geqslant \sum_1^N \epsilon\mu(F_n) = \sum_1^N \epsilon \int \chi_{F_n}\, d\mu$$
$$= \int \Big(\sum_1^N \epsilon\chi_{F_n}\Big) d\mu \geqslant \int (f - \epsilon 1)\, d\mu = \int f\, d\mu - \epsilon\mu(T).$$

Since $\epsilon = 1/N$, letting $N \to \infty$ gives

$$\phi(f) \geqslant \int f\, d\mu.$$

Replacing f by $1 - f$, and using the fact that $\phi(1) = \mu(T)$, is easily seen to yield the reverse inequality, and the proof is complete.

Finally we turn to the question of uniqueness.

If ν is *any* finite measure defined on a σ-algebra containing the class $\boldsymbol{B}$ then continuous functions are ν-measurable and integrable and it is clear that the map

$$\phi: f \mapsto \int f\, d\nu$$

is a positive linear functional on $\mathcal{C}(T)$. If now we construct the measure μ as in the previous pages, do we get the original ν back again? Not precisely, since μ and ν will in general be defined on different σ-algebras; but these σ-algebras have the class $\boldsymbol{B}$ in common, and for the purpose of integrating continuous functions **B** is the σ-algebra of most interest. In fact we have

15.18 Theorem With the above notation, $\mu(A) = \nu(A)$ for all $A \in \mathbf{B}$.

Proof By the main theorem, $\int f\, d\mu = \phi(f) = \int f\, d\nu \quad (f \in \mathcal{C}(T))$. First let E be any closed set. We can choose a decreasing sequence of continuous functions f_n such that $f_n \to \chi_E$ pointwise (for instance $f_n(t) = (1 + n\, d(t, E))^{-1}$, as the reader may verify). By the Monotone Convergence Theorem we conclude that

$$\mu(E) = \lim \int f_n\, d\mu = \lim \int f_n\, d\nu = \nu(E) \quad (E \text{ closed}).$$

Now for any $A \in \mathbf{B}$ and any closed $E \subset A$,

$$\mu(E) = \nu(E) \leqslant \nu(A) \tag{*}$$

Since μ is constructed from ϕ, we can supply Corollary 15.11 to μ to obtain from (*):

$$\mu(A) = \sup\{\mu(E): E \text{ closed}, E \subset A\} \leqslant \nu(A) \quad (A \in \mathbf{B}).$$

If in fact $\mu(A) < \nu(A)$ for any A we should have

$$\mu(T) = \mu(A) + \mu(T \sim A) < \nu(A) + \nu(T \sim A) = \nu(T)$$

$$= \int 1\, d\nu = \int 1\, d\mu = \mu(T)$$

a contradiction. Hence $\mu(A) = \nu(A) \quad (A \in \mathbf{B})$, as required.

Corollary Every finite measure ν on the Borel sets of a compact metric space is necessarily **regular**, that is

$$\nu(A) = \sup\{\nu(E): E \text{ closed}, E \subset A\}$$

or equivalently, by taking complements,

$$\nu(A) = \inf\{\nu(U): U \text{ open}, U \supset A\}.$$

15.19 We now give some examples of functionals and their associated measures.

Example 1 Let μ be the unit point mass ϵ_t at a point t of T (2.7 Example 1). It is simple to verify that $\int f\, d\mu = f(t)$ for any function f, so μ is associated with the *evaluation* functional $\hat{t}$ defined by

$$\hat{t}(f) = f(t) \quad (f \in \mathcal{C}(T)).$$

Similarly any finite linear combination of point masses,

$$\mu = \sum_1^k c_j\, \epsilon_{t_j}$$

with positive coefficients, is associated with the positive functional

$$\phi: f \mapsto \sum_{1}^{k} c_j\, f(t_j) \quad (f \in \mathcal{C}(T)).$$

Example 2 A more important fact is that Theorem 15.4 proves the existence of Lebesgue measure on any finite interval $[a, b]$ in $\mathbf{R}$: for the elementary *Riemann* integral $f \mapsto \int_b^a f(t)\,dt$ is a positive linear functional on $\mathcal{C}[a, b]$, and therefore there exists a measure λ such that $\int_a^b f(t)\,dt = \int f\ d\lambda \quad (f \in \mathcal{C}[a, b])$; it is left as an exercise to show that $\lambda(I)$ coincides with the ordinary idea of the length of I when I is a subinterval of $[a, b]$, and thus that λ is indeed Lebesgue measure.

15.20 Though Theorem 15.4 exemplifies the techniques needed in representing a functional on $\mathcal{C}(T)$ by a measure, some further remarks are called for. First, what about non-positive functionals? It can be proved (quite simply – see Problem 15C) that every $\phi \in \mathcal{C}(T)^*$ is a linear combination of positive functionals, and from this it is not hard to set up a one-to-one correspondence between $\mathcal{C}(T)^*$ and the space $\mathbf{M}(T)$ of bounded real or complex **charges** (countably additive scalar-valued set functions, also called **signed measures**) on the Borel sets of T; under a natural norm on $\mathbf{M}(T)$ this becomes a congruence.

Second, 15.19 Example 2 suggests that it would be desirable to relax the compactness assumption on T, so as to construct Lebesgue measure on the whole of $\mathbf{R}$ 'in one go'. We mention that the general setting for the Riesz Representation Theorem is in a locally compact Hausdorff topological space T; but the general proof presents many extra difficulties, both of topology and of measure theory.

Barycentres and Choquet's Theorem

15.21 Before turning to our next topic, a non-trivial application of the Riesz Representation Theorem to the construction of measures, we discuss briefly its physical background. One of the obvious physical interpretations of a measure is as a *distribution of mass*. Starting with a very simple case we recall that if one has:

particles of masses $m_1, \ldots, m_k$ at points with position vectors $u_1, \ldots, u_k$ (1)

in ordinary space, an important physical property is their **barycentre** (centroid, centre of mass, centre of gravity) given by the formula

$$u_0 = \frac{\sum_1^k m_j u_j}{\sum_1^k m_j} \tag{2}$$

Of course to compute a barycentre one represents vectors as $u = (x, y, z)$ in some coordinate system and the x coordinate of u_0 is then given by

$$x = \frac{\Sigma m_j x_j}{\Sigma m_j} \tag{3}$$

with similar formulae for the y and z coordinates. We can, and shall, interpret (3) as an integral: mathematically, the mass distribution (1) is represented by the measure m where

$$m = \sum_1^k m_j \, \epsilon_{u_j} \tag{4}$$

– the sum of m_j times the unit point mass at u_j – and one can write (3) as

$$x_0 = \frac{\int x dm}{\int dm} \text{ with } y_0, z_0 \text{ similarly.} \tag{5}$$

This form, which generalizes to continuous as well as discrete mass distributions, has probably been used by the reader at some stage in finding the barycentres of cones, hemispheres etc.

There remains one simplifying convention and one simple observation before we recast this in its final form. First, we reduce (5) to the form

$$x_0 = \int x dm, \text{ with } y_0, z_0 \text{ similarly} \tag{6}$$

by dealing only with **probability measures**, that is measures m whose total mass $\int dm$ is unity. Second, noting that $u \mapsto x$, $u \mapsto y$, $u \mapsto z$ are a spanning set of linear functionals on the linear space $\mathbf{U}$ of 3-dimensional vectors we see that (6) is equivalent to

$$f(x\ ) = \int f(u)\, dm(u) \text{ for all linear functionals } f \tag{7}$$

the integration being taken over all of $\mathcal{O}$, or any convenient subset whose complement is m-null. We now make the

15.22 Definition Let X be a normed linear space, A a bounded Borel subset of X, and μ a Borel probability measure on X such that $X \sim A$ is μ-null. A point x of X is a **barycentre** of μ if

$$f(x) = \int f\,d\mu \text{ for all } f \in X^*. \tag{8}$$

15.23 The restrictions imposed on μ ensure of course that members of X^* are μ-measurable, and bounded outside a μ-null set, so that they are μ-integrable. We shall say that μ **lives on** A if $X \sim A$ is μ-null, and denote by $\mathcal{P}(A)$ the set of all Borel probability measures on X living on the Borel subset A. The integral in (8) can be taken either over X or over A. Clearly we can – and often do – identify μ with its restriction to A (2.7 Example 4). Note that (8) includes our motivating example in 3-space as a special case. A barycentre is unique when it exists, for if x and y are both barycentres of μ then $f(x) = f(y)$ $(f \in X^*)$ so that $x = y$ by Corollary 8.11; but in general, μ may have no barycentre at all.

In order to use the separation theorems of §8 we shall assume that X is a real normed space; the results are generalized to the complex case by the simple expedient of restricting multiplication to be by real scalars.

It is clear from formula (2) of 15.21 that barycentres are closely linked with convex combinations and convex hulls. The next two results make this more precise.

15.24 Lemma If x_0 is the barycentre of a measure $\mu \in \mathcal{P}(A)$ then $x_0 \in \overline{\text{co}}\, A$.

Proof If $x_0 \notin \overline{\text{co}}\, A$ then by Corollary 8.15 there exist $f \in X^*$ and $\delta > 0$ such that $f(x_0) + \delta \leqslant \inf\{f(x)\colon x \in \overline{\text{co}}\, A\} \leqslant \inf\{f(x)\colon x \in A\}$, which gives

$$f(x_0) = \int_A f\,d\mu \geqslant \int_A (f(x_0) + \delta)\,d\mu = (f(x_0) + \delta)\int d\mu = f(x_0) + \delta,$$

a contradiction.

Apart from the idea of the barycentre x_0 being in some sense the 'average position' of the measure μ the reader may find another intuitive picture helpful: μ describes the contributions made by different parts of A in expressing x_0 as a limit of convex combinations of points of A. (Of course this can usually be done in different ways, i.e. different measures may have the same barycentres.) This is the idea underlying the proof of the two fundamental results below.

15.25 Theorem Let A be a compact subset of X and let $x_0 \in \overline{\text{co}}\, A$. Then x_0 is the barycentre of some $\mu \in \mathcal{P}(A)$.

This clearly follows, on taking T to be A and e to be the identity map, from the following more technical but in many ways more useful result.

15.26 Theorem Let T be a compact metric space, e a continuous map of T into X, $\tilde{T}$ the image $e(T)$ and x_0 a point of $\overline{\text{co}}\, \tilde{T}$. Then there is a probability Borel measure μ on T such that

$$f(x_0) = \int_T f(e(t))\, d\mu(t) \text{ (or } \int_T f \circ e\, d\mu, \text{ for short)}$$

for each $f \in X^*$.

Proof By the Riesz Representation Theorem it suffices to show that there exists a positive linear functional ϕ on $\mathcal{C}(T)$ such that $\phi(1) = 1$ and

$$\phi(f \circ e) = f(x_0) \quad (f \in X^*). \tag{9}$$

(Note that for each $f \in X^*$, $f \circ e$ is indeed a member of $\mathcal{C}(T)$.)

If x_0 is in co $\tilde{T}$ and thus is a convex combination $\sum_1^k \alpha_j\, e(t_j)$ of points $e(t_j)$ in $\tilde{T}$ this is easy, for the functional defined by

$$\phi(g) = \sum_1^k \alpha_j\, g(t_j) \quad (g \in \mathcal{C}(T))$$

is clearly linear and positive with $\phi(1) = 1$, and for each $f \in X^*$ one has, by the linearity of f,

$$\phi(f \circ e) = \sum_1^k \alpha_j\, f(e(t_j)) = f \sum_1^k \alpha_j\, e(t_j) = f(x_0).$$

Note that $\phi(1) = \|\phi\|$ for positive functionals, so $\|\phi\| = 1$.

A general $x_0 \in \overline{\text{co}}\, \tilde{T}$ is the limit of some sequence $\{x_n\}$ in $\overline{\text{co}}\, \tilde{T}$. By the last paragraph, for each n there is a positive linear functional ϕ_n on $\mathcal{C}(T)$ such that $\|\phi_n\| = \phi_n(1) = 1$ and $\phi_n(f \circ e) = f(x_n)$ for all $f \in X^*$. By continuity, $f(x_n) \to f(x_0)$ as $n \to \infty$; we deduce that whenever g belongs to the set M of functions in $\mathcal{C}(T)$ of the form $f \circ e + \lambda 1$ for some $f \in X^*$ and scalar λ, the limit $\phi_0(g) = \lim_n \phi_n(g)$ exists and one has

$$\phi_0(g) = \lim_n (\phi_n(f \circ e) + \lambda\phi_n(1)) = f(x_0) + \lambda. \tag{10}$$

Clearly M is a linear subspace and ϕ_0 a linear functional on M with $\|\phi_0\| = 1$. Extending ϕ_0 by the Hahn–Banach Theorem to a functional ϕ on $\mathcal{C}(T)$ with the same norm we clearly have $\phi(1) = 1$, and $\phi(f \circ e) = f(x_0)$ $(f \in X^*)$, by (10). Problem 8M shows that ϕ is positive and the proof is complete.

Problem 16J gives an alternative approach to this proof, which avoids the Hahn–Banach Theorem.

15.27 Lemma 15.24 and Theorem 15.25 imply that for a compact subset A of a normed space X, $\overline{\text{co}}\, A$ consists precisely of all barycentres of members of $\mathcal{P}(A)$. It is worth mentioning that when X is complete (though not in general) every $\mu \in \mathcal{P}(A)$ actually posseses a barycentre; this ties up $\mathcal{P}(A)$ and $\overline{\text{co}}\, A$ in a very pleasing way.

15.28 We now make a more subtle analysis of barycentres. Instead of starting with a compact set A of arbitrary shape and linking $\overline{\text{co}}\, A$ with the measures on A we start with a compact *convex* set C and investigate those subsets A of C for which $\overline{\text{co}}\, A = C$. Among such sets, which play roughly the same role for the operation 'closed convex hull' as fundamental sets do for 'closed linear span' we are especially interested in 'economical', that is small, ones. Motivated by the observation that many simple convex solids (squares, cubes, tetrahedra, etc.) are the convex hulls of their corners we make a definition. A point x of C is an **extreme point** of C if there do not exist distinct points $u, v \in C$ and $\alpha \in (0, 1)$ such that $x = \alpha u + (1 - \alpha)v$; that is, if no line segment in C has x as an interior point (Diagram 21).

15.29 Theorem (G. Choquet 1956) Let C be a compact convex set in a normed space X and x_0 a point of C. Then x_0 is the barycentre of some $\mu \in \mathcal{P}(C)$ such that μ lives on a subset of the extreme points of C.

Before going on to the proof we state an important result which follows at once from Choquet's Theorem, though one should add that the usual proof (Problem 15G) is by far more elementary arguments!

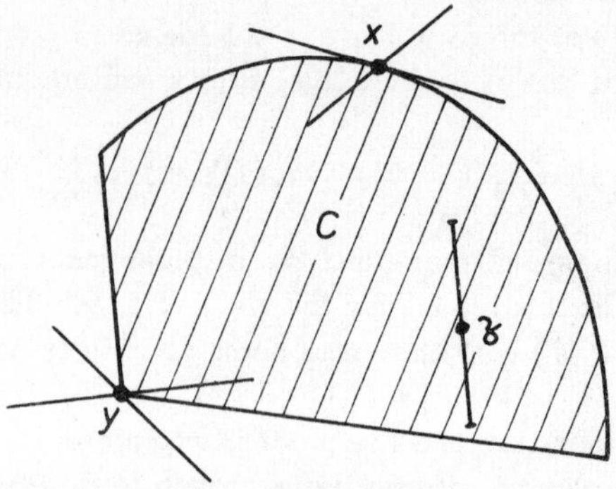

(x and y are extreme points; z is not)

Diagram 21.

15.30 Corollary (Krein–Milman Theorem) Every compact convex set C in a normed space is the closed convex hull of its set E of extreme points.

Proof The Theorem and Lemma 15.24 imply at once that $C \subset \overline{\mathrm{co}}\, E$ while the reverse inclusion is trivial.

This ties in with the remarks above about squares, cubes, etc., since it is simple to verify that the extreme points of such sets are just their corner points in the intuitive sense of the word.

15.31 Our proof of Choquet's Theorem (due essentially to Bonsall 1963) depends on the following notions. A real function f on C is **convex, concave,** or **affine** according as the relation $\leqslant$, $\geqslant$ or $=$ holds between

$$f(\alpha x + (1 - \alpha)y) \text{ and } \alpha f(x) + (1 - \alpha)\, f(y)$$

for any points $x, y \in C$ and any $\alpha \in [0, 1]$; *strictly convex* or *strictly concave* if $<$ or $>$ respectively hold whenever $x \neq y$ and $0 < \alpha < 1$. Clearly the affine functions are just those that are both convex and concave. The following simple properties are left as an exercise.

15.32 Lemma

(i) If $f \in X^*$ and $\lambda \in \boldsymbol{R}$ then $f + \lambda 1$ is affine.
(ii) The pointwise infimum of a sequence of affine functions (assumed bounded below) is concave.
(iii) f is strictly convex $\Leftrightarrow -f$ is strictly concave.
(iv) The sum of a concave and a strictly concave function is strictly concave.

(v) If f is strictly concave and $f \geqslant 0$ then $\{x \in C: f(x) = 0\}$ is a subset of the extreme points of C.

15.33 Lemma There exists a strictly convex continuous function on C.

Proof Since C is a compact metric space it is separable and one easily sees that there exists a countable set $\{f_n\}$ of members of X^*, such that $\|f_n\| \leqslant 1$ for all n, separating the points of C. Let f be any strictly convex continuous function on $\boldsymbol{R}$ to $\boldsymbol{R}$ (such as $F(t) = t^2$) and define

$$g(x) = \sum_1^\infty 2^{-n} F(f_n(x)), \; (x \in C).$$

Since C is bounded the series is easily seen to converge uniformly on C, so that g is continuous. Given $x, y \in C$ and $\alpha \in [0, 1]$ and writing $z = \alpha x + (1 - \alpha)y$ we have, for each n,

$$\begin{aligned} 2^{-n} F(f_n(z)) &= 2^{-n} F(\alpha f_n(x) + (1-\alpha) f_n(y)) \\ &\leqslant \alpha . 2^{-n} F(f_n(x)) + (1-\alpha) 2^{-n} F(f_n(y)); \end{aligned}$$

if $0 < \alpha < 1$ and $x \neq y$ then strict inequality holds for at least one n, namely an n for which $f_n(x) \neq f_n(y)$. Summing from 1 to ∞ shows that g is strictly convex.

15.34 Proof of Theorem 15.29 Let g be the function constructed in the last Lemma and, in the product space $X \times \boldsymbol{R}$, let $\tilde{C}$ be the graph of g (a sort of 'bowl-shaped surface'). Then $\tilde{C}$ is the image of C under the continuous map $e: x \mapsto (x, g(x))$ of C into $X \times \boldsymbol{R}$.

Now $\tilde{C}$ is compact and therefore bounded; therefore $\overline{\text{co}}\, \tilde{C}$ is a closed bounded set and it is clear that among the points of the form (x_0, t) in $\overline{\text{co}}\, \tilde{C}$, ($x_0$ being the given point in the statement of the Theorem) there is one, say (x_0, α), whose t coordinate is as large as possible. By Theorem 15.26 there is a probability Borel measure μ on C such that

$$G(x_0, \alpha) = \int_C G(e(x)) \, d\mu(x) \tag{11}$$

for every bounded linear functional G on $X \times \boldsymbol{R}$. We now show that μ is the measure we want. Given $f \in X^*$ the functional $(x, t) \mapsto f(x)$ is in $(X \times \boldsymbol{R})^*$; taking this for G and using the definition of e yields

$$f(x_0) = \int_C G(x, g(x)) \, d\mu(x) = \int_C f(x) \, d\mu(x) \quad (f \in X^*) \tag{12}$$

so that (back in the normed space X)

$$x_0 \text{ is the barycentre of } \mu. \tag{13}$$

The function $(x, t) \mapsto t$ is in $(X \times R)^*$; taking this for G yields

$$\alpha = G(x_0, \alpha) = \int_C G(x, g(x))\, d\mu(x) = \int_C g(x)\, d\mu(x). \tag{14}$$

Now, by the definition of α, for each $\epsilon > 0$ the point $(x_0, \alpha + \epsilon)$ lies outside $\overline{\text{co}}\, \tilde{C}$. By the Separation Theorem, Corollary 8.15, there exists $H \in (X \times R)^*$ whose value β at $(x_0, \alpha + \epsilon)$ is greater than its supremum on $\overline{\text{co}}\, \tilde{C}$ and in particular on $\tilde{C}$. It should be clear from a diagram, and is left to the reader to verify that the set $\{(x, t): H(x, t) = \beta\}$ is the graph of a function h of the form $f + \lambda 1$ where $f \in X^*$ such that

$$h(x_0) = \alpha + \epsilon,\ h \geqslant g \text{ on } C. \tag{15}$$

Thus using (12),

$$\int f\, d\mu = \int f\, d\mu + \lambda \int d\mu = f(x_0) + \lambda = h(x_0) = \alpha + \epsilon. \tag{16}$$

Do this for a sequence of $\epsilon_n \to 0$ and let k be the pointwise infimum of the resulting h_n; then $k \leqslant h_n$ for all n, so by (16), $\int k\, d\mu \leqslant \alpha$. Thus the function $k - g$ satisfies

$$k - g \geqslant 0 \text{ on } C,\ \int (k - g)\, d\mu \leqslant \alpha - \alpha = 0 \text{ by (14)},$$

and therefore vanishes μ-almost everywhere, in other words μ lives on the set $N = \{x \in C: (k - g)(x) = 0\}$. The various parts of Lemma 15.32 yield: k is concave; $k - g$ is strictly concave; N is a subset of the extreme points of C. With (13) this completes the proof.

Though a discussion of the uses of Choquet's Theorem would take us too far afield into the realms of complex analysis and potential theory, the following deduction from Theorem 15.26 shows quite well the sort of representation theorems that can be deduced by these techniques. Let f be a continuous scalar function on the square $0 \leqslant s \leqslant 1$, $0 \leqslant t \leqslant 1$, and for each s let f_s denote the 'section' of f at s,

$$f_s(t) = f(s, t).$$

The functions f_s form a subset A of the normed space $\mathcal{C}[0, 1]$.

15.35 Proposition $\overline{\text{co}}\, A$ consists precisely of all functions g of the form

$$g(t) = \int_0^1 f(s, t)\, d\mu(s) \tag{17}$$

where μ is a Borel probability measure on $[0,1]$.

Proof By Problem 6F, $e: s \mapsto f_s$ is a continuous map of $[0,1]$ into $\mathcal{C}[0,1]$, whose image is the set A. Theorem 15.26, with appropriate change of notation, asserts that for each $g \in \overline{\text{co}}\, A$ there is a measure μ on $[0,1]$ such that

$$\phi(g) = \int_0^1 \phi(f_s)\, d\mu(s) \quad (\phi \in \mathcal{C}[0,1]^*).$$

Taking ϕ to be the functional $g \mapsto g(t)$ gives the required equation (17). That, conversely, every function of the form (17) is in $\overline{\text{co}}\, A$ is left as an exercise.

Problems

Unless otherwise stated T denotes a compact metric space and the scalars are real. $\mathcal{C}^+(T)$ denotes $\{f \in \mathcal{C}(T): f \geqslant 0\}$.

15A Verify, from the definitions in 15.1, that the measure λ constructed in 15.19 Example 2 is indeed Lebesgue measure on $[a, b]$. (The usual properties of the Riemann integral of a continuous function, as in elementary calculus, may be assumed.)

15B Let M denote the space obtained from the class $\mathbf{M}$ of measurable sets by counting two sets as the same if their ρ-distance apart is zero. (Notation of 15.1 – 15.4; make the above informal description rigorous before proceeding!) Show that ρ makes M into a complete metric space.

15C The Decomposition Lemmas (DLs) For $f, g \in \mathcal{C}(T)$ with $f \leqslant g$ define $[f; g] = \{h \in \mathcal{C}(T): f \leqslant h \leqslant g\}$.

(i) DL for $\mathcal{C}(T)$. Show that for all $f, g \in \mathcal{C}^+(T)$, $[0; f + g] = [0; f] + [0; g]$ in the sense of vector addition of subsets.

(ii) DL for $\mathcal{C}(T)^*$. Given $\phi \in \mathcal{C}(T)^*$ define

$$\phi_+(f) = \sup\{\phi(g): g \in [0; f]\} \quad (f \in \mathcal{C}^+(T)).$$

Show that $\phi_+(\alpha f) = \alpha\phi_+(f)$, $\phi_+(f + g) = \phi_+(f) + \phi_+(g)$ whenever $f, g \in \mathcal{C}^+(T)$, $\alpha \geqslant 0$ and deduce that ϕ_+ has a unique extension (also called ϕ_+) to a positive linear functional on $\mathcal{C}(T)$. Show that

$$\phi = \phi_+ - \phi$$

where $\phi_- = (-\phi)_+$.

(iii) Deduce that every $\phi \in \mathcal{C}(T)^*$ is of the form $f \mapsto \int f\,d\mu - \int f\,d\nu$ where μ, ν are finite regular Borel measures on T.

15D Let $\{f_n\}$ be a bounded sequence in $\mathcal{C}(T)$, let $f \in \mathcal{C}(T)$ and suppose $f_n \to f$ pointwise. Show there is a sequence of finite convex combinations of the f_n which converges to f uniformly.

[If not, then $f \notin \overline{\text{co}}\,\{f_1, f_2, \ldots\}$. Separate f from the latter set by a functional ϕ, apply Problem 15C (iii) and the Dominated Convergence Theorem to get a contradiction. This result lies much deeper than it appears to!]

Show by an example that 'bounded' cannot be dropped from the hypothesis.

15E Density Theorem Let μ be a finite regular Borel measure on T, and let $1 \leqslant p < \infty$. Show that $\mathcal{C}(T)$ is dense in $L_p(\mu)$. [Adapt the arguments in Section 9.]

15F Let T be a countably infinite compact metric space. Show that $\mathcal{C}(T)^*$ is congruent to ℓ_1.

Barycentres

15G Fill in the details of the following direct proof of the **Krein–Milman Theorem** 15.30: (Notation as used there.)

If $\overline{\text{co}}\,E$ is a proper subset of C then by Corollary 8.15 there exist $f_1 \in X^*$, $\alpha \in \boldsymbol{R}$ such that $\{x \in X: f_1(x) \leqslant \alpha\}$ contains points of C but not of E. Let $\{f_2, f_3, \ldots\} \subset X^*$ separate the points of C (see 15.33) and, inductively, define $C_0 = C$ and C_{n+1} to be the set of points of C_n on which f_n attains its infimum. The C_n are compact convex sets and $\bigcap C_n$ consists of exactly one point e. Show e is an extreme point and $f_1(e) \leqslant \alpha$, a contradiction.

15H Let C be the set of continuous real functions f on $[0,1]$ which are convex and increasing and satisfy $f(0) = 0, f'(1) \leqslant 1$ ($f'(1)$ here means the left-hand derivative at 1–it necessarily exists for convex f (why?)).

Show that C is a compact convex subset of $\mathcal{C}[0,1]$ whose extreme points are precisely the functions f_s $(0 \leqslant s \leqslant 1)$ where $f(t) = (t - s) \vee 0$.

Deduce that $f \in C$ iff f has the form

$$f(t) = \int_0^t (t - s)\, d\mu(s) \quad (0 \leqslant t \leqslant 1)$$

for some probability Borel measure μ on $[0,1]$.

15I Clearly an important problem is: given two normed spaces, how to tell if they are congruent? A useful method is to look at the extreme points of their unit balls, since it is easy to see that if T is a congruence from X to Y then T maps the set $E(X)$ of extreme points of the closed unit ball of X onto the corresponding set in Y.

(i) Show that $E(X)$ lies on the surface $S_X = \{x: \|x\| = 1\}$ of the ball; that $E(X) = S_X$ for a Hilbert space, but $E(X) = \emptyset$ for the spaces c_0 and $L_1[0,1]$.

(ii) The following are worth investigating for extreme points: c; ℓ_∞; ℓ_1; real $\mathcal{C}[0,1]$; complex $\mathcal{C}[0,1]$; any L_p space, $1 < p < \infty$; any L_∞ space; the dual of $\mathcal{C}[0,1]$.

6

INFINITE PRODUCTS AND RELATED TOPICS

16. Infinite products and related topics

This chapter covers several topics connected with infinite products of topological spaces. The ideas developed in the first few sections are so important in modern analysis as to warrant detailed treatment, and though they seem to have tenuous contact with normed spaces, they lead up to the **Bourbaki–Alaoglu Theorem** which shows that there is a natural topology under which the closed unit ball in the dual of a normed space becomes a compact Hausdorff space. It should be emphasized that there is only one difficult proof in the chapter, that of Tychonov's Theorem 16.13; the ideas, however, are difficult in that they demand from the reader a flexibility of viewpoint, for instance the ability to regard the dual X^* of a normed space X both as a normed space in its own right and also as a subset of a certain infinite product. Most of the proofs consist of an appropriate change of viewpoint coupled with routine manipulation.

Weak topologies

One of the reasons for the emergence of topological spaces as a fruitful generalization of metric spaces was the result below, whose importance earns it the title of theorem in spite of the utter triviality of its proof. It has to do with the question: how is the topology $\boldsymbol{T}$ on a space T related to the set of all continuous functions from T to some given topological space Z? In view of the open set characterization of continuity, it is clear that the larger $\boldsymbol{T}$ is, the more continuous functions there are; to put it the other way round, one cannot reduce $\mathbf{T}$ beyond a certain point without certain functions becoming discontinuous. Our first result makes this precise.

16.1 Theorem Let T be a nonempty set and let $\{f_i\}_{i \in I}$ be a family of mappings defined on T, each f_i mapping into a topological space Z_i. Then there is a unique smallest topology $\mathbf{T}$ on T which makes each f_i continuous.

Proof Any topology on T which makes each f_i continuous must contain the family $\mathcal{S}$ of all sets $f_i^{-1}(V_i)$ where i runs over I and V_i runs over all open sets in Z_i. Clearly the smallest topology $\mathbf{T}$ containing $\mathcal{S}$ – that is the one for which $\mathcal{S}$ is a subbase (Proposition 1.19) – is the one we require.

It is clear that the proof depends on the fact that *any* family of subsets of T gives rise to a topology in a natural way; there is no similar way of generating metrics on T.

$\mathbf{T}$ is called the **weak topology** generated by the f_i. The following simple but important properties are left as exercises, the notation being as in the theorem except where otherwise stated.

16.2 Proposition

(1) The family of all sets of the form $\bigcap_{r=1}^{n} f_{i_r}^{-1}(V_r)$ where V_r is open in Z_{i_r} $(r = 1, \dots, n)$ is a base for the weak topology.

(2) It is sufficient to let V_r run over a base for the open sets in Z_{i_r} in (1).

(3) If all the Z_i are the scalar field $\mathbf{R}$ or $\mathbf{C}$ then a base of 'weak neighbourhoods' of a point a is formed by all sets of the form $\{t : |f_{i_r}(t) - f_{i_r}(a)| < \epsilon,\ r = 1, \dots, n\}$ with $\epsilon > 0$, $i_1, \dots, i_n \in I$, n arbitrary.

(4) If $S \subset T$ then the weak topology generated by the (restrictions of the) f_i on S is the same as the restriction to S of the weak topology on T.

(5) A sequence $\{t_n\}$ in T converges weakly (i.e. in the weak topology) to $t \Leftrightarrow f_i(t_n) \to f_i(t)$ for every i. (This fact must be treated with caution, for sequences are usually not enough to describe the weak topology. However, weak convergence of sequences is very important in applications.)

(6) A function g *from* some topological space *into* T is weakly continuous $\Leftrightarrow f \circ g$ is continuous for each i.

(7) If all the Z_i are Hausdorff and the f_i separate the points of T then the weak topology is Hausdorff.

16.3 Example To illustrate (3) let T be a subset of $\mathbf{C}$ not meeting the negative real axis, let $Z_1 = Z_2 = \mathbf{R}$, let $f_1(z) = |z|$ and $f_2(z) = \text{ph}(z)$ – that is, $f_2(z)$ is the unique θ such that $z = |z|(\cos\theta + i\sin\theta)$ and $-\pi \leqslant \theta < \pi$. The weak topology generated by f_1, f_2 has a base of open sets that are easiest described in polar coordinates, being the intersections of T with sets of the form $r_0 - \epsilon < r < r_0 + \epsilon$, $\theta_0 - \epsilon < \theta < \theta_0 + \epsilon$ (Diagram 22), and reduces in this case to the usual topology.

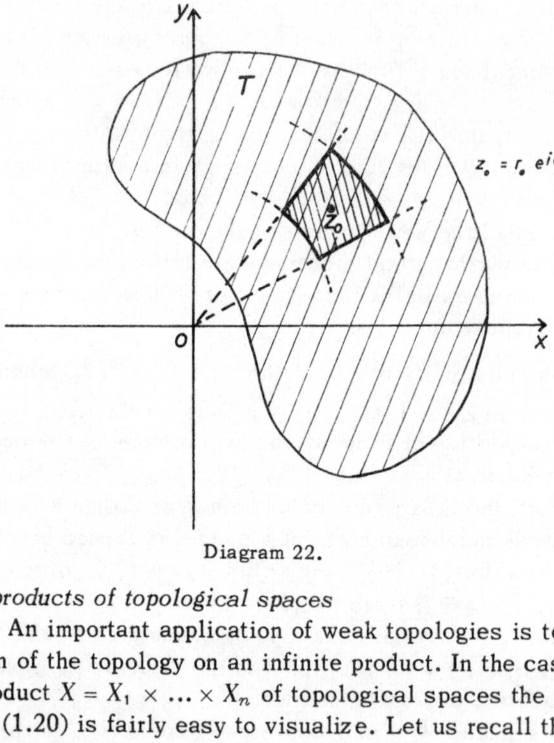

Diagram 22.

Infinite products of topological spaces

16.4 An important application of weak topologies is to the definition of the topology on an infinite product. In the case of a *finite* product $X = X_1 \times \ldots \times X_n$ of topological spaces the product topology (1.20) is fairly easy to visualize. Let us recall that:

(1) A base is formed by all sets $U_1 \times \ldots \times U_n$ where for each i U_i is open in X_i.

(2) The coordinate projections $p_i : x = (x_1, \ldots, x_n) \mapsto x_i$ are continuous.

(3) When each X_i is metrizable, so is X, and a sequence $\{x_k\}$ in X converges to $a \Leftrightarrow$ each coordinate x_{ki} converges to a_i as $k \to \infty$.

These facts suggest analogous properties that ought to be possessed by an infinite product of topological spaces; the snag is that they are not in general compatible, as we now show.

Formally, the product $X = \underset{i \in I}{\times} X_i$ of an arbitrary family of sets X_i is defined as the set of all functions x defined on I such that $x(i) \in X_i$ for each i; to make this accord with our intuitive ideas we usually, but not always, write $x(i)$ as x_i and visualize x as a family of coordinates $\{x_i\}_{i \in I}$. If each X_i is a topological space, property 1 above suggests the obvious course of taking, as a base for a topology on X, all products $U = \underset{i}{\times} U_i$, where U_i is open in X_i (Problem 16B shows we have a bonafide base). The result, called the **box topology** from the shape of the basic open sets, turns out to have more undesirable properties than useful ones.

16.5 Example Let each $X_i = [0, 1]$ $(i = 1, 2, \ldots)$ and think of $X = \underset{i=1}{\overset{\infty}{\times}} X_i$ as the set of all scalar sequences $x = (x(1), x(2), \ldots)$ with $0 \leqslant x(i) \leqslant 1$ – a subset, in fact, of ℓ_∞. Let x_n $(n = 1, 2, \ldots)$ be the constant sequence $(1/n, 1/n, 1/n, \ldots)$, and let x be the zero sequence. Then $x_n(i) = 1/n \to 0 = x(i)$ for each i as $n \to \infty$ (in fact uniformly over i), but for no n does x_n lie in the neighbourhood of x defined by $U = \underset{j=1}{\overset{\infty}{\times}} [0, 1/j)$, so that $x_n \not\to x$ in the box topology, and therefore properties 1 and 3 are incompatible for infinite products.

16.6 The accepted **product topology** is a much smaller one: it is by definition the weak topology generated by the coordinate projections $p_i : x \mapsto x_i$. Thus a sub-base for the product topology consists of all sets $p_k^{-1}(U_k)$ where U_k is open in X_k. Since $x \in p_k^{-1}(U_k) \Leftrightarrow x_k \in U_k$, all the other coordinates x_i being unrestricted, the reader can, if he finds such pictures helpful, visualize sub-basic sets as slabs stretching all the way across X in all coordinate directions but one (Diagram 23). Such a 'slab' is a particular sort of 'box', so that the box topology is larger than the product topology. For a finite product $X_1 \times \ldots \times X_n$ both topologies coincide, for clearly every box $U_1 \times \ldots \times U_n$ is the intersection of slabs $p_1^{-1}(U_1) \cap \ldots \cap p_n^{-1}(U_n)$. For an infinite product, a base is formed by all *finite* intersections of sub-basic open sets and thus a typical basic open set can be imagined as a slab or cylinder, stretching all the way across X in *all but finitely many* coordinate directions. From the properties of weak topologies we at once deduce several facts which put flesh and blood on the

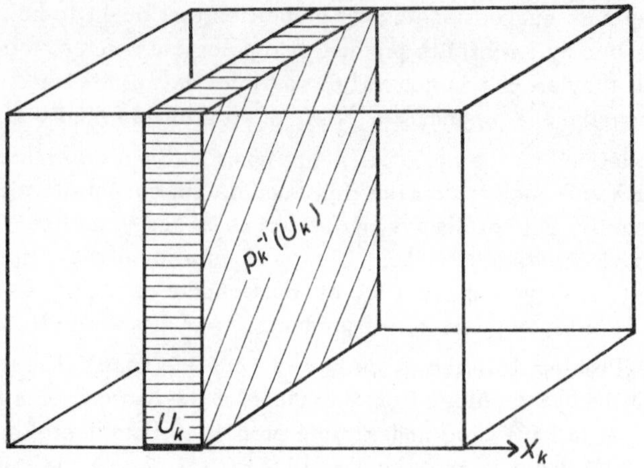

Diagram 23.

bare bones of definition and are extremely useful. Here are some: again we leave the proof to the student.

16.7 Proposition (Notation as in the previous paragraph)

(1) The basic open sets can be described as $\{x : x_{i_r} \in U_r,\ r = 1, \ldots, n\}$ where U_r are open sets in a finite number of coordinate spaces X_{i_r} $(r = 1, \ldots, n)$.

(2) It is sufficient to let U_r run over a base for the open sets in X_{i_r} in (1); while if U_r runs over a sub-base in X_{i_r} the sets in (1) form a sub-base for the product topology.

(3) If all the X_i are $\boldsymbol{R}$ or $\mathbf{C}$, a basic neighbourhood of a point a looks like $\{x : |x_{i_r} - a_{i_r}| < \epsilon,\ r = 1, \ldots, n\}$ with $\epsilon > 0$, $i_1, \ldots, i_n \in I$, n arbitrary.

(4) If $Y_i \subset X_i$ for each i, the product topology on $Y = \underset{i}{\mathsf{X}}\, Y_i$ is the same as the restriction to Y of the product topology on X.

(5) A sequence $\{x_n\}$ in X converges to $x \Leftrightarrow x_n(i) \to x(i)$ for each i. (See the caution in Proposition 16.2 (5).)

(6) A function g *from* some topological space T *into* a product space is continuous $\Leftrightarrow$ each of its **components (coordinate functions)** g_i defined by $g_i(t) = (g(t))_i$ is continuous.

(7) The product of Hausdorff spaces is Hausdorff.

The pleasantest spaces are metric spaces, and we noted as a pleasant property of *finite* products the fact that a finite product of metric spaces is metrizable. We shall now see that this generalizes to *countable* products (definitely not to arbitrary ones).

16.8 Theorem Let $X_1, X_2, \ldots$ be metric spaces. Then $\mathop{\mathsf{X}}_{n=1}^{\infty} X_n$, with the product topology, is metrizable.

To form a visual idea of the proof, it is probably easier to prove the following more general theorem about weak topologies. Clearly it contains the previous theorem as a special case, and it has other uses.

16.9 Theorem In the notation of 16.1 suppose that:

(1) the index set I is finite or countably infinite;
(2) the f_i separate points of X;
(3) each Z_i is a metric space.

Then X in the weak topology is metrizable.

Proof The case where I is finite is treated by an obvious variant of the proof below, so we shall assume I is countably infinite and so can be taken as $\mathbf{N}$ without loss of generality. To keep notation simple let us denote the metric in each Z_i by d, and write $f_i x$ instead of $f_i(x)$ (for $i \in \mathbf{N}$, $x \in X$).

Given $x, y \in X$ define

$$\rho(x, y) = \sum_{i=1}^{\infty} \min\{2^{-i}, d(f_i x, f_i y)\}.$$

It is easy to see ρ is a metric on X (we need the f_i to separate points in order that $\rho(x, y) = 0$ should imply $x = y$). Fix attention on a particular f_n: if $x, y \in X$ and $\epsilon > 0$ are given it is clear that the situation $d(f_n x, f_n y) \geqslant \epsilon$ implies $\rho(x, y) \geqslant \min\{2^{-n}, \epsilon\}$.

Thus if $\delta = \min\{2^{-n}, \epsilon\}$ then $\rho(x, y) < \delta$ implies $d(f_n x, f_n y) < \epsilon$, proving that with respect to ρ each f_n is continuous (indeed uniformly continuous). Thus by definition the weak topology is smaller than the ρ-topology. Conversely given a ball $V = \{x : \rho(x, a) < r\}$ choose n so that $2^{-n} < \frac{1}{2} r$. Then the set $U = \{x : d(f_i x, f_i a) < r/2^{i+1} \text{ for } i = 1, \ldots, n\}$, which is a weak neighbourhood of a, lies inside V because if $x \in U$ then

$$\rho(x, a) \leqslant \sum_{i=1}^{n} d(f_i x, f_i a) + \sum_{i=n+1}^{\infty} 2^{-i}$$

$$< \sum_{i=1}^{n} \frac{r}{2^{i+1}} + 2^{-n}$$

$$= r(\tfrac{1}{4} + \tfrac{1}{8} + \ldots) + 2^{-n} < \tfrac{1}{2}r + \tfrac{1}{2}r = r.$$

This proves the weak topology is larger than the ρ-topology, and therefore the two topologies coincide.

16.10 Corollary A metric (out of many possible ones) for the product topology on a countable product X of metric spaces $X_1, X_2, \ldots$ is

$$\rho(x, y) = \sum_{i=1}^{\infty} \min\{2^{-i}, d(x_i, y_i)\}.$$

Exercise Prove that $\sup_i \min\{2^{-i}, d(x_i, y_i)\}$ is also a metric on X but does not generate the product topology.

Products of compact spaces

16.11 It is generally agreed that the most important single result in topology is Tychonov's celebrated Theorem that the product of any family of compact topological spaces is itself compact in the product topology. Since the proof, however one does it, is almost totally un-visualizable, it may help to start less ambitiously with a look at finite and countably infinite products of compact metric spaces, where elementary arguments will suffice.

Consider first the product of two compact metric spaces X, Y, and recall that a metric space is compact iff each sequence in it has a convergent subsequence. For any sequence $\{(x_n, y_n)\}$ in $X \times Y$ one can find a sequence $\{n_k\}$ of indices such that $\{x_{n_k}\}$ converges in X; one can then refine $\{n_k\}$ further to a sequence $\{n_k{}'\}$ such that $\{y_{n_k{}'}\}$ converges in Y; then $\{(x_{n_k{}'}, y_{n_k{}'})\}$ converges in $X \times Y$ (see 1.15). This technique of successive refinement clearly extends to any *finite* product; the only snag, notation apart, in generalizing it to a countable product is that after infinitely many refinements one may have no subsequence left at all (consider the effect of starting with $\{x_1, x_2, x_3, \ldots\}$, extracting $\{x_2, x_4, x_6, \ldots\}$, then extracting $\{x_4, x_8, x_{12}, \ldots\}$ and so on) so that the **diagonal process** has to be used in the following proof.

16.12 Theorem The product X of a sequence of compact metric spaces $X_1, X_2, \ldots$ is compact.

Proof Since X is metrizable by Theorem 16.8 it is sufficient to show that each sequence in X has a convergent subsequence.

It will be convenient to use the notation $x(j)$ for the jth coordinate of a point x in X, and subscripts to denote sequences.

Let $\{x_n\}$ be a sequence in X. By the successive refinement process described above we produce a whole array of subsequences of indices

$$S_1 = \{n_{11}, n_{12}, n_{13}, \ldots\}$$
$$S_2 = \{n_{21}, n_{22}, n_{23}, \ldots\}$$
$$\ldots\ \ldots\ \ldots,$$

where each S_j is a subsequence of the previous one, chosen to make the jth coordinates $\{x_{n_{j1}}(j), x_{n_{j2}}(j), x_{n_{j3}}(j), \ldots\}$ converge in the compact metric space X_j. We may, and do, demand that:

The first j terms of S_{j+1} and S_j coincide.

This ensures that (considering S_j simply as a subset of $\mathbf{N}$) the intersection of all the S_j is an infinite set $S = \{m_1, m_2, \ldots\}$ – in fact the first j terms of S and S_j coincide, and one has $m_k = n_{kk}$, whence the name 'diagonal process'. Clearly $\{m_k\}$ is a subsequence of *all* our subsequences, so that the jth coordinates of the x_{m_k} converge for every j, in other words $\{x_{m_k}\}$ converges in the product topology, and the proof is complete.

For general products of topological spaces one cannot describe the topology by sequences, nor can one count off the factor spaces $X_1, X_2, X_3, \ldots$, so successive refinement will not work. Nevertheless the 'refinement' idea is present in a more sophisticated form in the following, which is Bourbaki's proof of the Tychonov theorem. A fimily $\mathfrak{F}$ of subsets of a set X has the **finite intersection property** if every finite family $F_1, \ldots, F_n$ of members of $\mathfrak{F}$ has nonempty intersection. By taking complements in the definition of compactness one easily sees (cf. 1.13) that a topological space X is compact iff each family of closed sets in X with the finite intersection property has nonempty intersection.

16.13 Tychonov's Theorem The product $X = \mathsf{X}_{\alpha} X_\alpha$ of a family of compact topological spaces X_α is compact.

Proof Let $\mathfrak{F}_0$ be a family of closed sets in X with the finite intersection property. We shall show that there is a point common to all the sets in $\mathfrak{F}_0$. The first step is to extend $\mathfrak{F}_0$ to a larger family $\mathfrak{F}$ of (not necessarily closed) sets which is *maximal* with respect to the finite intersection property. To show this can be done involves setting up a situation in which Zorn's lemma can be applied, which we do by considering the class of *all* families $\mathfrak{F}$, $\mathfrak{G}$, ... of sets in X having the finite intersection property – call them FIP families. Partially order the FIP families by inclusion: that is, $\mathfrak{F} \leqslant \mathfrak{G}$ means that every F in $\mathfrak{F}$ is also in $\mathfrak{G}$. Now, if $\mathbf{C}$ denotes a *chain* of FIP families, (that is if for any $\mathfrak{F}, \mathfrak{G} \in \mathbf{C}$, either $\mathfrak{F} \leqslant \mathfrak{G}$ or $\mathfrak{G} \leqslant \mathfrak{F}$), then (this is the vital point) the union of all the $\mathfrak{F}$'s in $\mathbf{C}$ is also a FIP family. For suppose $A_1, \ldots, A_n$ are sets belonging to this union. Each A_r has come from one of the constituent $\mathfrak{F}$'s, call it $\mathfrak{F}_r$, for $r = 1, 2, \ldots, n$. Since $\mathbf{C}$ is a chain and there are only finitely many $\mathfrak{F}_r$ one of these, say $\mathfrak{F}_{r_0}$, must be larger than all the others. Then *all* the A_r belong to $\mathfrak{F}_{r_0}$; since $\mathfrak{F}_{r_0}$ is a FIP family we have $A_1 \cap \ldots \cap A_n \neq \emptyset$ which is what we wished to show.

The existence of a maximal FIP family $\mathfrak{F}$ containing the original family $\mathfrak{F}_0$ of closed sets now follows easily from Zorn's lemma. The projections of the sets of $\mathfrak{F}$ on the coordinate space X_α form a family $\mathfrak{F}^\alpha$ of sets in that space, also having the finite intersection property, as the reader can easily verify. Since X_α is assumed compact, there is a point x_α which is in the closure of every set of $\mathfrak{F}^\alpha$. Let x be the point of X whose αth coordinate is x_α for each α. We shall show x is in the closure of each set of $\mathfrak{F}$, and therefore is in every set of $\mathfrak{F}_0$, which will finish the proof. Accordingly let U be any open set in X containing x. Then there exists, by the definition of the product topology, a finite set of indices $\alpha_1, \ldots, \alpha_n$ and open sets $U_{\alpha_i} \subset X_{\alpha_i}$, $i = 1, \ldots, n$ such that

$$x \in \bigcap_{i=1}^{n} p_{\alpha_i}^{-1}(U_{\alpha_i}) \subset U,$$

where p_α is the αth coordinate projection. This implies in particular that $x_{\alpha_i} \in U_{\alpha_i}$ and hence (since x_{α_i} is in the closure of every set in $\mathfrak{F}^{\alpha_i}$) that U_{α_i} meets every set in $\mathfrak{F}^{\alpha_i}$. But then $p_{\alpha_i}^{-1}(U_{\alpha_i})$ meets every set of $\mathfrak{F}$ and so belongs to $\mathfrak{F}$ by maximality (if it didn't, we could add it to $\mathfrak{F}$ to get a strictly larger FIP family). Then again $\bigcap_{i=1}^{n} p_{\alpha_i}^{-1}(U_{\alpha_i})$ is in $\mathfrak{F}$ because $\mathfrak{F}$ is a FIP family, and this implies U intersects every member of $\mathfrak{F}$. Since U was an arbitrary open set in X containing x it follows that x is in the closure of every set in $\mathfrak{F}$, as required.

Function spaces and the pointwise topology

16.14 We now turn to what, though it does not at first sight look like an infinite product, is in fact the most important example of such a thing. Basically, a **function space** means simply a set of functions from one set A to another set B, or, as it will be more convenient to say in this section, a set of B-valued functions on A. (This language emphasizes that B is the important space, while A plays a passive role.) In practice we are interested in sets of functions with extra structure: continuity, as in the space $\mathcal{C}(T)$, T a compact space; linearity, as with the space X^* of bounded linear functionals on a normed space X.

One of the functional analyst's basic notions is that of treating F as a space in its own right, on which further functions can be defined, and an important class of such mappings is formed by the **evaluations** at points a of A, that is the B-valued functions

$$\hat{a}: f \mapsto f(a) \quad (f \in F). \tag{1}$$

We have met these in 14.2–6, and the student may find that the situation there – A is a normed space X, B is the scalar field, F is X^* – may help to fix his ideas.

Consider for the moment the space {all B-valued functions on A}. Comparing it with the definition of a product $\bigtimes_{i \in I} X_i$, we see it is precisely the same thing as that product, if we take A as the index set I and let each coordinate space X_i be B. Thus we can visualize it as $B \times B \times \ldots \times B$ (A times) and for this reason it is denoted B^A and called a **Cartesian power**. Thus we can write

$$F \text{ is a family of } B\text{-valued functions on } A \Leftrightarrow F \subset B^A. \tag{2}$$

Moreover the evaluation $\hat{a}: f \mapsto f(a)$ is nothing other than the coordinate projection onto the ath coordinate space. The following examples may help the student grasp the mathematical sleight of hand whereby we have thus identified 'function space' with 'subset of a Cartesian power'. As usual $\mathbf{F}$ denotes either $\mathbf{R}$ or $\mathbf{C}$.

16.15 Example 1 Let T be a compact space. $\mathcal{C}(T)$ is a subset of the set $\mathbf{F}^T$ of all scalar functions on T – indeed a linear subspace of it if $\mathbf{F}^T$ is given the obvious ('**coordinatewise**' or '**pointwise**') linear space structure. Evaluations on $\mathcal{C}(T)$ are the maps $\hat{t}(f) = f(t)$ and are bounded linear functionals on $\mathcal{C}(T)$.

Example 2 If X is a normed space, X^* is a linear subspace of $\mathbf{F}^X$; evaluations on X^* have been discussed above. Similarly,

$\mathcal{B}(X)$ is a linear subspace of X^X.

Example 3 By considering an n-tuple $x = (x_1, \ldots, x_n)$ as the function $i \mapsto x_i$ on $\{1, \ldots, n\}$ one identifies $\mathbf{F}^n$ with $\mathbf{F}^{\{1,\ldots,n\}}$.

Example 4 In the same way the sequence spaces ℓ_p can be considered as linear subspaces of $\mathbf{F}^N$ – as was done in Example 16.5. In this, and the last example, the evaluations are what we have called the **standard coordinate functionals** $f_n(x) = x_n = x(n)$ $(x \in \ell_p;\ n = 1, 2, \ldots)$.

16.16 The simplest and most useful function spaces are those where the functions are scalar-valued and we assume henceforth that this is so. Thus let F be a family of scalar functions on a set X, that is

$$F \subset \mathbf{F}^X.$$

Since $\mathbf{F}^X$ is a product of copies of the topological space $\mathbf{F} = \mathbf{R}$ or $\mathbf{C}$, F has a ready-made topology: the restriction to F of the product topology, invariably called the **pointwise topology** on the function space F for reasons that will soon be clear. Combining 16.2(4), the definition 16.6 of the product topology, the identification of evaluations with coordinate projections, 16.7(3, 5, 7) and 16.8, one obtains the following properties of the pointwise topology $\mathbf{P}$ on F. Any of statements 1, 2, 3 could be taken as the definition of $\mathbf{P}$.

16.17 Proposition (Notation as above)

(1) $\mathbf{P}$ is the restriction to F of the product topology on $\mathbf{F}^X$.
(2) $\mathbf{P}$ is the weak topology generated by the evaluations at points of X, that is the smallest topology making all the functions $\hat{x} : f \mapsto f(x)$ into continuous functions on F.
(3) Basic $\mathbf{P}$-neighbourhoods of a function f_0 are sets of the form $\{f \in F : |f(x_r) - f_0(x_r)| < \epsilon,\ r = 1, \ldots, n\}$ with $\epsilon > 0$, $x_1, \ldots, x_n \in X$, n arbitrary.
(4) A sequence of functions f_n in F is $\mathbf{P}$-convergent to f iff $f_n(x) \to f(x)$ for all $x \in X$, that is iff $f_n \to f$ pointwise on X.
(5) $\mathbf{P}$ is Hausdorff; if X is countable then $\mathbf{P}$ is metrizable.

We have, incidentally, thus given the answer Yes to a very natural question, namely 'Does there exist a topology $\mathbf{T}$ on a set of scalar functions, such that for sequences, $\mathbf{T}$-convergence is just the same as pointwise convergence?' In view of the inadequacies of sequences in non-metrizable topologies this question is not trivial: it is worth mentioning that there is, for instance, no topology on the Lebesgue-measurable functions on [0, 1] that corresponds to 'pointwise convergence a.e.' in this fashion.

Pointwise compactness – the Bourbaki–Alaoglu Theorem

It is now very easy to write down large numbers of sets of scalar functions that are compact in the pointwise topology; for a simple and not very useful example we have:

16.18 Proposition The set F of all scalar functions f on a set A, such that $|f(x)| \leqslant 1$ $(x \in A)$ is compact in the pointwise topology.

Proof F is just D^A where $D = \{\alpha : |\alpha| \leqslant 1\}$ in $\mathbf{R}$ or $\mathbf{C}$. Since D is compact, D^A is compact by Tychonov's Theorem.

16.19 The next theorem, which is the most important single application of Tychonov's Theorem to functional analysis, uses a slightly more subtle version of the same idea. Let X be a normed space with dual X^*. For reasons of history and convenience the product, alias pointwise, topology on X^* is called the **weak* topology** (or **vague topology** – both terms indicate that it finds difficulty in telling one functional from another). Recall that one can visualize it as the topology in which $f_n \to f$ means $f_n(x) \to f(x)$ $(x \in X)$.

The closed unit ball B^* in X^* is a subset of the space $\mathbf{F}^X$ of all scalar functions on X. In order for an $f \in \mathbf{F}^X$ to qualify for membership of B^* it is clearly necessary and sufficient that (i) f be linear and (ii) $|f(x)| \leqslant \|x\|$ for every $x \in X$. If we show that the first condition restricts f to a closed subset of $\mathbf{F}^X$ and the second restricts f to a compact subset of $\mathbf{F}^X$ we shall have shown B^* to be the intersection of a closed set with a compact set and (since $\mathbf{F}^X$ is Hausdorff) derived:

16.20 Theorem The closed unit ball B^* in X^* with the weak* topology is a compact Hausdorff space.

Completion of proof Clearly an f in $\mathbf{F}^X$ is linear iff it obeys all possible equations of the form

$$\alpha f(x) + \beta f(y) = f(z) \tag{3}$$

where z denotes $\alpha x + \beta y$. By definition the point-evaluations $\hat{x} : f \mapsto f(x)$ are continuous scalar functions on $\mathbf{F}^X$, hence so is any linear combination of them. But clearly the set N of all f satisfying (3) can be written as $N = \{f : (\alpha\hat{x} + \beta\hat{y} - \hat{z})(f) = 0\}$

and so, being the zero set of a continuous function, is closed in $\mathbf{F}^X$. Thus the set of *linear* f is the intersection of a number of closed sets of the form of N, and so is closed.

Secondly the restriction $|f(x)| \leqslant \|x\|$ means that for each x the 'xth component', namely $f(x)$, of f lies in the compact set $D_x = \{\alpha : |\alpha| \leqslant \|x\|\}$ in $\mathbf{F}$. That is, the set of f that are restricted in this way are precisely the elements of the product $\underset{x \in X}{\times} D_x$, a 'box' in $\mathbf{F}^X$, which is compact by Tychonov's theorem. Together with the last paragraph this completes the proof.

This result has many applications, and in fact compactness in the weak*-topology plays a much more important role in functional analysis than compactness in the norm topology. Many of the geometric facts about norm compactness proved in 4.15 – for instance that the vector sum of two compact sets is compact – are true also of weak*-compactness.

An immediate application of the Bourbaki–Alaoglu theorem is

16.21 Theorem Every normed space X is congruent to a linear subspace of $\mathcal{C}(T)$ for some compact Hausdorff space T.

Proof Let T be the compact Hausdorff space B^*, in the weak*-topology $\mathbf{W}$. For each $x \in X$ the evaluation

$$\hat{x} : f \mapsto f(x) \qquad (f \in B^*)$$

is, by the definition of $\mathbf{W}$, a continuous scalar function on B^*, so that the map $e : x \mapsto \hat{x}$ maps X into $\mathcal{C}(B^*)$. That e is linear follows exactly as with the similar embedding of X in X^{**} (Theorem 14.2). Note that the $\hat{x}$'s used here are simply the restrictions to B^* of the $\hat{x}$'s in 14.2. By Corollary 8.12,

$$\|\hat{x}\|_\infty = \sup\{|\hat{x}(f)| : f \in B^*\} = \sup\{|f(x)| : f \in B^*\} = \|x\|$$

so that e is a congruence of X with the subspace $e(X)$ of $\mathcal{C}(B^*)$.

General theory of evaluation maps

The 'evaluation map' $e : x \mapsto \hat{x}$, used in Theorems 16.21 and 14.2, is one of the most powerful and flexible tools available for the analysis of mathematical structures of all kinds. A systematic account of its diverse uses is out of place here, but we describe some of the basic ideas involved, and two typical applications (more are given in the problems). The arguments, except when Tychonov's Theorem enters, are trivial computations in set theory and elementary algebra.

16.22 Let E be a nonempty set, and F a nonempty collection of scalar functions on E. (Purely to avoid confusion between F and $\mathbf{F}$ we take the scalar field to be $\mathbf{R}$, but $\mathbf{C}$, and indeed more general spaces, will do equally well.) For each $x \in E$ the function $\hat{x} : f \mapsto f(x)$ (an **evaluation**) is a scalar function on F; the **evaluation map** is the map e that associates to each x the function $\hat{x}$; the range $e(E)$ we shall call $\hat{E}$. We thus have a symmetrical situation:

$$F \text{ is a set of scalar functions on } E,$$

$$\hat{E} = e(E) \text{ is a set of scalar functions on } F;$$

or in symbols,

$$F \subseteq \mathbf{R}^E \text{ and } \hat{E} \subset \mathbf{R}^F.$$

16.23 Lemma

(i) e is one-to-one $\Leftrightarrow F$ separates points of E.

(ii) E always separates points of F.

Proof By the definition of equality for functions, $\hat{x} \neq \hat{y} \Leftrightarrow \hat{x}(f) \neq \hat{y}(f)$ for some $f \in F \Leftrightarrow f(x) \neq f(y)$ for some $f \in F$. Thus

$$\begin{aligned} &e \text{ is one-to-one} \Leftrightarrow x \neq y \text{ implies } \hat{x} \neq \hat{y} \\ \Leftrightarrow\ & x \neq y \text{ implies that for some } f \in F,\ f(x) \neq f(y) \\ \Leftrightarrow\ & F \text{ separates points,} \end{aligned}$$

proving (i); we leave (ii) as an exercise.

The *algebraic* rationale behind this is that E will be a structure of some kind – a normed linear space, a group, etc. – such that the scalar field is a particularly simple structure of the same sort; F will be a family of mappings on E that preserve the structure. It then usually turns out that $\hat{E}$ has a natural structure of the same kind. If F separates points, the map e is thus a **faithful representation** (i.e. structure-preserving one-to-one map) of E as a set of functions, which is often very enlightening, as 14.2 and 16.21 indicate, and as we see in 16.27 and 16.28. The student should note that the evaluation $\hat{x}$ (being the *restriction* to F of the corresponding $\hat{x}$ on all of $\mathbf{R}^E$) becomes a different function, often with quite different algebraic properties, if one varies the collection of functions F. Compare 14.2, where $\hat{x}$ was a bounded linear functional on $F = X^*$, with 16.27, where $\hat{x}$ was a continuous function on $F = B^*$, a subset of X^*.

From the topological point of view, evaluation maps show that weak topologies and product topologies are really the same thing in disguise; and when F is a finite set they provide a method of embedding topological spaces in Euclidean space $\boldsymbol{R}^n$, and a way to visualize weak topologies (Example 16.25). This is made precise by the next result, whose proof follows easily by considering sub-basic open sets (cf. 1.18), and is left as an exercise.

16.24 Proposition

(i) Let E be given the weak topology $\boldsymbol{W}$ generated by F and let $\hat{E}$ carry the pointwise topology $\boldsymbol{P}$ for functions on F; its topology, in other words, as a subset of $\boldsymbol{R}^F$. Then e sets up a one-to-one correspondence between open sets in $(E, \boldsymbol{W})$ and those in $(\hat{E}, \boldsymbol{P})$ – that is, e is continuous and open. In particular if F separates points of E then e is a homeomorphism.

(ii) Suppose all the functions in F are continuous with respect to some given topology $\boldsymbol{T}$ on E. Then e is a continuous map of $(E, \boldsymbol{T})$ onto $(\hat{E}, \boldsymbol{P})$.

When F is a finite set of functions $\{f_1, \ldots, f_n\}$ one can regard $\boldsymbol{R}^F$ as $\boldsymbol{R}^n$ and the evaluation map is easily visualized as the embedding

$$x \mapsto (f_1(x), \ldots, f_n(x))$$

of E into $\boldsymbol{R}^n$, which, assuming F to separate points and E to carry the weak topology, is a homeomorphism.

16.25 Example Let E be $\mathbf{C} = \boldsymbol{R}^2$ with the weak topology $\boldsymbol{W}$ generated by the functions Re(z), Im(z), ph(z). Thus e is the homeomorphic mapping $(x, y) \mapsto (x, y, \text{ph}(x + iy))$ of $\boldsymbol{R}^2$ into $\boldsymbol{R}^3$ whose range is the graph of ph (a spiral surface). $\boldsymbol{W}$ is now easy to describe: it is roughly the same as the usual topology, except for a 'break' caused by the discontinuity of ph along the negative real axis.

16.26 The results below show some algebraic implications of the evaluation method. The first is really the easy half of the fundamental **Gelfand Representation Theorem** for commutative Banach algebras. Suppose that A, B are normed algebras over the same scalars. A map $T : A \to B$ is an **algebra-homomorphism** if it satisfies $T(x + y) = Tx + Ty$, $T(\alpha x) = \alpha Tx$, $T(xy) = Tx \,.\, Ty$; an **isomorphism** if also it is one-to-one and onto (its inverse is then clearly also an isomorphism). Clearly an algebra homomorphism is in particular a linear map: hence it is continuous iff it is bounded in the sense of 7.2. A **character** on a unital normed algebra A is a nonzero bounded algebra-homomorphism ϕ from A

into the scalar field – note that in particular ϕ belongs to A^* (considering A as a normed linear space) and is often called a **multiplicative functional.**

16.27 Theorem A commutative unital normed algebra A, for which there exist enough characters to separate the points of A, is algebra-isomorphic to a subalgebra of $\mathcal{C}(T)$ for some compact Hausdorff space T.

Proof It is easy to prove (Exercise 16L) that $\phi(1) = 1$ and $\|\phi\| \leqslant 1$ for any character ϕ; a straightforward extension of the proof of the Bourbaki–Alaoglu theorem shows that the set

$$\Phi = \{\phi : \phi \text{ is a character on } A\}$$

– called the **carrier space** of A – is a compact subset of A^* in the weak* topology. For each $x \in A$ define the evaluation

$$\hat{x}(\phi) = \phi(x) \qquad (\phi \in \Phi).$$

It is routine to verify that the map $e : x \mapsto \hat{x}$ is an algebra isomorphism of A onto a subalgebra $\hat{A}$ of $\mathcal{C}(\Phi)$.

The difficult part of the Gelfand Theorem is a subtle argument, intertwining algebra and analysis, which shows among other things that a commutative Banach algebra A has enough characters to separate points if and only if $\limsup \|x^n\|^{1/n} > 0$ for each nonzero $x \in A$. These subtleties however are not needed for the next result which is a special case of the preceding. A **closed sup-norm algebra** means any closed subalgebra of the space $B(S)$ of all bounded scalar functions on a nonempty set S; $B(S)$ is clearly a Banach algebra with the usual pointwise operations and the sup-norm $\|f\|_\infty$. Examples are the space of all bounded continuous functions on $\boldsymbol{R}$ or all bounded Borel-measurable functions on $[0, 1]$, or the disc algebra 6.23, or any closed subalgebra of these.

16.28 Theorem Every closed – and in the case of complex scalars, self-adjoint – sup-norm algebra A containing 1 is isometrically algebra isomorphic to $\mathcal{C}(T)$ for some compact Hausdorff space T.

Proof The functions $\hat{s} : f \to f(s)\,(f \in A)$ where s runs over the underlying set S, clearly form a subclass $\hat{S}$ of the set Φ of characters on A which separates the points of A. In the notation of the last theorem it follows that $e : f \mapsto \hat{f}$ where $\hat{f}(\phi) = \phi(f)$

($\phi \in \Phi$) is an algebra isomorphism of A with a sub-algebra $\hat{A}$ of $\mathcal{C}(\Phi)$. Since $\|\phi\| \leqslant 1$ ($\phi \in \Phi$) one has

$$|\hat{f}(\phi)| = |\phi(f)| \leqslant \|f\|_\infty \quad (\phi \in \Phi) \text{ so that } \|\hat{f}\|_\infty \leqslant \|f\|_\infty;$$

while since every $\hat{s}$ is a ϕ one has

$$|f(s)| = |\hat{s}(f)| = |\hat{f}(\hat{s})| \leqslant \|\hat{f}\|_\infty \qquad (s \in S)$$

giving the reverse inequality. Thus $f \mapsto \hat{f}$ is a congruence; since A is closed and thus complete, $\hat{A}$ is a complete and therefore closed subalgebra of $\mathcal{C}(\Phi)$. Clearly $\hat{A}$ separates the points of Φ (16.23 (ii)) and contains 1 since one easily sees that $\hat{1} = 1 \in \mathcal{C}(\Phi)$. If the scalars are real, the Stone–Weierstrass Theorem immediately gives $\hat{A} = \mathcal{C}(\Phi)$, completing the proof. The extension to complex scalars is left as an exercise.

The proof above made use of a typical technique – what one might term 'leap-frogging' of evaluation maps. The following shows the various sets involved:

$$A \subset \mathbf{R}^S,$$

$$\hat{S} \subset \Phi \subset A^* \subset \mathbf{R}^A,$$

$$\hat{A} \subset \mathcal{C}(\Phi) \subset \mathbf{R}^\Phi.$$

Problems

Weak topologies and infinite products

16A Prove parts 5 and 7 of Proposition 16.2 (on basic properties of weak topologies).

16B Let X be the product of a family of topological spaces X_i ($i \in I$) and let $\mathbf{B}$ be the class of all boxes $U = \mathop{\times}\limits_{i \in I} U_i$ where U_i is open in X_i. Verify that the class $\mathbf{T}$ of arbitrary unions of members of $\mathbf{B}$ is indeed a topology on X.

16C Prove that a product of discrete spaces is in general not discrete in the product topology. What about the box topology?

16D Prove the following statements about a compact Hausdorff space T are equivalent, thus strengthening Theorem 9.18. (Assume that $\mathcal{C}(T)$ separates the points of T, i.e. Urysohn's Lemma. That (c) implies (a) is an important and often used **metrization lemma.**)

(a) T is metrizable;

(b) $\mathcal{C}(T)$ is separable;

(c) There exists a countable family of continuous scalar functions separating the points of T.

16E The **weak topology** on a normed space X is by definition the weak topology generated by the members of X^* – thus a sequence $\{x_n\}$ **converges weakly** to x (written $x_n \rightharpoonup x$) iff $f(x_n) \to f(x)$ for all f in X^*.

Let x_n ($n = 1, 2, \ldots$) be the functions on $[0, 1]$ defined by $x_n(t) = \sin nt$. Show that $\{x_n\}$ is weakly convergent when considered as a sequence in $L_2[0, 1]$ but not when considered as a sequence in $\mathcal{C}[0, 1]$; in neither space is it norm-convergent.

16F Two interesting properties of the weak topology on a normed space are the following. The second shows that sometimes 'sequences are adequate' in a nonmetrizable topology.

(i) Show that a convex subset of X is closed in the norm topology iff it is closed in the weak topology.

(ii) Let $T : X \to X$ be a linear map. Show that the following are equivalent:

(a) $Tx_n \rightharpoonup 0$ for any sequence $\{x_n\}$ such that $x_n \rightharpoonup 0$;

(b) T is continuous when X has the weak topology;

(c) T is bounded.

[(a) $\Rightarrow$ (c) uses Uniform Boundedness in the form of Theorem 10.8; for (c) $\Rightarrow$ (b) 16.2 part 6 may help.]

Tychonov's theorem and the weak topology*

16G Let $\mathbf{S}$ be the class of all subsets of a given set S, and for each s in S define $f_s : \mathbf{S} \to \mathbf{R}$ by

$$f_s(A) = \begin{cases} 1 & (s \notin A) \\ 0 & (s \notin A), \end{cases}$$

where $A \in \mathbf{S}$. Show that $\mathbf{S}$, with the weak topology generated by the functions f_s, is a compact Hausdorff space.

16H Let $1 \leqslant p < \infty$, and identify ℓ_p^* with ℓ_q as in Theorem 14.11. Show that a bounded sequence $\{y_n\}$ in ℓ_p^* is weak*-convergent iff it is coordinatewise convergent, that is iff the ith coordinates $\{y_n(i)\}_{n=1,2,\ldots}$ form a convergent sequence for each i.

16I Show that a normed space X is separable iff the unit ball of X^* is metrizable in the weak* topology. [Use Theorem 16.21 and Problem 16D.]

16J Let X be a separable normed space, B^* the unit ball in X^* and $\{x_n\}$ a dense set in the unit ball of X. By a direct argument not relying on Tychonov's Theorem or anything else that needs Zorn's Lemma, show that the metric

$$\rho(f, g) = \sum_{n=1}^{\infty} \min\{2^{-n}, |f(x_n) - g(x_n)|\}$$

(a) gives the weak* topology on B^* and (b) makes B^* into a compact metric space.

Finally, use this result to give a 'Zorn-free' proof of Theorem 15.26 (at present it uses it via the Hahn–Banach Theorem). [Show, in the notation of the theorem, that $\{\phi_n\}$ has a weak*-convergent subsequence.]

Evaluation maps

16K Show that a topological space T is compact and metrizable iff it is homeomorphic to a closed subset of $[0, 1]^{\mathbf{N}}$.

[For ⇒, construct a countable fundamental subset F of $\mathcal{C}(T)$ such that $0 \leqslant f \leqslant 1$ ($f \in F$) and consider the evaluation map of T into $\mathbf{R}^F$; the latter can clearly be identified with $\mathbf{R}^{\mathbf{N}}$.]

16L In the proof of Theorem 16.27 verify that:

(i) A character ϕ satisfies $\phi(1) = 1$ and $\|\phi\| \leqslant 1$. [For the latter show that $|\phi(a)|^n \leqslant \|\phi\| \; \|a\|^n$.]

(ii) Φ is weak*-compact.

(iii) e is an algebra-isomorphism.

16M In Theorem 16.28 prove that the set S of evaluations is dense in Φ. Hence, when S is compact and A is a closed subalgebra of $\mathcal{C}(S)$, give an elementary proof of the theorem not using any of the big theorems of this chapter.

16N Let T be a compact space and F a subset of $\mathcal{C}(T)$ containing 1. Define an equivalence relation on T by writing $s \sim_F t$ if $f(s) = f(t)$ for all f in F. The equivalence classes of F containing more than one point are the **non-trivial level sets** of F. Prove the following **Generalized Stone–Weierstrass Theorem**: the closed (in the complex case self-adjoint) subalgebra generated by F consists precisely of those $f \in \mathcal{C}(T)$ which are constant on each nontrivial level set of F. [Use the last problem.]

Use this to deduce Corollary 6.28.

16O Find, with their appropriate topology, the carrier spaces Φ of the following sup-norm algebras: (i) The space of functions $f + \alpha 1$ where f is a continuous function on $\boldsymbol{R}$ vanishing at infinity; (ii) The space of **regulated** functions on $[0, 1]$ (the functions f such that the left and right hand limits $\lim_{t \to \alpha+} f(t)$, $\lim_{t \to \alpha-} f(t)$ exist at each point).

16P This problem sketches a proof that every separable normed space is congruent to a subspace of $\mathcal{C}[0,1]$ (see p. 46). The key idea is the topological result below. A *path* in a compact metric space Z means a continuous map α from $I = [0,1]$ into Z. A path α is *universal* if it goes through each point of Z, i.e. if $\operatorname{im} \alpha = Z$.

Theorem Suppose Z has the following property:
(*) Any two points in Z can be joined by a path; and given $\epsilon > 0$ there is $\delta > 0$ such that any two points x,y with $d(x,y) \leqslant \delta$ can be joined by a path lying wholly inside the ϵ-ball round x. Then Z has a universal path.

The proof is in (i) to (iii) below, and (iv), (v) derive the (routine) consequences.

(i) Show that the set $\mathcal{C}(I, Z)$ of paths in Z is a complete metric space under the *uniform metric* $\rho(\alpha, \beta) = \sup_{t \in I} d(\alpha(t), \beta(t))$.

(ii) Let a δ-*tour* mean a path passing within δ of every point of Z – that is $d(x, \operatorname{im} \alpha) \leqslant \delta$ $(x \in Z)$. Suppose (*) holds, let ϵ, δ be as in (*), let α be a $\delta/2$-tour, and let $x_1, \dots x_n \in Z$. Show that one can 'adjust' α to produce a path β passing through each x_j, such that $\rho(\alpha, \beta) \leqslant \epsilon$.

(iii) Using (ii), inductively construct a ρ-Cauchy sequence of δ_n tours α_n in Z where $\delta_n \to 0$, and show that the ρ-limit path α is universal. (It is necessary to check stage $n = 1$ of the induction!)

(iv) Let X be a separable normed space and let Z denote the unit-ball of X^* in the weak*-topology, which is a com-m-pact metrizable space. Verify that Z has property (*).

(v) Let α be a universal path in Z and combine the canonical map (16.21) of X into $\mathcal{C}(Z)$ with the map $f \mapsto f \circ \alpha$ of $\mathcal{C}(Z)$ into $\mathcal{C}[0,1]$ to get the required embedding of X in $\mathcal{C}[0,1]$.

7

OPERATORS

17. Operators on a Banach space

The most important linear mappings on a normed space X are those into the scalar field – the functionals – and those from X into itself, which form the subject of this chapter. Eventually we are going to specialize X to a Hilbert space H, and we prove very few of the properties of the spectra and adjoints of operators on X except what is directly relevant to this Hilbert space situation or else makes it more lucid by placing it in its proper perspective. In the finite-dimensional case the link between linear operators and such mundane things as arithmetic – which is too important to be omitted or left to the exercises but is not central to the theme of the book – is achieved by matrix algebra. Rather than devote a whole section to it we give a brief discussion of the finite-dimensional case of each concept as it occurs.

17.1 The structure of the space $\mathcal{B}(X) = \mathcal{B}(X, X)$ of operators on a normed space X is immensely enriched by the presence of a natural **multiplication**, and the whole of the chapter is devoted to exploiting its properties. The **product** of two operators on X is defined to be their composition:

$$ST = S \circ T$$

Note that in general $ST \neq TS$. If $ST = TS$ we say that S and T **commute.** More generally we say that a family of operators $\{T_j\}$ commute if any two of them commute.

If we grant for the moment that this deserves its name as a bonafide multiplication, we see that it certainly has a unit, for clearly the **identity operator** I defined by

$$I(x) = x \qquad (x \in X)$$

is in $\mathcal{B}(X)$ and satisfies

$$IT = TI = T \qquad (T \in \mathcal{B}(X)).$$

The verification of the following basic properties of the multiplication is very easy and will be left as an exercise.

17.2 Proposition Let $S, T, U \in \mathcal{B}(X)$ and let λ, μ be scalars. Then

(1) ST is a linear map of X into itself;
(2) ST is bounded with $\|ST\| \leqslant \|S\|\,\|T\|$ (and so the product of elements of $\mathcal{B}(X)$ is indeed in $\mathcal{B}(X)$);
(3) $(ST)U = S(TU)$ (immediate, since composition of maps is always associative);
(4a) (4b) $S(\lambda T + \mu U) = \lambda ST + \mu SU$ and $(\lambda T + \mu U)S = \lambda TS + \mu US$;
(5) $\|I\| = 1$.

The only not quite obvious fact is (4a), which follows from $(S(\lambda T + \mu U))x = S((\lambda T + \mu U)x) = S(\lambda Tx + \mu Ux) = \lambda S(Tx) + \mu S(Ux) = (\lambda ST + \mu SU)x$. When the properties listed above are combined with Theorem 7.12 and the definitions at the beginning of §6 they can be summarized in the following:

17.3 Theorem With the above product, the normed space $\mathcal{B}(X)$ is a unital (in general very non-commutative) normed algebra with I as its unit. If X is complete then $\mathcal{B}(X)$ is a Banach algebra.

This theorem forms the starting point on the road to the higher realms of operator theory, a road along which this book takes only a few steps.

17.4 If follows at once from the above properties that operator multiplication is jointly continuous, that is that if $S_n \to S$ and $T_n \to T$, in norm, then $S_nT_n \to ST$ in norm. The proof is exactly as in 6.3.

The finite dimensional case

17.5 Recall from linear algebra that when X is of finite dimension n, A a linear map of X into itself (automatically bounded, by Corollary 7.10) and $\{e_1, \ldots, e_n\}$ a basis for X, the **representing matrix** for A under this basis is the $n \times n$ array $a = (a_{ij})$ where a_{ij} $(1 \leqslant i \leqslant n, 1 \leqslant j \leqslant n)$ is the coefficient of e_i in the unique expression of the vector Ae_j in terms of the basis. We assume the reader knows the rules for addition, scalar multiplication and multiplication of matrices. If linear maps A, B are represented (under the same basis) by matrices $a = (a_{ij})$, $b = (b_{ij})$ and if λ is a scalar then it is trivial that $A + B$, λA are represented by $a + b$, λa; and routine, though less trivial, to show that the product $C = AB$ is represented by the matrix product $c = ab$ where the matrix $c = (c_{ij})$ is given by

$$c_{ik} = \sum_{j=1}^{n} a_{ij}\, b_{jk};$$

which is of course the reason for defining matrix multiplication in this way. This amounts to saying that the correspondence, let us call it $T \mapsto m_T$, which associates to each $T \in \mathcal{B}(X)$ its matrix under the basis $\{e_j\}$, is an *algebra-isomorphism* of $\mathcal{B}(X)$ with the algebra $\mathcal{M}_n(\boldsymbol{R})$ or $\mathcal{M}_n(\mathbf{C})$ of all real or complex $n \times n$ matrices. The reader should realize however that changing the basis will result in each T having in general a quite different m_T and thus produce a quite different isomorphism. Since no basis has any preference over any other there is no 'natural' isomorphism of $\mathcal{B}(X)$ with the algebra of matrices.

The inverse of an operator

17.6 Ignoring topology for a moment, let us consider an arbitrary linear map T of a normed space X into itself. T has an **inverse map** T^{-1} iff it is one-to-one and onto – equivalently, iff $\ker(T) = \{0\}$, $\operatorname{im}(T) = X$. By definition, of course, $T^{-1}y$ is the unique vector x such that $Tx = y$, and it is trivial to verify that T^{-1} is linear when it exists.

When considerations of continuity enter, it is more convenient to use a different definition. An operator $T \in \mathcal{B}(X)$ is **invertible** (or **regular**) if there exists an $S \in \mathcal{B}(X)$ (the **inverse** of T) such that

$$ST = TS = I. \qquad (*)$$

The connexion of this innocent-looking definition with the underlying normed space X is not so straightforward as one might think, and is a frequent source of error to the beginner, so we examine it more closely. The equation $ST = I$ – that is, $S(Tx) = x$ ($x \in X$) – implies that S is onto, because every x is the image under S of some vector, namely Tx; and that T is one-to-one, because $Tx = Ty \Rightarrow x = S(Tx) = S(Ty) = y$. Conversely, $TS = I$ implies that T is onto and S is one-to-one, so that (*) implies that both S and T are one-to-one maps of X onto itself. It also follows from (*) that S is the inverse map T^{-1} of T and so is unique when it exists. The difference from the purely algebraic situation shows itself here, in that in general T^{-1} may exist but not be in $\mathcal{B}(X)$ – in other words not be bounded – and in this case clearly T cannot be invertible.

17.7 It is worth listing, without proof, some of the simplifications that occur when X is finite-dimensional: mainly so that the reader, who is probably more familiar with this case, can rid himself of preconceived ideas by comparing it with the more complicated general case. When $\dim X < \infty$ all linear maps T on X are automatically continuous and one has the *rank and nullity theorem* of linear algebra which says that $\dim(\ker T) + \dim(\operatorname{im} T) = \dim X$. In particular $\dim(\ker T) = 0 \Leftrightarrow \dim(\operatorname{im} T) = \dim X$ and therefore either of 'one-to-one' and 'onto' implies the other, with the result that the following assertions are equivalent:

(a) T is invertible;
(b) T^{-1} exists;
(c) T is one-to-one;
(d) T is onto.

In a general normed space, *all* of these equivalences fail, but it is an important consequence of the Open Mapping Theorem that (a) $\Leftrightarrow$ (b) is valid when X is a Banach space. The next two results summarize the situation.

17.8 Proposition For an operator T on a normed space X the following are equivalent:

(i) T is invertible;
(ii) T^{-1} exists and is bounded;
(iii) T is a topological isomorphism of X onto itself;
(iv) T is onto, and there exists $c > 0$ such that $\|Tx\| \geqslant c\|x\|$ for all $x \in X$; equivalently, such that $\|Tx\| \geqslant c$ whenever $\|x\| = 1$.

When T is invertible the operator S such that $ST = TS = I$ is unique, and is the inverse map T^{-1} of T.

Proof (i) $\Leftrightarrow$ (ii), and the last assertion, are contained in the discussion in 17.5. (ii) $\Leftrightarrow$ (iii) $\Leftrightarrow$ (iv) are immediate from the definition of a topological isomorphism and Proposition 7.6.

17.9 Theorem An operator T on a Banach space is invertible iff T^{-1} exists, that is iff T is one-to-one and onto.

Proof Immediate from Banach's Isomorphism Theorem 10.11.

17.10 Example 1 We introduce the **shift operators,** whose simple definition belies some quite complex properties which are examined further in the problems.

Let X be any of the standard sequence spaces, say ℓ_2 to fix ideas. The **right shift** R is defined by

$$R(x_1, x_2, ...) = (0, x_1, x_2, ...)$$

for $x = (x_1, x_2, ...)$ in X. The **left shift** L is defined by

$$L(x_1, x_2, ...) = (x_2, x_3, ...).$$

It is clear that R and L are linear and bounded with $\|R\| = \|L\| = 1$. It is simple to verify that

$$LR(x_1, x_2, ...) = (x_1, x_2, ...)$$

$$RL(x_1, x_2, ...) = (0, x_2, ...)$$

so that $LR = I \neq RL$. Thus L is onto and R is one-to-one; clearly R is not onto since $\text{im } R = \{x \in X : x_1 = 0\}$, while L is not one-to-one since $(1, 0, 0, ...) \in \ker L$. Neither of the two is invertible.

Example 2 Let X be the space ℓ_0 of sequences $x = (x_1, x_2, ...)$ with only finitely many nonzero terms, with the ℓ_2 norm, and define $T \in \mathcal{B}(X)$ by

$$T(x_1, x_2, x_3, ...) = (\tfrac{1}{1}x_1, \tfrac{1}{2}x_2, \tfrac{1}{3}x_3, ...).$$

It is clear that T is linear and bounded with $\|T\| = 1$. Moreover T^{-1} exists, being given by the formula

$$T^{-1}(x_1, x_2, x_3, ...) = (x_1, 2x_2, 3x_3, ...).$$

If e_n is the standard nth basis vector, $\|T^{-1}e_n\| = n\|e_n\|$, so that T^{-1} is unbounded. Thus T maps X one-to-one onto itself but is not invertible. It is clear that this construction is only possible because ℓ_0 is a very 'meagre and incomplete' space.

Manipulation of operators

We start with some purely algebraic facts; the aim is to give the reader an idea of what one is and isn't allowed to do with operators (see also Problem 17G). Algebraic manipulation of elements in a commutative algebra like $\mathcal{C}(T)$ differs little from that in the fields $\boldsymbol{R}$ and $\boldsymbol{C}$, provided that one remembers division to be a more restricted operation in that not all nonzero elements have inverses. (In $\mathcal{C}(T)$, T a compact space, a function f is clearly invertible iff it never vanishes, for the function $f^{-1}(t) = 1/f(t)$ is then also in $\mathcal{C}(T)$.) In a non-commutative algebra the differences are far greater, which is one reason why a systematic study of $\mathcal{B}(X)$ is beyond the scope of this book. An important instance is that, in $\mathcal{C}(T)$, if fg is invertible so are f and g (and see Lemma 17.14 (vi)). This fails in the non-commutative case: see Example 1 of 17.10 where the operators L, R are non-invertible but $LR = I$. We shall avoid such snags by dealing mainly with commuting sets of operators, in particular with **polynomials** in a single operator, which we now define.

17.11 If $T \in \mathcal{B}(X)$ one defines powers of T in the natural way: $T^0 = I$, $T^1 = T$, $T^2 = T \cdot T$ and inductively $T^n = TT^{n-1}$. It is easy to check the **index law**

$$T^m T^n = T^n T^m = T^{m+n} \tag{1}$$

for all integers $m, n \geqslant 0$; (if T is invertible one defines negative powers by $T^{-n} = (T^{-1})^n$ and it is clear that the index law holds for *any* integers m, n – but we do not need this fact). By induction from the relation $\|ST\| \leqslant \|S\| \, \|T\|$ one has the important inequality

$$\|T^n\| \leqslant \|T\|^n \tag{2}$$

Now let $p(t) = c_0 + c_1 t + \dots + c_n t^n$ be a polynomial function in the real or complex variable t. As in 6.7 one defines the corresponding **polynomial in** T by the formula

$$p(T) = c_0 I + c_1 T + \dots + c_n T^n. \tag{3}$$

Note in particular that for any operator

$$1(T) = I, \quad u(T) = T$$

where, as usual, u is the polynomial $t \mapsto t$.

The next result is neither surprising nor hard, but extremely useful; its extensions however, to functions more complicated than polynomials, form an important part of the theory of operators – see 18.12 and Problem 17G.

17.12 Lemma The mapping $p \mapsto p(T)$ is an algebra homomorphism of the algebra $\mathcal{P}$ of polynomial functions into $\mathcal{B}(X)$, that is

(i) $(p+q)(T) = p(T) + q(T)$
(ii) $(\alpha p)(T) = \alpha p(T)$
(iii) $(pq)(T) = p(T)\, q(T)$

or all p, q in $\mathcal{P}$ and all scalars α. Moreover one has

(iv) $(p \circ q)(T) = p(q(T))$

Proof (i) and (ii) are trivial; (iii) is obvious, but a bit tedious to write out rigorously. To prove (iv), regard q as fixed and define $\mathcal{A} = \{p \in \mathcal{P} : (p \circ q)(T) = p(q(T))\}$. Then $\mathcal{A}$ contains 1 and u, because $(1 \circ q)(T) = 1(T) = I = 1(q(T))$ and $(u \circ q)(T) = q(T) = u(q(T))$. In view of the identities $(p_1 + p_2) \circ q = p_1 \circ q + p_2 \circ q$, $(\alpha p) \circ q = \alpha p \circ q$ and $(p_1 p_2) \circ q = p_1 \circ q \cdot p_2 \circ q$ it is clear that $\mathcal{A}$ is a subalgebra of $\mathcal{P}$; since it contains 1 and u it must be all of $\mathcal{P}$.

17.13 Two simple corollaries will show the reader the uses of these properties. First, since $pq = qp$, (iii) shows that any two polynomials in the same operator commute. Second, by (iii) and induction, or by taking $p(t) = t^n$ and $q(t) = 1 + t$ in (iv), we see that for any integer $n \geqslant 0$ the usual binomial expansion

$$(I+T)^n = \sum_{r=0}^{n} \binom{n}{r} T^r$$

is valid.

The next lemma summarizes some elementary and frequently used properties of operators on X.

17.14 Lemma

(i) The product of commuting operators $T_1, \ldots, T_n$ is independent of the order of the factors.
(ii) If T is invertible, so is T^{-1}, and $(T^{-1})^{-1} = T$.
(iii) If T is invertible so is αT for any $\alpha \neq 0$, and $(\alpha T)^{-1} = \alpha^{-1} T^{-1}$.
(iv) If S commutes with T and T is invertible then S commutes with T^{-1}.
(v) If S and T are invertible so is ST, and $(ST)^{-1} = T^{-1} S^{-1}$.
(vi) The product of commuting operators $T_1, \ldots, T_n$ is invertible iff each T_j is invertible.

Proof We prove (iv) and (vi) and leave the rest as an exercise. Note that (ii) and (v) amount to the assertion that the set of invertible operators is a group under multiplication – a basic fact in ring theory. For (iv), note that if $ST = TS$ then $T^{-1}STT^{-1} = T^{-1}TST^{-1}$. On simplifying this becomes $T^{-1}S = ST^{-1}$.

(vi) If $T_1, \ldots, T_n$ are invertible then so is their product, by (v) and induction. If they commute and their product $S = T_1 T_2 \ldots T_n$ is invertible then by (i) it suffices to prove T_1 is invertible. T_1 commutes with S and hence with S^{-1} by (iv); it follows easily that $S^{-1}(T_2 \ldots T_n)$ is the inverse of T_1.

17.15 Example The following result, while unrelated to the rest of the book, illustrates what one can achieve by exploiting the similarity between ordinary algebra and the algebra of linear maps. No considerations of continuity enter here – the manipulation is purely algebraic – but the reader should have no trouble convincing himself that 17.12 and 17.13 are valid for an arbitrary linear map of a linear space into itself. Let X be the linear space of all scalar sequences, written as $f = (f(0), f(1), \ldots)$ and let L be the *left shift operator* on X defined in 17.10. The **difference operator** $\Delta = L - I$ applied to f produces the **difference sequence** $\Delta f = (f(1) - f(0), f(2) - f(1), \ldots)$, and successive applications produce a family of sequences $f, \Delta f, \Delta^2 f$ which when written beneath each other form a **difference table**, for instance:

$$\begin{array}{lccccccccccc}
f & 2 & & 1 & & 4 & & 11 & & 22 & & 37\ldots \\
\Delta f & & -1 & & 3 & & 7 & & 11 & & 15\ldots & \\
\Delta^2 f & & & 4 & & 4 & & 4 & & 4\ldots & & \\
\Delta^3 f & & & & 0 & & 0 & & 0\ldots & & &
\end{array} \tag{4}$$

The numbers $\Delta^r f(0)$ – that is, the numbers $2, -1, 4, 0, \ldots$ at the left of the table – are the *leading differences* of f. In the formula below, whose uses in numerical analysis are hinted at in Problem 17E, the choice of notation emphasizes that it can be thought of as a discrete version of the Taylor–Maclaurin series. For any real x define

$$x^{(r)} = x(x-1)\ldots(x-r+1) \qquad (r = 1, 2, \ldots)$$

– thus $x \mapsto x^{(r)}$ is a polynomial of degree r.

Proposition For any scalar sequence $f = (f(0), f(1), \ldots)$ and $n = 0, 1, 2, \ldots$ one has

$$f(n) = f(0) + \frac{\Delta f(0)}{1!} n + \frac{\Delta^2 f(0)}{2!} n^{(2)} + \ldots + \frac{\Delta^n f(0)}{n!} n^{(n)}$$

that is

$$f(n) = \sum_{r=0}^{n} \frac{c_r}{r!} n^{(r)}$$

where the c_r are the leading differences of f.

Proof Clearly L^n is the mapping that shifts a sequence n places to the left so that the binomial expansion 17.13 gives

$$f(n) = L^n f(0) = (I + \Delta)^n f(0) = \sum_{r=0}^{n} \binom{n}{r} \Delta^r f(0)$$

$$= \sum_{r=0}^{n} \frac{n(n-1) \ldots (n-r+1)}{r!} \Delta^r f(0)$$

whence the result.

For example, assuming that the sequence f in table (4) continues to follow the pattern shown in the table, it is clear that it is given by the explicit formula $f(n) = 2 + \frac{(-1)}{1!} n + \frac{4}{2!} n^{(2)} = 2 - n + 2n(n-1) = 2 - 3n + 2n^2$ – as the reader can verify.

The existence of inverses

Since many mathematical problems can be cast in the form of an equation $Tx = a$ where T is a linear operator and x is the unknown (vector, function, etc.) to be determined, it is clear that one of the basic practical problems in linear analysis is that of finding the inverse of an operator – if possible by an explicit formula – that is, of replacing the formula $Tx = a$ by the formula $x = T^{-1}a$. The next result, which shows that when X is complete the identity operator is 'surrounded' by invertible operators, is important both as a computational technique and in the structure theory of $\mathcal{B}(X)$. It can be regarded as the extension to the case $n = -1$ of the binomial expansion 17.13, the proof being lifted directly from the elementary series $(1 - x)^{-1} = 1 + x + x^2 + \ldots$ ($|x| < 1$).

17.16 Theorem Let T be an operator on a Banach space X. If $\|T\| < 1$ then $I - T$ is invertible; equivalently if $\|I - T\| < 1$ then T is invertible.

Proof We work with the first of these assertions. By Theorem 7.16, $\mathcal{B}(X)$ is a Banach space, so the series

$$I + T + T^2 + \ldots,$$

which is absolutely convergent because $\sum_0^\infty \|T^n\| \leqslant \sum_0^\infty \|T\|^n < \infty$ by 17.11(2), converges to some operator $S \in \mathcal{B}(X)$. Clearly for each n,

$$(I + T + \ldots + T^{n-1})(I - T) = (I - T)(I + T + \ldots + T^{n-1}) = I - T^n.$$

Letting $n \to \infty$ and using the continuity of multiplication 17.4 gives (since $\|T^n\| \to 0$ on the right hand side)

$$S(I - T) = (I - T)S = I,$$

whence the result.

The spectrum

17.17 Throughout this section X will denote a complex Banach space; the importance of having the scalars complex and the space complete will emerge as we proceed. The **spectrum** of an operator $T \in \mathcal{B}(X)$ is the subset of the complex plane defined by

$$sp(T) = \{\lambda : T - \lambda I \text{ fails to be invertible}\}.$$

17.18 Though the notion of the spectrum has become a fundamental tool in most branches of linear analysis, it originated with the following problem on a finite-dimensional space X:

> Given an operator T on X, find a basis $e_1, \ldots, e_n$ of X with respect to which T is represented by a matrix (17.5) of especially simple form.

and a quick look at this will motivate the definition. First, suppose $e_1, \ldots, e_n$ is a *given* basis for X (we assume $n \geqslant 1$) and T is an operator whose matrix (a_{ij}) under this basis has **diagonal** form $\operatorname{diag}(\lambda_1, \ldots, \lambda_n)$, this being short for

$$(a_{ij}) = \begin{bmatrix} \lambda_1 & & & & 0 \\ & \lambda_2 & & & \\ & & \cdot & & \\ & & & \cdot & \\ 0 & & & \cdot & \lambda_n \end{bmatrix} \qquad (5)$$

It is quickly verified that the effect of T on a vector $x = \sum_1^n c_j e_j$ is to map it to the vector $\sum_1^n \lambda_j c_j e_j$ and in particular the effect of T on e_j is just to multiply it by λ_j:

$$Te_j = \lambda_j e_j \tag{6}$$

The geometrical effect of T is vividly brought out: it consists of a 'squash' or 'stretch' in each of the coordinate directions determined by the basis $\{e_j\}$.

Conversely when an operator T is given one would like to find a basis with respect to which T has a matrix of the form (5). It is easy to see (Problem 17H) that the existence of n linearly independent e_j's and corresponding λ_j's such that (6) holds is precisely what is needed for the purpose. We say a complex number λ is an **eigenvalue** of T if there is a nonzero vector e such that $Te = \lambda e$, the vector e being an **eigenvector** of T belonging to the eigenvalue λ. (Obviously there is exactly one eigenvalue corresponding to each eigenvector, but there will in general be many eigenvectors for a given eigenvalue.) Since $Te = \lambda e$ can be written as $(T - \lambda I)e = 0$, to say that λ is an eigenvalue is the same as saying that $T - \lambda I$ fails to be one-to-one; by 17.7 this is the same as saying that $T - \lambda I$ fails to be invertible, so that we have proved the

17.19 Proposition When $\dim X < \infty$, the set of all the eigenvalues of T is precisely the set of all λ for which $T - \lambda I$ fails to be invertible, in other words $sp(T)$.

To continue the finite-dimensional theory one makes use of two further facts: that eigenvectors $e_1, \ldots, e_k$ belonging to *distinct* eigenvalues $\lambda_1, \ldots, \lambda_k$ are necessarily linearly independent (Problem 17H) and that an operator S is invertible $\Leftrightarrow$ its **determinant** $\det S$ is nonzero. The determinant has the property that $\det(T - \lambda I)$ is a polynomial $p(\lambda)$ of degree n, called the **characteristic polynomial** of T, and it follows that λ is an eigenvalue $\Leftrightarrow \lambda$ is a zero of $p(\lambda)$. Now $p(\lambda)$ cannot have more than n zeros, but by the fundamental theorem of algebra (here *complex* scalars are essential) all polynomials of degree $n \geqslant 1$ have a zero and 'most' (in a sense that can be made precise) have a full complement of n distinct zeros. By combining these facts one easily derives the following

17.20 Theorem Let $\dim X = n < \infty$. Then for every $T \in \mathcal{B}(X)$, $sp(T)$ is a nonempty set with at most n points. For 'most' T (in a sense that can be made precise) $sp(T)$ consists of n points. In this case (and in certain others as we shall see) it is possible to choose a basis $e_1, \ldots, e_n$ consisting of eigenvectors of T, and with respect to this basis T takes the diagonal form (5) of 17.18.

In other words 'most' operators T on a finite dimensional space are **diagonalizable.** The exceptional T, which can be shown to form a nowhere dense subset of $\mathcal{B}(X)$, can be represented by an almost-diagonal matrix, the **Jordan canonical form**, which however does not concern us here.

17.21 In a general, infinite-dimensional Banach space, the situation is more complicated. Recall from Theorem 17.9 that $T - \lambda I$ is invertible $\Leftrightarrow$ $T - \lambda I$ is one-to-one and onto, and that *both* conditions are needed. Those λ for which the first condition is violated are, as in the finite-dimensional case, the **eigenvalues** of T. In general there will exist λ for which the second condition is violated but not the first (Problem 17I indicates some of the possibilities), in other words the eigenvalues constitute part but not all of the spectrum.

17.22 Theorem $sp(T)$ is a compact set lying entirely inside the closed disc $\{\lambda : |\lambda| \leqslant \|T\|\}$.

Proof If $|\lambda| > \|T\|$ then $\|\lambda^{-1} T\| < 1$; by Theorem 17.16,

$$T - \lambda I = -\lambda^{-1}(I - \lambda^{-1} T)$$

is invertible, so that $\lambda \notin sp(T)$. Thus $sp(T)$ lies in the disc $|\lambda| \leqslant \|T\|$.

We prove that $sp(T)$ is closed, by showing that if $\lambda \in \mathbf{C} \sim sp(T)$ there is a disc round λ lying in $\mathbf{C} \sim sp(T)$. $T - \lambda I$ has inverse $S \in \mathcal{B}(X)$, and for any $\alpha \in \mathbf{C}$ one has

$$T - \lambda I - \alpha I = (T - \lambda I)(I - \alpha S).$$

The first factor on the right is invertible; if $|\alpha| < \|S\|^{-1}$ then $\|\alpha S\| < 1$ so that the second factor is also invertible by Theorem 17.16. By Lemma 17.14 (v), $T - \lambda I - \alpha I$ is then invertible; in other words the disc of radius $\|S\|^{-1}$ round λ lies in $\mathbf{C} \sim sp(T)$. Thus $sp(T)$ is closed and bounded, and therefore compact.

17.23 Conceivably, the spectrum of T might be empty and in fact this can happen in a real Banach space, an example being $T : (x, y) \mapsto (-y, x)$ on $\mathbf{R}^2$. (One then defines $sp(T)$, of course, as the set of all *real* λ for which $T - \lambda I$ is not invertible.) In a complex Banach space one has the important

Theorem $sp(T) \neq \emptyset$.

We do not stop to give a proof, since in the cases which concern us later the existence of points of $sp(T)$ can be established by elementary methods, but in Problem 17O one of the standard proofs, which relies on *Liouville's Theorem* from complex analysis, is sketched for the interested reader. If we grant the truth of the theorem it makes sense to define the **spectral radius** of an operator T by

$$\rho(T) = \sup\{|\lambda| : \lambda \in sp(T)\}.$$

Clearly $0 \leqslant \rho(T) \leqslant \|T\|$.

The next result forms a basic computational tool for spectra (e.g. Problem 17J). It, also, relies in an essential way on the scalar field being complex.

17.24 Spectral Mapping Theorem Let $T \in \mathcal{B}(X)$ and p be a polynomial. Then

$$sp(p(T)) = p(sp(T)),$$

the latter set denoting of course $\{p(\lambda) : \lambda \in sp(T)\}$.

Proof First we show that $0 \notin sp(p(T)) \Leftrightarrow 0 \notin p(sp(T))$. By the fundamental theorem of algebra p decomposes into linear factors $p(z) = (z - \mu_1) \ldots (z - \mu_n)$ where the μ_j run (possibly with repetitions) over *all* the zeros of p. Thus $p(T) = (T - \mu_1) \ldots (T - \mu_n)$, the factors all commuting, and we have

$$\begin{aligned} 0 \notin sp(p(T)) &\Leftrightarrow p(T) \text{ is invertible} \\ &\Leftrightarrow \text{each } T - \mu_j \text{ is invertible, by Lemma 17.14(vi)} \\ &\Leftrightarrow \text{none of the zeros of } p \text{ lies in } sp(T) \\ &\Leftrightarrow 0 \notin p(sp(T)). \end{aligned}$$

Replacing p by $p - \lambda$ we deduce that for any $\lambda \in \mathbf{C}$,

$$\begin{aligned} \lambda \notin sp(p(T)) &\Leftrightarrow (p - \lambda)(T) = p(T) - \lambda I \text{ is invertible} \\ &\Leftrightarrow 0 \notin (p - \lambda)(sp(T)) \\ &\Leftrightarrow \lambda \notin p(sp(T)) \end{aligned}$$

which is equivalent to the assertion of the theorem.

The adjoint operator

17.25 We now show that to each operator T on a normed space X there corresponds in a natural way an operator T^* on the dual space X^*, called the **adjoint** of T. We define

$$T^* : f \mapsto f \circ T \qquad (f \in X^*),$$

in other words T^*f is the functional whose value at x is $f(Tx)$. The arguments used in 17.1, 17.2 to justify the definition of the multiplication $ST = S \circ T$ of operators carry through here with S replaced by f, and we leave the reader to verify that, given f, $g \in X^*$ and scalars α, β:

(1) $f \circ T$ is a linear scalar-valued function on X;
(2) $f \circ T$ is bounded with $\|f \circ T\| \leqslant \|f\| \; \|T\|$ (thus T^* does indeed map X^* into X^*);
(3) $(\alpha f + \beta g) \circ T = \alpha(f \circ T) + \beta(g \circ T)$ (thus T^* is linear).

From property 2, T^* is bounded with $\|T^*\| \leqslant \|T\|$, and therefore $T^* \in \mathfrak{B}(X^*)$. The basic algebraic properties of the adjoint are:

17.26 Proposition
(i) $(\alpha S + \beta T)^* = \alpha S^* + \beta T^*$.
(ii) The adjoint I^* of the identity I on X is the identity I on X^*.
(iii) $(ST)^* = T^* S^*$.

Proof (i) amounts to saying that $f \circ (\alpha S + \beta T) = \alpha f \circ S + \beta f \circ T$ – see 4a of Proposition 17.2; (ii) is trivial, and (iii) follows from $T^*(S^*f) = (S^*f) \circ T = (f \circ S) \circ T = f \circ (S \circ T) = (ST)^*f$.

The reader may find it instructive to see how the same manipulations can prove quite different facts simply by a change of viewpoint. The argument which showed multiplication to be associative in 17.2(3) is used here to show that $(ST)^* = T^* S^*$; the right-hand distributive law 17.2(4a) corresponds here to the fact that $T \mapsto T^*$ is a linear mapping, and so forth.

The next result is by no means trivial, being based on the Hahn–Banach Theorem.

17.27 Theorem $\|T^*\| = \|T\|$.

Proof Given $x \in X$ we can choose, by Theorem 8.10, an $f \in X^*$ such that $\|f\| = 1$ and $f(Tx) = \|Tx\|$. Then

$$\begin{aligned}\|Tx\| = |f(Tx)| &= |(T^*f)(x)| \leqslant \|T^*f\| \; \|x\| \leqslant \|T^*\| \; \|f\| \; \|x\| \\ &= \|T^*\| \; \|x\|.\end{aligned}$$

This holds for all x, so $\|T\| \leqslant \|T^*\|$. We already know $\|T^*\| \leqslant \|T\|$ and the result follows.

An immediate corollary is that if $T_n \to T$ in norm then $T_n^* \to T^*$ in norm (for $\|T_n^* - T^*\| = \|T_n - T\|$).

To summarize the discussion so far we can say that the map $T \mapsto T^*$ is a congruence of $\mathcal{B}(X)$ into $\mathcal{B}(X^*)$ which reverses products and maps the identity to the identity.

The finite dimensional case

17.28 Recall that to each basis $e_1, \ldots, e_n$ in a space X of finite dimension n corresponds the dual basis of coefficient functionals $f_1, \ldots, f_n$ (Theorem 8.4) characterized by saying that $f_i(x)$ is the coefficient of e_i in the expansion $x = \Sigma \lambda_j e_j$, or equivalently by the formula $f_i(e_j) = \delta_{ij}$.

Proposition Let $A \in \mathcal{B}(X)$ have the matrix $a = (a_{ij})$ under the basis $\{e_j\}$. Then the matrix $\alpha = (\alpha_{ij})$ of $A^* \in \mathcal{B}(X^*)$ under the dual basis $\{f_i\}$ is the transpose of a,

$$\alpha_{ij} = a_{ji} \qquad (1 \leqslant i \leqslant n,\ 1 \leqslant j \leqslant n).$$

Proof The definition (17.5) of the matrix a can be expressed as

$$a_{ij} = f_i(Ae_j),$$

while that of the matrix α amounts to saying that $A^*f_j = \sum_i \alpha_{ij} f_i$. We then have, using $f_j(e_i) = \delta_{ji}$,

$$\alpha_{ij} = \sum_i \alpha_{ij} f_j(e_i) = (A^*f_j)(e_i) = f_j(Ae_i) = a_{ji}.$$

Thus the product-reversing map $T \mapsto T^*$ of $\mathcal{B}(X)$ into (obviously in this case onto) $\mathcal{B}(X^*)$ is mirrored by the product-reversing map $a \mapsto \alpha = a^t$ of *transposition* of matrices.

The $\langle x, f \rangle$ notation

17.29 Finding a convenient representation for the adjoint of an operator on a concrete, but infinite dimensional, space (like a sequence space) looks a daunting task. The notation we now introduce helps a lot, by disposing of the jumble of canonical maps which tend to confuse the essential properties in such cases; but its main merit is that it shows, better than the formula $T^*f = f \circ T$, the underlying symmetry that exists between T and T^*. Given $f \in X^*$, $x \in X$ let us write $\langle x, f \rangle$ instead of $f(x)$, rather as if X were an inner-product space. Then the definition of T^* gives $T^*f(x) = f(Tx)$, or in this notation the pleasantly symmetrical relation

$$\langle x, T^*f \rangle = \langle Tx, f \rangle. \tag{7}$$

The map $(x, f) \mapsto \langle x, f \rangle$ is a scalar-valued function on $X \times X^*$ with the following properties:

(1) $\langle \alpha x + \beta y, f \rangle = \alpha \langle x, f \rangle + \beta \langle y, f \rangle$
(2) $\langle x, \lambda f + \mu g \rangle = \lambda \langle x, f \rangle + \mu \langle x, g \rangle$
(3) $|\langle x, f \rangle| \leqslant \|x\| \, \|f\|$

The reason for the next definition should be clear to the student if he glances at the canonical identifications of dual spaces in 14.10 to 14.15. A **pairing** of two normed spaces X, Y is a scalar function ϕ on $X \times Y$, $\phi(x, f)$ being denoted by $\langle x, f \rangle$, such that (1, 2, 3) hold. (We denote elements of Y by letters f, g ... to aid in thinking of them as functionals on X.) For each $f \in Y$ the mapping

$$\tilde{f} : x \mapsto \langle x, f \rangle \qquad (x \in X)$$

is clearly a bounded linear functional on X with $\|\tilde{f}\| \leqslant \|f\|$ (by 1, 3) and (by 2) the mapping $f \mapsto \tilde{f}$ is a bounded linear map of Y into X^*. If

(4) The map $f \mapsto \tilde{f}$ is one-to-one, onto and norm-preserving,

in other words a congruence of Y with X^*, we can clearly think of Y as 'being' X^*; we say that Y is then the **dual of X under the pairing $\langle x, f \rangle$**. It is then natural to think of the adjoint T^* of an operator T on X as 'being' an operator on Y. Rather than spell out this identification process we make a new definition taking Equation (7) as the basic property:

17.30 Theorem For each $T \in \mathcal{B}(X)$ there is a *unique* $T^* \in \mathcal{B}(Y)$ such that

$$\langle x, T^*f \rangle = \langle Tx, f \rangle \qquad (x \in X, f \in Y).$$

T^* is called the **adjoint of T under the pairing.** One has

$$(\alpha S + \beta T)^* = \alpha S^* + \beta T^*;$$

$$I^* = I;$$

$$(ST)^* = T^* S^*;$$

$$\|T^*\| = \|T\|.$$

Proof The reader should have no difficulty in adapting the proofs of 17.25, 26 and 27 to this case. We give a sample argument:

$$\langle STx, f \rangle = \langle Tx, S^*f \rangle = \langle x, T^* S^*f \rangle$$

shows, by uniqueness, that $(ST)^* = T^* S^*$.

In the case where $Y = X^*$ and $\langle x, f \rangle$ means $f(x)$ (this is called the 'natural pairing' of X and X^*), T^* is of course the 'ordinary' adjoint. The point of all this is, of course, that 'new T^*' and 'old T^*' have identical properties – for instance sp(new T^*) = sp(old T^*); im (new T^*) is dense in $Y \Leftrightarrow$ im (old T^*) is dense in X^*, and so on – whereas 'new T^*' may be much more convenient to work with. In practice one usually ignores the distinction between them.

17.31 Example Theorem 14.12 asserts, in our present notation, that ℓ_1 is the dual of c_0 under the pairing defined by

$$\langle x, f \rangle = \sum_1^\infty x_n f_n$$

where $x = (x_1, x_2, \ldots) \in c_0$ and $f = (f_1, f_2, \ldots) \in \ell_1$.

Proposition The adjoint of the left shift operator L on c_0 (17.10 Example 1) under this pairing is the right shift operator R on ℓ_1.

Proof This follows at once from the computation

$$\begin{aligned}\langle Lx, f \rangle &= \langle (x_2, x_3, \ldots), (f_1, f_2, \ldots) \rangle \\ &= x_2 f_1 + x_3 f_2 + x_4 f_3 + \ldots \\ &= \langle (x_1, x_2, \ldots), (0, f_1, f_2, \ldots) \rangle = \langle x, Rf \rangle.\end{aligned}$$

The Hilbert space adjoint

The theory of linear operators on a Hilbert space H is given a quite different flavour from the general Banach space theory by the fact that, since the dual of H can be regarded as H itself, the adjoint of an operator on H can be identified with an operator on H. On a *real* Hilbert space the inner product $\langle x, y \rangle$ obeys the rules 1 to 4 of 17.29 so that Theorem 17.30 applies; but in the more important complex case the inner product is conjugate-linear in the second argument, which entails a small but significant change in the properties of T^*. We therefore prove the next theorem separately.

17.32 Theorem For each $T \in \mathcal{B}(H)$ there is a unique $T^\bullet \in \mathcal{B}(H)$ such that

$$\langle Tx, y \rangle = \langle x, T^\bullet y \rangle \qquad (x, y \in H) \tag{8}$$

$T^\bullet$ is called the **Hilbert space adjoint** of T. One has

$$T^{\bullet\bullet} = T,$$

$$(\alpha S + \beta T)^* = \bar{\alpha} S^* + \bar{\beta} T^*,$$
$$I^* = I,$$
$$(ST)^* = T^* S^*,$$
$$\|T^*\| = \|T\|.$$

Proof By the Cauchy–Schwarz inequality, $\langle Tx, y\rangle \leqslant \|T\|\,\|x\|\,\|y\|$ so that for each fixed $y \in H$ the map $x \mapsto \langle Tx, y\rangle$ $(x \in H)$ is easily seen to be a bounded linear functional of norm at most $\|T\|\,\|y\|$. By the Riesz–Fréchet Theorem 12.10 there is thus a unique element of H, which we designate by T^*y, such that $\langle Tx, y\rangle = \langle x, T^*y\rangle$ for all $x \in H$; moreover T^*y has norm equal to the bound of the linear functional, so

$$\|T^*y\| \leqslant \|T\|\,\|y\|. \tag{9}$$

That the resulting map T^* is a linear map of H into H follows from

$$\begin{aligned} \langle x, T^*(\alpha y + \beta z)\rangle &= \langle Tx, \alpha y + \beta z\rangle \\ &= \bar{\alpha}\langle Tx, y\rangle + \bar{\beta}\langle Tx, z\rangle \\ &= \bar{\alpha}\langle x, T^*y\rangle + \bar{\beta}\langle x, T^*z\rangle \\ &= \langle x, \alpha T^*y + \beta T^*z\rangle. \end{aligned}$$

(Note the appearance and disappearance of the complex conjugates $\bar{\alpha}, \bar{\beta}$: one might have expected T^* to be a 'conjugate-linear' operator.) By (9), T^* is bounded with $\|T^*\| \leqslant \|T\|$; the uniqueness of T^* is clear from the construction. That $T^{**} = T$, and consequently $\|T^*\| = \|T\|$, follows from uniqueness and

$$\begin{aligned} \langle x, T^{**}y\rangle &= \langle T^*x, y\rangle \qquad \text{(definition of } T^{**}) \\ &= \overline{\langle y, T^*x\rangle} = \overline{\langle Ty, x\rangle} = \langle x, Ty\rangle. \end{aligned}$$

The remaining properties are left as an exercise.

17.33 Corollary

(i) T is invertible $\Leftrightarrow T^*$ is invertible, and in that case $(T^*)^{-1} = (T^{-1})^*$.

(ii) $sp(T^*) = \overline{sp(T)}$ (the latter denotes the set $\{\bar{\lambda} : \lambda \in sp(T)\}$).

Proof (i) is clear, and (ii) follows by applying (i) to the operator $T^* - \lambda I = (T - \bar{\lambda} I)^*$.

17.34 In 17.28 we saw that when A is a linear map on a finite-dimensional space, the matrix of A^* is the transpose of the matrix of A, when these are computed with respect to a basis $\{e_j\}$ and the dual basis $\{f_i\}$. In a finite-dimensional Hilbert space H it is natural to take an *orthonormal* basis $\{e_j\}$. In view of the relation

$$\langle e_i, e_j \rangle = \delta_{ij}$$

one can think of $\{e_j\}$ as being its own dual basis: it is simple to verify that the formula $a_{ij} = f_i(Ae_j)$, in 17.28, for the matrix of A can be written

$$a_{ij} = \langle Ae_j, e_i \rangle. \tag{10}$$

Once again, conjugate linearity makes a small but important difference:

Proposition Let $A \in \mathcal{B}(H)$ have matrix $a = (a_{ij})$ under the orthonormal basis $\{e_j\}$. Then the matrix $\alpha = (\alpha_{ij})$ of the Hilbert-space adjoint $A^{\bullet}$ under the same basis is the **conjugate transpose** of a, that is $\alpha_{ij} = \bar{a}_{ij}$.

Proof Using (10), applied to $A^{\bullet}$ instead of A, one has

$$\alpha_{ij} = \langle A^{\bullet} e_j, e_i \rangle = \langle e_j, Ae_i \rangle = \overline{\langle Ae_i, e_j \rangle} = \overline{a_{ji}}.$$

Problems

Unless otherwise specified X is a nonzero complex Banach space and T is a bounded linear operator on X.

Algebraic manipulation with operators

17A The linear maps J, D, M, T_h $(h \in \mathbf{R})$ on the space $\mathcal{P}$ of polynomial functions on $\mathbf{R}$ are defined by

$$Jp(x) = \int_0^x p(t)\,dt \qquad Dp(x) = p'(x)$$

$$Mp(x) = x\,p(x) \qquad T_h p(x) = p(x-h)$$

Let $[S, T]$ denote $ST - TS$. Show that

(i) $[D, M] = I$

and similarly compute

(ii) $[D, T_h]$

(iii) $[M, T_h]$

(iv) $[M, J]$

(v) $[D, J]$.

17B Prove that there cannot exist U, V in $\mathcal{B}(X)$ such that $[U, V] = I$. [Show $[U, V] = I$ implies $[U, V^n] = nV^{n-1}$, and this contradicts the boundedness of U, V as $n \to \infty$.]

17C

(i) Given nonzero $u \in X$, $h \in X^*$ one defines the operator $u \otimes h$ on X (called a **rank one operator**; often pronounced 'u bun h') by $(u \otimes h)x = h(x)u$ $(x \in X)$. Show that $u \otimes h$ is a linear operator of norm $\|u\| \, \|h\|$.

(ii) An operator $T \in \mathcal{B}(X)$ has **finite rank** if $\dim(\operatorname{im} T) < \infty$; $\dim(\operatorname{im} T)$ is called the **rank** of T. Show that if T has rank n it can be written (non-uniquely) as the sum of n rank one operators.

17D We have seen that one can have $S, T \in \mathcal{B}(X)$ such that ST is invertible but TS is not. Show however that if $I - ST$ is invertible with inverse U, so is $I - TS$, with inverse $I + TUS$.

17E In the notation of 17.15 suppose $f(0), f(1), \dots, f(N)$ denote tabulated values of a function $F\colon \mathbf{R} \to \mathbf{R}$ at equally spaced points $a, a + h, \dots, a + Nh$. Show that 17.15 leads to the (formal) expansion

$$F(a + xh) \simeq \sum_{r=0}^{N} \frac{c_r}{r!} x^{(r)}$$

(the **forward Newton interpolation formula**), which is exact if F is a polynomial of degree $\leqslant N$.

Use it to estimate sin 45°, given sin x tabulated for $x = 0°$, 30°, 60°, 90° and compare with the true value (to 5 decimal places).

17F Show that if T is invertible so is any S such that $\|S - T\| < \|T^{-1}\|^{-1}$, and deduce that the set of invertible operators is open in $\mathcal{B}(X)$.

17G Given a rational function $f(z)$ represented as $p(z)/q(z)$ where p, q are polynomials, define $f(T)$ to be $p(T)\,q(T)^{-1}$ provided $q(T)$ is invertible. (By 17.24 this is the same as asking that q should have no zeros on $sp(T)$.) Show that the mapping $f \mapsto f(T)$ (f a rational function whose denominator is nonzero on $sp(T)$) has the algebra-homomorphism properties (i) to (iii) of Lemma 17.12 and the extra property

$$f(T)^{-1} = (1/f)(T)$$

provided either side of this equation is defined.

(Note: This problem justifies the use and manipulation of expressions like $\dfrac{I + T}{I - T}$, just like ordinary fractions, provided one knows the denominator to be invertible.)

The spectrum

17H Let T be an operator on a finite-dimensional space.

(i) Show that T is diagonalizable iff there exists a basis consisting of eigenvectors of T.
(ii) Show that any set of eigenvectors $e_1, \ldots, e_k$ whose eigenvalues $\lambda_1, \ldots, \lambda_k$ are all different, is linearly independent.
(iii) Show that the operator $T: (x_1, \ldots, x_n) \mapsto (0, x_1, \ldots, x_{n-1})$ on $\mathbf{C}^n$ is not diagonalizable. What is $sp(T)$?

17I Classifying the spectrum The points $\lambda \in sp(T)$ are customarily divided into three classes (call them P, C, and R). Those λ for which $\ker(T - \lambda I) \neq \{0\}$ – that is, the eigenvalues – form the **point spectrum.** Of the rest, those λ for which $\operatorname{im}(T - \lambda I)$ is a proper dense subspace of X form the **continuous spectrum,** while those for which $\operatorname{im}(T - \lambda I)$ is not dense form the **residual spectrum.**

(i) For the operator L of 17.10 (acting on ℓ_2) show that $sp(L) = \{\lambda : |\lambda| \leqslant 1\}$ with $P = \{\lambda : |\lambda| < 1\}$, $C = \{\lambda : |\lambda| = 1\}$, $R = \emptyset$.
(ii) Investigate similarly the operator R of 17.10.
(iii) Investigate similarly the operator T on ℓ_1 defined by $Tx = y$ where $y_n = \alpha_n x_n$, $\{\alpha_n\}$ being a given bounded scalar sequence.
(iv) Let A be a compact subset of $\mathbf{C}$ and investigate the operator M on $\mathcal{C}(A)$ where $Mf(z) = zf(z)$ $(z \in A,\ f \in \mathcal{C}(A))$.

Spectral mapping theorem

17J Show that $sp(\Delta^2)$, where $\Delta = L - I$ and L is as in the last problem, is the region inside the *cardioid curve* whose equation in polars is $r = 2 + 2\cos\theta$.

17K Show that if $T \in \mathcal{B}(X)$ and f is a rational function such that $f(T)$ is defined, then $sp(f(T)) = f(sp(T))$.

The adjoint operator

17L Show that $\ker(T) = \operatorname{im}(T^*)_\perp$, $\ker(T^*) = \operatorname{im}(T)^\perp$, and $\operatorname{im}(T)^- = \ker(T^*)_\perp$.

Deduce that T^* is invertible $\Leftrightarrow$ T is invertible. [For $\Rightarrow$, suppose T^* invertible. $\operatorname{im}(T)$ is dense in X by the first part; use the formula $\langle x, f\rangle = \langle Tx, (T^*)^{-1}f\rangle$ to show that $\|Tx\| \geqslant \|(T^*)^{-1}\|^{-1}\|x\|$.]

Hence show that $sp(T^*) = sp(T)$. (Cf. Corollary 17.33.)

17M Let $T = \sum_{r=1}^{n} u_r \otimes h_r$ be a finite-rank operator (Problem 17C). Find T^*.

17N Differential equation existence theorem This illustrates the great power of Theorem 17.16, and the 'renorming' method introduced in Problem 7P. Consider the differential equation

$$\frac{d^2u}{dt^2} + a(t)\frac{du}{dt} + b(t)u = g(t) \tag{1}$$

or $u'' + au' + bu = g$ for short, with initial conditions

$$u(0) = c_0, \quad u'(0) = c_1 . \tag{2}$$

Here the unknown function u is defined on a compact interval I containing 0, and a, b, g are continuous on I.

(i) Show (1) takes the form

$$\left(\frac{1}{p}u'\right)' - qu = f \tag{3}$$

on division by $p(t) = \exp(-\int_0^t a(s)\,ds)$.

(ii) Let X be the Banach space $\mathcal{C}(I, \mathbf{C}^2)$ of continuous $\mathbf{C}^2$-valued functions $\mathbf{u}(t) = (u(t), v(t))$ on I, with norm $\|\mathbf{u}\| = \|u\|_\infty + \|v\|_\infty$, and define $T \in \mathcal{B}(X)$ by

$$T\mathbf{u}(s) = \left(\int_0^s p(t)\,v(t)\,dt, \int_0^s q(t)\,u(t)\,dt\right).$$

Show that by setting $v = \dfrac{1}{p}u'$ one reduces the problem above to the form

$$\mathbf{u} = \mathbf{f} + T\mathbf{u} \tag{4}$$

with $\mathbf{u} = (u, v)$, $\mathbf{f}(s) = (c_0, c_1 + \int_0^s f(t)\,dt)$.

(iii) Show that for any k the norm on X given by

$$\|\mathbf{u}\| = \sup_{t \in I} e^{-k|t|}(|u(t)| + |v(t)|)$$

is equivalent to the usual one and for sufficiently large k it makes $\|T\| < 1$. Deduce that (4) has a unique solution and hence that there is a unique twice differentiable function u on I satisfying equations (1), (2).

17O Let u, v be continuous real functions on a compact space X, and T a positive operator on $\mathcal{C}(X)$ in the sense of 13.11, such that one can renorm $\mathcal{C}(X)$ with an equivalent norm to make $\|T\| < 1$. Show that

$$u \leqslant v + Tu$$

implies

$$u \leqslant (I - T)^{-1} v.$$

Deduce **Gronwall's inequality** (useful in differential equation theory): If u, w are continuous on $[0, b]$ with $w \geqslant 0$ and, for some real α,

$$u(s) \leqslant \alpha + \int_0^s w(t)u(t)\,dt \qquad (0 \leqslant s \leqslant b)$$

then

$$u(s) \leqslant \alpha \exp \int_0^s w(t)\,dt \qquad (0 \leqslant s \leqslant b).$$

(The next two problems need some knowledge of complex analysis.)

17P Suppose $T \in \mathcal{B}(X)$ and $T - \lambda I$ is invertible for all $\lambda \in \mathbf{C}$. Show that for each x in X and f in X^* the function

$$F(z) = f((T - zI)^{-1} x)$$

is analytic on the whole plane and tends to 0 at infinity. By Liouville's Theorem, $F = 0$; choosing z, x and f suitably, derive a contradiction and deduce that $sp(T)$ must be nonempty.

17Q Let $T \in \mathcal{B}(X)$ be invertible. Show that T^{-1} is the limit of polynomials in T iff 0 belongs to the unbounded component of $\mathbf{C} \sim sp(T)$. [Look at the proof of Theorem 8.16.]

18. Self-adjoint operators on a Hilbert space

There is no space in this small book to justify a discussion of the various interesting types of operator on a Hilbert space; and no lack of good texts such as Berberian (**2**), Halmos (**6**) which give a leisurely introduction to the wide perspectives of Hilbert space theory. Since self-adjoint operators are the most important in applications we have chosen to concentrate almost entirely on proving the **functional calculus theorem** which, when T is self-adjoint, extends the elementary concept of a polynomial $p(T)$ of T by making sense of '$f(T)$' for any complex function f continuous on the spectrum of T. Throughout the section H will be a complex Hilbert space; $H \neq \{0\}$ to avoid trivialities.

Hilbert space theory depends, more than most analysis, on ingenious manipulations. There is little point in trying to motivate them: we hope they will commend themselves to the reader by their results.

18.1 An operator T on H is **self-adjoint** (or **Hermitian**) if $T^* = T$, and **positive** (we write $T \geqslant 0$) if $\langle Tx, x \rangle \geqslant 0$ for all $x \in H$.

Clearly, T is self-adjoint iff $\langle Tx, y \rangle = \langle x, Ty \rangle$ for all $x, y \in H$, so that $\langle Tx, x \rangle = \langle x, Tx \rangle = \overline{\langle Tx, x \rangle}$; thus for a self-adjoint operator $\langle Tx, x \rangle$ is always real. In Problem 18B the reader is asked to show that conversely if $\langle Tx, x \rangle$ is always real then T is self-adjoint: in particular a positive operator is necessarily self-adjoint, but we do not need this fact.

In finite-dimensional linear algebra the 'canonical examples' of these two types of operator are those that with respect to some orthonormal basis are represented by real diagonal, and positive diagonal, matrices respectively. For instance

$$\begin{pmatrix} -3 & 0 \\ 0 & 2 \end{pmatrix}, \begin{pmatrix} 2 & 0 \\ 0 & 1 \end{pmatrix}$$

are examples of a self-adjoint and a positive operator on $\mathbf{C}^2$. Self-adjoint operators exist in large numbers, as the next two propositions show. The proofs are simple and left to the reader.

18.2 Proposition

(i) Let S, T be self-adjoint and α a real scalar. Then αS, $S + T$ are self-adjoint.

(ii) The norm limit of a sequence of self-adjoint operators is self-adjoint.

(iii) Any $T \in \mathcal{B}(H)$ is uniquely of the form $A + iB$ where A, B are self-adjoint. One then has $T^* = A - iB$, $A = \frac{1}{2}(T + T^*)$ and $B = -\frac{1}{2} i(T - T^*)$.

To summarize this: the set $\mathcal{S} = \{T \in \mathcal{B}(H): T^* = T\}$ is a closed real linear subspace of $\mathcal{B}(H)$ such that $\mathcal{B}(H) = \mathcal{S} \oplus i\mathcal{S}$.

18.3 Proposition

(i) The product of self-adjoint operators S, T is self-adjoint iff they commute.

(ii) For any $T \in \mathcal{B}(H)$, T^*T and TT^* are self-adjoint, and positive. In particular $T^2 \geqslant 0$ if T is self-adjoint.

18.4 Proposition Let $T \in \mathcal{B}(H)$. Then $\|T^*T\| = \|T\|^2$.

Proof By Theorem 17.27 $\|T^*\| = \|T\|$ so that $\|T^*T\| \leqslant \|T^*\| \, \|T\| = \|T\|^2$, while the reverse inequality follows from

$$\|Tx\|^2 = \langle Tx, Tx \rangle = \langle T^*Tx, x \rangle \leqslant \|T^*Tx\| \, \|x\| \leqslant \|T^*T\| \, \|x\|^2.$$

18.5 The subset $\{\langle Tx, x\rangle : \|x\| = 1\}$ of the complex plane plays an important role in the study of $\mathcal{B}(H)$; it is called the **numerical range** of the operator T and denoted by $W(T)$. The quantity

$$w(T) = \sup\{|\lambda| : \lambda \in W(T)\} = \sup\{|\langle Tx, x\rangle| : \|x\| = 1\}$$

is called the **numerical radius** of T. The Cauchy–Schwarz inequality is easily seen to imply that

$$w(T) \leqslant \|T\|.$$

By the remarks in 18.1, T is self-adjoint $\Rightarrow W(T) \subset \boldsymbol{R}$. Clearly also, T is positive $\Rightarrow W(T) \subset \boldsymbol{R}^+$.

18.6 Lemma

(i) If $\langle Tx, x\rangle \geqslant 0$ for all unit vectors x – equivalently, if $W(T) \subset \boldsymbol{R}^+$ – then T is positive.

(ii) If the numerical range of a self-adjoint operator T lies in the real interval $[\alpha, \beta]$ then $T - \alpha I$ and $\beta I - T$ are positive.

Proof (i) Any $y \in H$ is a multiple λx of some unit vector x; then $\langle Ty, y\rangle = \langle \lambda Tx, \lambda x\rangle = |\lambda|^2 \langle Tx, x\rangle \geqslant 0$, so $T \geqslant 0$.

(ii) If $\|x\| = 1$ then $\langle Tx, x\rangle \in [\alpha, \beta]$, so

$$\langle (T - \alpha I)x, x\rangle = \langle Tx, x\rangle - \alpha\langle x, x\rangle = \langle Tx, x\rangle - \alpha \geqslant 0,$$

proving $T - \alpha I \geqslant 0$. Similarly $\beta I - T \geqslant 0$.

18.7 Theorem

(i) Let A be a positive self-adjoint operator. Then $\|Ax\|^2 \leqslant w(A)\langle Ax, x\rangle$ for all x.

(ii) Let T be self-adjoint. Then $\|T\| = w(T)$.

Proof We derive these as special cases of:

(*) Let T be an operator whose numerical range lies in the real interval $[\alpha, \beta]$. Then $(\beta I - T)(T - \alpha I) \geqslant 0$.

To prove (*) we may assume $\alpha < \beta$ since the case $\alpha = \beta$ (trivial in any case by Problem 18B(iii)) can be dealt with by considering $[\alpha - \epsilon, \beta + \epsilon]$ and letting $\epsilon \to 0$. Using the positivity of $\beta I - T$ and $T - \alpha I$ we have, for any x,

$$\begin{aligned} 0 &\leqslant (\beta - \alpha)^{-1}\{\langle (\beta I - T)(T - \alpha I)x, (T - \alpha I)x\rangle \\ &\qquad + \langle (T - \alpha I)(\beta I - T)x, (\beta I - T)x\rangle\} \\ &= (\beta - \alpha)^{-1}\langle (\beta I - T)(T - \alpha I)x, (T - \alpha I + \beta I - T)x\rangle \\ &= \langle (\beta I - T)(T - \alpha I)x, x\rangle, \end{aligned}$$

as required.

Now take $T = A \geqslant 0$, $\alpha = 0$, $\beta = w(A)$ in (*) to get

$$\begin{aligned} 0 \leqslant \langle (\beta I - A)Ax, x\rangle &= \beta\langle Ax, x\rangle - \langle AAx, x\rangle \\ &= \beta\langle Ax, x\rangle - \langle Ax, Ax\rangle \\ &= w(A)\langle Ax, x\rangle - \|Ax\|^2, \end{aligned}$$

which proves (i).

To prove (ii) let $c = w(T)$ and take $\alpha = -c$, $\beta = c$ in (*) to get

$$\begin{aligned} 0 &\leqslant \langle (cI - T)(T + cI)x, x\rangle = \langle (c^2I - T^2)x, x\rangle \\ &= c^2\|x\|^2 - \|Tx\|^2 \text{ (using } \langle T^2x, x\rangle = \langle Tx, Tx\rangle = \|Tx\|^2). \end{aligned}$$

Thus $\|Tx\| \leqslant c\|x\|$ for all x, so $\|T\| \leqslant c = w(T)$. Since we know that $w(T) \leqslant \|T\|$ the result follows.

18.8 Theorem Let T be self-adjoint and let $\alpha = \inf W(T)$, $\beta = \sup W(T)$. Then $\alpha, \beta \in sp(T)$. Moreover $sp(T)$ lies wholly in the real interval $[\alpha, \beta]$.

Proof By Lemma 18.6(ii), $T - \alpha I \geqslant 0$. Choose a sequence $\{x_n\}$ in H with $\|x_n\| = 1$ and $\langle Tx_n, x_n\rangle \to \alpha$. Applying Theorem 18.7(i) gives

$$\begin{aligned} \|(T - \alpha I)x_n\|^2 &\leqslant w(T - \alpha I)\langle (T - \alpha I)x_n, x_n\rangle \\ &= w(T - \alpha I)(\langle Tx_n, x_n\rangle - \alpha) \\ &\to 0 \end{aligned}$$

proving that $T - \alpha I$ cannot be invertible, and therefore $\alpha \in sp(T)$. A similar argument shows that $\beta \in sp(T)$.

We now show that $sp(T) \subset [\alpha, \beta]$. Choose any $\mu \in \mathbf{C} \sim [\alpha, \beta]$: then μ lies at a positive distance from $[\alpha, \beta]$, say δ. For any unit vector x, since $\langle Tx, x\rangle \in [\alpha, \beta]$ one has

$$\|(T - \mu I)x\| \geqslant |\langle (T - \mu I)x, x\rangle| = |\langle Tx, x\rangle - \mu| \geqslant \delta.$$

It follows by Proposition 7.6 that $T - \mu I$ is a topological isomorphism of H onto the range R of $T - \mu I$; thus $T - \mu I$ will be invertible if we can show $R = H$. Now by Corollary 7.8, R is complete and therefore closed in H; if $R \neq H$ then by Theorem 12.19 there exists a unit vector $y \in R^{\perp}$. But $(T - \mu I)y \in R$ by the definition of R so that by the reasoning used above,

$$0 = |\langle (T - \mu I)y, y\rangle| = |\langle Ty, y\rangle - \mu| \geqslant \delta,$$

a contradiction. Hence $R = H$, $T - \mu I$ is invertible, $\mu \notin sp(T)$, $sp(T) \subset [\alpha, \beta]$ and the proof is complete.

18.9 Corollary If T is self-adjoint then $sp(T)$ and $W(T)$ have the same infimum and supremum; in particular $T \geqslant 0$ if and only if $sp(T) \subset \boldsymbol{R}^+$.

Since the supremum of the absolute values of numbers in a bounded subset A of $\boldsymbol{R}$ is easily seen to be $\max\{\sup A, -\inf A\}$, Corollary 18.9 and Theorem 18.7(ii) yield immediately one of the most important properties of self-adjoint operators:

18.10 Theorem Let T be self-adjoint. Then the spectral radius $\rho(T)$, the numerical radius $w(T)$, and the norm $\|T\|$ are equal.

For *any* $T \in \mathcal{B}(H)$ it is true that $\rho(T) \leqslant w(T) \leqslant \|T\|$, the second inequality having been noted already, and the first being essentially contained (see Problem 18F) in the proof of Theorem 18.8. In general both inequalities are strict (Problem 18E).

Corollary 18.9 is a remarkable result, for $W(T)$ depends only on the *geometry* of H, whereas $sp(T)$ depends only on the *algebraic* structure of $\mathcal{B}(H)$. The next theorem, though a digression from our argument, is interesting in its own right and illustrates the consequences of this knitting together of apparently unrelated concepts. Note that the product of *noncommuting* positive self-adjoint operators will not even be self-adjoint and so cannot be positive (18.1).

18.11 Theorem Let A, B be commuting positive self-adjoint operators. Then AB is positive.

Proof Let $\alpha = \inf W(A)$, $a = \sup W(A)$. Assume first of all that A is **strictly positive**, by which is meant that $\alpha > 0$. It then follows easily that the function ${}_A\langle\,,\,\rangle$ defined by

$$ {}_A\langle x, y\rangle = \langle Ax, y\rangle $$

is an inner product on H giving a norm $|x|$ equivalent to the original norm (in fact $\alpha^{\frac{1}{2}}\|x\| \leqslant |x| \leqslant a^{\frac{1}{2}}\|x\|$). Moreover B is self-adjoint also under this new inner product, for since $AB = BA$,

$$ {}_A\langle Bx, y\rangle = \langle ABx, y\rangle = \langle BAx, y\rangle = \langle Ax, By\rangle = {}_A\langle x, By\rangle . $$

Clearly $\mathcal{B}(H)$ consists of exactly the same operators under either norm so that any purely algebraic property like invertibility and spectrum is the same in either case. Since $B \geqslant 0$ with respect to the old inner product, Corollary 18.9 shows that $sp(B) \subset \boldsymbol{R}^+$; applied in reverse it shows that $B \geqslant 0$ with respect to the new inner product, in other words

$$0 \leqslant_A \langle Bx, x \rangle = \langle ABx, x \rangle \quad (x \in H)$$

so that AB is indeed positive.

If A is not strictly positive, we obtain the result by considering the strictly positive operator $A + \epsilon I$ ($\epsilon > 0$) and letting $\epsilon \to 0$; this is left as an exercise.

18.12 Before we turn to the main result of the chapter some explanation is in order. For any operator T, to define 'the square of T' and more generally 'p-of-T' for any polynomial p is an elementary and purely algebraic process. But to ask whether, for instance, T has a **square root** $T^{\frac{1}{2}}$ is essentially to ask whether one can sensibly define $f(T)$ for a class $\mathcal{A}$ of functions f which is wider than the class $\mathcal{P}$ of polynomials. Of course one wants $(T^{\frac{1}{2}})^2 = T$, so $f(T)$ is to be defined in such a way that the *algebra homomorphism* properties, $f(T)\, g(T) = (fg)(T)$ and so on, of Lemma 17.12 continue to hold. Thus it is natural to look for:

(a) An algebra $\mathcal{A}$ of functions, such that $\mathcal{A} \supset \mathcal{P}$; and
(b) an algebra homomorphism $\theta : f \mapsto f(T)$ defined on $\mathcal{A}$, which *extends* the elementary homomorphism $p \mapsto p(T)$.

This task (we hope the sequel persuades the reader that it is worth doing) involves analysis as well as algebra.

It turns out, rather surprisingly, that the definition of $f(T)$ depends only on the behaviour of f on $sp(T)$. We mention, for the interested reader, that for *any* operator T on *any* Banach space it is possible to take $\mathcal{A} = \{$all functions analytic in some open neighbourhood of $sp(T)\}$ – for a treatment of this important piece of complex analysis see Schechter (**12**). For a self-adjoint operator on a Hilbert space the role of $sp(T)$ is shown up even more sharply by the next theorem (also called the **Gelfand–Naimark Spectral Theorem**).

18.13 Functional Calculus Theorem Let T be a self-adjoint operator on H, and let $\mathcal{A}$ denote the (norm-) closed subalgebra of $\mathcal{B}(H)$ generated by I and T. Then there is a unique isometric algebra isomorphism θ of the complex Banach algebra $\mathcal{C}(sp(T))$ onto $\mathcal{A}$, extending the map $p \mapsto p(T)$ of polynomials.
We denote θf by $f(T)$. Thus one has

FC1 $(f + g)(T) = f(T) + g(T)$
FC2 $(\alpha f)(T) = \alpha f(T)$

FC3 $(fg)(T) = f(T)g(T)$

FC4 $\|f(T)\| = \|f\|$

for all $f, g \in \mathcal{C}(sp(T))$ and all scalars α. In particular since $\|f(T) - g(T)\| = \|f - g\|$ by FC1, 4 one has for any $f, g \in \mathcal{C}(\mathbf{R})$,

FC5 $f(T) = g(T) \Leftrightarrow f = g$ on $sp(T)$.

18.14 The **closed subalgebra generated** by a subset of $\mathcal{B}(H)$ is defined, of course, to be the smallest closed algebra containing that subset. The arguments concerning subalgebras of the algebra $\mathcal{C}(T)$ in 6.2 to 6.7 carry over without change to $\mathcal{B}(H)$, as the reader should verify. In particular it is easily seen that the closed subalgebra generated by I, T is the closure of the set of all polynomials in T.

Here and later, Σ will denote $sp(T)$ and u will be the function $u(t) = t$. Note that $\Sigma \subset R$ by Theorem 18.8.

We ignore the distinction between a polynomial p on $\mathbf{R}$ (or $\mathbf{C}$) and its restriction to Σ; but the student should note that the properties of p as an element of the algebra $\mathcal{C}(\Sigma)$ depend on Σ. For instance the function u, as an element of $\mathcal{C}[1, 2]$, is positive and invertible; as an element of $\mathcal{C}[-1, 1]$ it is neither.

Proof of theorem Let p be a polynomial with real coefficients: then by Propositions 18.2(i), 18.3(i), and induction, $p(T)$ is self-adjoint. Combining Theorem 18.10 with the Spectral Mapping Theorem 17.24 gives

$$\begin{aligned}\|p(T)\| &= \rho(p(T)) = \sup\{|\mu| : \mu \in sp(p(T))\} \\ &= \sup\{|p(\lambda)| : \lambda \in sp(T)\} = \|p\|,\end{aligned}$$

the latter denoting the norm of p as an element of $\mathcal{C}(\Sigma)$. Now let p be any polynomial. We can write $p = q + ir$ where the polynomials q, r have real coefficients. Then

$$p(T)^{*} = (q(T) + ir(T))^{*} = q(T)^{*} + \bar{i}r(T)^{*} = q(T) - ir(T)$$

so that by Proposition 18.4,

$$\begin{aligned}\|p(T)\|^2 &= \|p(T)p(T)^{*}\| = \|(q(T) + ir(T))(q(T) - ir(T))\| \\ &= \|(q^2 + r^2)(T)\| = \|q^2 + r^2\| = \|p\|^2,\end{aligned}$$

since $q^2 + r^2$ has real coefficients and (for real t, in particular for $t \in \Sigma$) equals $|p|^2$.

Hence the map $p \mapsto p(T)$ is an isometric algebra isomorphism of the set $\mathcal{P}(\Sigma)$ of polynomial functions on Σ onto the set $\mathcal{P}(T)$ of all polynomials in T. It was noted above that $\mathcal{P}(T)$ is dense in $\mathcal{A}$, and by the Stone–Weierstrass Theorem $\mathcal{P}(\Sigma)$ is dense in $\mathcal{C}(\Sigma)$. (This latter fact depends on having $\Sigma \subset \boldsymbol{R}$.) By Problem 7L, the map $p \mapsto p(T)$ extends uniquely to an isometric algebra isomorphism $\theta : f \mapsto f(T)$ of $\mathcal{C}(\Sigma)$ onto $\mathcal{A}$, and the proof is complete.

18.15 It should be emphasized that $f(T)$ is not the result of f 'acting on' T in any explicit sense, but is the image of f under a map constructed in a rather complicated way. However it is possible to work out $f(T)x$ for any $f \in \mathcal{C}(\Sigma)$ and any $x \in H$, in principle, by choosing a sequence of polynomials p_n such that $p_n \to f$ uniformly on Σ, and evaluating $\lim p_n(T)x$, for

$$\|f(T)x - p_n(T)x\| \leqslant \|f(T) - p_n(T)\| \; \|x\| = \|f - p_n\| \; \|x\| \to 0.$$

The map $f \mapsto f(T)$ has many further properties that make it a powerful computational tool. The first one strengthens the 'uniqueness' part of the theorem.

FC6 θ is the only continuous map of $\mathcal{C}(\Sigma)$ into $\mathcal{B}(H)$ obeying FC1, 2, 3 and

$$\theta 1 = I, \qquad \theta u = T.$$

Proof Let $\theta' : \mathcal{C}(\Sigma) \to \mathcal{B}(H)$ be any other map with these properties and let $A = \{f \in \mathcal{C}(\Sigma) : \theta'(f) = \theta(f)\}$. Clearly $1, u \in A$. The continuity of θ', θ implies A is closed; since θ' and θ both obey FC1, 2, 3, A is easily seen to be a subalgebra. By the Stone–Weierstrass Theorem, $A = \mathcal{C}(\Sigma)$, so $\theta' = \theta$.

FC7 $f(T)$ and $g(T)$ commute for any $f, g \in \mathcal{C}(\Sigma)$.

Proof $f(T)\,g(T) = (fg)(T) = (gf)(T) = g(T)\,f(T)$.

FC8 An operator $S \in \mathcal{B}(H)$ which commutes with T commutes with every $f(T)$.

Proof It is readily verified that if S commutes with U, V then it commutes with $U + V$, UV and αU for any scalar α; and that if S commutes with U_n for each n and $U_n \to U$ then S commutes with U. Thus $\{U \in \mathcal{B}(H) : SU = US\}$ is a closed subalgebra. Clearly it contains I and T, so it contains the subalgebra $\mathcal{A}$ which consists precisely of all the operators $f(T)$.

FC9 $f(T)^* = \bar{f}(T)$.

FC10 $f(T)$ is self-adjoint $\Leftrightarrow f$ is real.

Proofs When f is a polynomial, FC9 is contained in the proof of the main theorem, being equivalent to the formula $(q(T) + ir(T))^* = q(T) - ir(T)$. It extends to arbitrary f by continuity, using the remark after Theorem 17.27. An alternative proof, which the reader may find instructive, is to note that the map $f \mapsto (\bar{f}(T))^*$ satisfies the conditions of FC6 and so must coincide with $f \mapsto f(T)$. Property FC10 now follows from $f(T) = f(T)^* \Leftrightarrow f(T) = \bar{f}(T) \Leftrightarrow f = \bar{f}$, using FC5.

FC11 $f(T)$ is invertible $\Leftrightarrow f$ is never zero on Σ.

FC12 $sp(f(T)) = f(sp(T))$ – the generalized Spectral Mapping Theorem

FC13 $f(T)$ is a positive self-adjoint operator $\Leftrightarrow f \geqslant 0$ on Σ.

Proofs If f never vanishes on Σ then $g = 1/f$ is in $\mathcal{C}(\Sigma)$ and $g(T)$ is obviously the inverse of $f(T)$. The converse is not immediate because conceivably $f(T)^{-1}$ might exist but not lie in the algebra $\mathcal{A}$. Suppose that $f(T)$ has inverse S but that f vanishes at some $t_0 \in \Sigma$. It is easy to see that the functions $g_n = 1/(1 + n|f|)$ have norm 1 (attained at $t = t_0$) but that $\|fg_n\| \leqslant 1/n \to 0$. Thus using FC3, 4 we have

$$1 = \|g_n\| = \|g_n(T)\| = \|Sf(T)\,g_n(T)\| = \|S\,(fg_n)(T)\| \leqslant \|S\|\,\|fg_n\| \to 0,$$

a contradiction which proves that f cannot vanish on Σ. This proves FC11, from which FC12 follows easily and is left as an exercise. Using FC10, FC12 and Corollary 18.9 we then have: $f(T)$ is self-adjoint and positive $\Leftrightarrow f$ is real and $sp(f(T)) \subset \boldsymbol{R}^+ \Leftrightarrow f(\Sigma) = f(sp(T)) \subset \boldsymbol{R}^+ \Leftrightarrow f \geqslant 0$ on Σ, proving FC13.

FC14 Let f be a real-valued function, continuous on Σ, and g a complex-valued function, continuous on $f(\Sigma)$. Then

$$g(f(T)) = (g \circ f)(T) \tag{1}$$

Note Since f is real the operator $S = f(T)$ is self-adjoint. The left hand side of (1) means the image of g under the functional calculus map $g \mapsto g(S)$ associated with S. Note that it makes sense because g is continuous on $sp(S) = f(sp(T)) = f(\Sigma)$. The right hand side makes sense because $g \circ f$ is continuous on Σ.

Proof of FC14 Regard f as fixed and consider the map

$$\phi : g \mapsto (g \circ f)(T)$$

of $\mathcal{C}(sp(S))$ – in other words $\mathcal{C}(f(\Sigma))$ – into $\mathcal{B}(H)$. It is routine, though slightly tedious, to verify that ϕ obeys the conditions of FC6, with T replaced by S. Consequently ϕ must coincide with the map $g \mapsto g(S) = g(f(T))$.

18.16 A few simple applications of the functional calculus complete the chapter.

(1) Any positive self-adjoint operator T has a positive self-adjoint **square root** (an operator R such that $R^2 = T$).

To prove this note that the spectrum Σ of T lies in $\mathbf{R}^+$ by Corollary 18.9. The non-negative real function $g(t) = t^{\frac{1}{2}}$ is thus defined and continuous on Σ. Let $R = g(T)$; that R is positive self-adjoint follows from FC13, and that $R^2 = T$ is clear.

In general T may have many non-positive square roots (Problem 18H) but

(2) This positive self-adjoint square root is unique.

For let S be any other positive self-adjoint square root of T. Define $f(t) = t^2$, so that $T = f(S)$. Since (by Corollary 18.9 again) $sp(S) \subset \mathbf{R}^+$ we have $g(f(t)) = (t^2)^{\frac{1}{2}} = |t| = t$ $(t \in sp(S))$, in other words $g \circ f = u$ on $sp(S)$ – note that this would be false if $sp(S)$ extended into the negative real axis.

Thus FC14 applies to give $R = g(T) = g(f(S)) = (g \circ f)(S) = u(S) = S$.

We give a second proof of Theorem 18.11:

(3) The product of commuting positive self-adjoint operators A, B is positive.

For let $A^{\frac{1}{2}}$ denote the positive self-adjoint square root of A constructed above. Since B commutes with A it commutes with $A^{\frac{1}{2}}$ by FC8, so that $AB = A^{\frac{1}{2}}BA^{\frac{1}{2}}$. Thus

$$\langle ABx, x \rangle = \langle A^{\frac{1}{2}}BA^{\frac{1}{2}}x, x \rangle = \langle B(A^{\frac{1}{2}}x), A^{\frac{1}{2}}x \rangle \geqslant 0$$

for any x, showing that $AB \geqslant 0$.

Problems

Unless otherwise specified H is a complex Hilbert space and

$T \in \mathcal{B}(H)$; $\mathcal{S}$ denotes the set of self-adjoint operators.

18A Of the operators A, B, C, D on $L_2[0, 1]$ defined below, which are self-adjoint?

$$Af(s) = \int_0^s f(t)\,dt \qquad Bf(s) = \int_0^{1-s} f(t)\,dt$$

$$Cf(s) = \int_s^1 f(t)\,dt \qquad Df(s) = \int_{1-s}^1 f(t)\,dt$$

18B Prove that

(i) If $\langle Tx, x\rangle = 0$ for all x then $T = 0$. (*Not* true in a real space!)

(ii) If $\langle Tx, x\rangle$ is real for all x then $T \in S$.

[Express $\langle Ty, z\rangle$ in terms of inner products of the form $\langle Tx, x\rangle$.]

(iii) Show that if $W(T)$ reduces to a single point, then T is a scalar multiple of the identity operator.

18C Show (not using Theorem 18.8) that the eigenvalues of a self-adjoint operator are real.

18D Prove that any mapping $T: H \to H$ with the property $\langle Tx, y\rangle = \langle x, Ty\rangle$ for all x, y is necessarily a self-adjoint bounded linear operator.

[Uniform boundedness principle.]

Numerical range and radius

18E Show that for the operator $T: (x_1, x_2) \mapsto (x_1, x_1 + x_2)$ on $\mathbf{C}^2$ one has $\rho(T) < w(T) < \|T\|$. Find $W(T)$.

18F

(i) Show that for any $T \in \mathcal{B}(H)$ one has $sp(T) \subset W(T)^-$, so $\rho(T) \leqslant w(T)$.

(ii) On an arbitrary complex Banach space X one defines the numerical range of $T \in \mathcal{B}(X)$ to be $W(T) = \{f(Tx): x \in X, f \in X^*, \|x\| = \|f\| = f(x) = 1\}$. Show this reduces to the original definition when $X = H$, and extend (i) to this case.

Functional calculus

18G Show that if $T \in \mathcal{S}$ and the series $\Sigma\, a_n z^n$ converges absolutely on an interval I to $f(z)$, then $f(T) = \Sigma\, a_n T^n$ provided $sp(T) \subset I$.

18H Show that the identity operator on $\mathbf{C}^2$ has many self-adjoint square roots and many non-self-adjoint square roots.

18I Given $T \in \mathcal{S}$, $x \in H$, show that there exists a unique, positive, Borel measure μ on $sp(T)$ such that

$$\langle T^n x, x \rangle = \int_{sp(T)} t^n \, d\mu(t) \qquad (n = 0, 1, 2, \ldots).$$

18J Show that an isolated point α of the spectrum of a self-adjoint operator T is an eigenvalue. [Consider $f(T)$ where f is 1 at α, 0 elsewhere.]

18K **Other types of operator** T is **normal** if $T^*T = TT^*$ and **unitary** if $T^* = T^{-1}$. The sets of normal and unitary operators on H will be denoted by $\mathcal{N}$ and $\mathcal{U}$. (Note $\mathcal{U} \subset \mathcal{N}$, $\mathcal{S} \subset \mathcal{N}$.) Prove:

(i) $N \in \mathcal{N} \Rightarrow \|N^2\| = \|N\|^2$.
(ii) $U \in \mathcal{U} \Rightarrow \|U\| = 1$ and $sp(U) \subset \{z : |z| = 1\}$.
(iii) $T \in \mathcal{S} \Rightarrow \exp(iT)$ (that is $f(T)$ where $f(t) = e^{it}$) and $\dfrac{I + iT}{I - iT}$ are in $\mathcal{U}$.
(iv) P is an orthogonal projection (Problem 12F) $\Leftrightarrow P \in \mathcal{S}$ and $sp(P) \subset \{0, 1\}$.
(v) $N \in \mathcal{N}$ iff $\|N^*x\| = \|Nx\|$ for all x, while $U \in \mathcal{U}$ iff $\|U^*x\| = \|Ux\| = \|x\|$ for all x.

18L Given $T \in \mathcal{B}(H)$ define δ_T as in Problem 7E. Prove

(i) If $T \in \mathcal{S}$ then T is invertible iff $\delta_T > 0$.
(ii) If $T \in \mathcal{N}$ then T is invertible iff $\delta_T > 0$.

18M Let $\{T_n\} \subset \mathcal{S}$ be such that $\sup_n \|T_n\| < \infty$ and $T_{n+1} - T_n$ is a positive operator for each n. Show there exists $T \in \mathcal{S}$ such that $T_n x \to Tx$ for all x in H. [Use Theorem 18.7.]

18N Show that every $T \in \mathcal{B}(H)$ is a linear combination of unitary operators. [Reduce the problem to the case $T \in \mathcal{S}$, $\|T\| \leqslant 1$; consider $T \pm i(I - T^2)^{\frac{1}{2}}$.]

18O I cannot resist sketching a recent proof of the interesting inequality

$$w(T^n) \leqslant w(T)^n \qquad (T \in \mathcal{B}(H); \quad n = 1, 2, \ldots)$$

because it is a beautiful illustration of a **roots of unity argument**.

(i) Verify that suffices to show that $w(T) \leqslant 1$ implies $w(T^n) \leqslant 1$, and that $w(T) \leqslant 1$ is equivalent to

$$\operatorname{Re} \langle (I - e^{i\theta} T)x, x \rangle \geqslant 0 \text{ for all } \theta \in \mathbf{R},\ x \in H.$$

(ii) Let $\alpha = \exp(2\pi i/n)$ be a primitive nth root of unity and define

$$p_k(z) = 1 + \alpha^k z + \alpha^{2k} z^2 + \dots + \alpha^{(n-1)k} z^{n-1}$$

for $k = 0, 1, \dots, n-1$. Show that $(1 - \alpha^k z) p_k(z) = 1 - z^n$ and $\sum_{k=0}^{n-1} p_k(z) = n$.

(iii) If $w(T) \leqslant 1$ then by (i)

$$0 \leqslant \operatorname{Re}\langle (I - \alpha^k T) p_k(T) x, p_k(T) x \rangle .$$

Add up from 0 to $n-1$ and deduce $0 \leqslant \operatorname{Re}\langle (I - T^n) x, x \rangle$. Now replace T by $e^{i\theta/n} T$ and deduce the result using (i).

19. Compact self-adjoint operators and Sturm–Liouville theory

19.1 A linear map T from a normed space X into itself is called **compact** if whenever $\{x_n\}$ is a bounded sequence in X the sequence $\{Tx_n\}$ has a convergent subsequence. A compact linear map is necessarily bounded (hence an 'operator'), for if it were not there would be a bounded sequence $\{x_n\}$ with $\|Tx_n\| \to \infty$, and $\{Tx_n\}$ could not then have a convergent subsequence. The name 'compact' derives from the following fact – since we do not need it the proof is left as an exercise.

19.2 Lemma T is compact $\Leftrightarrow$ The image $T(B)$ of the unit ball B has compact closure $\Leftrightarrow$ The image $T(A)$ of each bounded set A in X has compact closure.

Compact operators are important because they are the simplest ones to which one can generalize the analysis in terms of eigenvectors and eigenvalues (**spectral analysis**) described for finite dimensional operators in 17.18–17.20, and because the resulting theory has many deep consequences in classical analysis. This section has the twofold purpose of introducing the reader to the simplest and most satisfying form of the theory, that of a compact self-adjoint operator on a Hilbert space; and of applying the results obtained to give a brief account of Sturm–Liouville theory and eigenfunction expansions. In view of the importance of the latter we begin and end the section with them, sandwiching the abstract theory in the middle; this arrangement points the lesson that the elegant and general techniques of functional analysis need to be supplemented by *ad hoc* methods in order to extract all the information about a non-trivial problem.

Sturm–Liouville systems

19.3 We start by defining a **regular Sturm–Liouville system** and describing the technique by which it can be recast as a special case of the theory of compact self-adjoint operators. The recasting is by no means trivial, but as it belongs rather to the theory of differential equations we refer the reader to a text such as Hille (**8**) for a more leisurely and motivated treatment.

The ingredients of the problem are:

SL1 A second order differential equation of the form

$$\frac{d}{dt}\left(p(t)\frac{dy}{dt}\right) + (\lambda - q(t))\,y(t) = 0 \qquad (a \leqslant t \leqslant b)$$

$$\text{or} \qquad (py')' + (\lambda - q)y = 0 \text{ for short,}$$

where $p \in \mathcal{C}^1[a, b]$, $q \in \mathcal{C}[a, b]$, p is strictly positive and q real, and λ is a scalar parameter, possibly complex. (Recall $\mathcal{C}^n[a, b]$ is the set of n times continuously differentiable functions on $[a, b]$.)

SL2 Boundary conditions of the form

$$\begin{array}{ll} \text{2a} & a_1 y(a) + a_2 y'(a) = 0, \\ \text{2b} & b_1 y(b) + b_2 y'(b) = 0, \end{array} \qquad (a_1, a_2, b_1, b_2 \text{ are real})$$

which are nontrivial, i.e. $(a_1, a_2) \neq (0, 0) \neq (b_1, b_2)$.

Just this situation, or generalizations of it, crops up when one attempts to solve boundary value problems involving the classical partial differential equations of physics by the method of **separation of variables.**

The **Sturm–Liouville problem** is to find those values of λ for which non-trivial solutions of SL1 satisfying the conditions SL2 exist, and the corresponding solution functions y. (Not surprisingly these are called the **eigenvalues** and **eigenfunctions** of the system.) In any specific instance one has to tackle this problem by *ad hoc* methods which do not concern us (but see the Examples in 19.30–32). What we are interested in is some of the **Sturm–Liouville theory**: namely that body of results which shows separation of variables to be a fruitful way to study partial differential equations. Briefly, eigenfunctions exist in large numbers, and choosing a suitably normalized one for each eigenvalue produces a complete orthonormal set in $L_2[a, b]$. By expanding functions in their Fourier series relative to this orthonormal set one obtains a powerful technique for solving boundary value problems.

We shall assume the basic fact (a special case of the *existence and uniqueness theorem* for linear differential equations; see Problem 17N for a proof):

19.4 Theorem For any $c \in [a, b]$, any scalars c_0, c_1 and any $f \in \mathcal{C}[a, b]$ there exists a unique twice differentiable function y (and in fact y'' is continuous so that $y \in \mathcal{C}^2[a, b]$) such that

$$(py')' + (\lambda - q)y = f,$$

and

$$y(c) = c_0, \; y'(c) = c_1.$$

If all the quantities concerned are real, so is the function y.

19.5 We set the Sturm–Liouville problem up as a question about linear operators as follows:

Define the **linear differential operator**

$$L_0 : y \mapsto (-py')' + qy \qquad (y \in \mathcal{C}^2[a, b]). \tag{1}$$

We are interested in the operator L, defined to be the *restriction* of L_0 to the subset (clearly a linear subspace)

$$Y = \{y \in \mathcal{C}^2[a, b] : y \text{ satisfies the conditions SL2}\} \tag{2}$$

Since $L_0 y$ is always a continuous function we shall *temporarily* take the range space as $\mathcal{C}[a, b]$, to fix ideas.

With this notation, SL1 can be written as $-Ly + \lambda y = 0$, or $(L - \lambda I)y = 0$, and since conditions SL2 are already written into the definition of L, the Sturm–Liouville problem is seemingly reduced to that of finding the eigenvalues and eigenvectors of L. The alert reader will have seen the flaw: L is not an operator on Y into itself; nor, if we were to put the sup norm on Y and $\mathcal{C}[a, b]$, would L be bounded. We overcome this difficulty by considering not L but its inverse, and we therefore put on one extra assumption:

19.6 Assumption $L = L_0|_Y$ is one to one. (Problem 19E shows this to be not really a restriction at all.)

We are now going to construct an explicit representation for L^{-1} by means of a *Green's function*. We need

19.7 Lagrange's Lemma For any $u, v \in \mathcal{C}^2[a, b]$,

$$u \cdot L_0(v) - v \cdot L_0(u) = (p \cdot (uv' - vu'))'$$

[Dash denotes differentiation, dot is pointwise product.]

Proof Work it out.

Theorem 19.4 implies that for each $c \in [a, b]$ and each pair of scalars c_0, c_1 there is a unique $y \in \mathcal{C}^2[a, b]$ such that $L_0 y = 0$ and

$$\begin{pmatrix} y(c) \\ y'(c) \end{pmatrix} = \begin{pmatrix} c_0 \\ c_1 \end{pmatrix}.$$

In other words if Z is the subspace $\{y : L_0 y = 0\}$ of $\mathcal{C}^2[a, b]$, the linear map $T_c : y \mapsto \begin{pmatrix} y(c) \\ y'(c) \end{pmatrix}$ is an isomorphism of Z with $\mathbf{C}^2$ (in particular Z is two dimensional). Hence if $L_0 u = L_0 v = 0$ then:

u, v are linearly independent $\Leftrightarrow \begin{pmatrix} u(c) \\ u'(c) \end{pmatrix}, \begin{pmatrix} v(c) \\ v'(c) \end{pmatrix}$ are linearly independent

$$\Leftrightarrow u(c)\, v'(c) - v(c)\, u'(c) \neq 0,$$

and since c was arbitrary this proves

19.8 Lemma If u, v are linearly independent solutions of $L_0 y = 0$ then the function $uv' - vu'$ is never zero on $[a, b]$.

Construction of Green's function

19.9 By Theorem 19.4 it is easy to see that there exist non-zero, real-valued solutions of $L_0 y = 0$, say u, v, such that u satisfies boundary condition SL2a of 19.3 and v satisfies SL2b. The functions u, v must be linearly independent, for if either is a multiple of the other – say $u = cv$ – then u satisfies both SL2a and SL2b and so is in Y; but $L_0 u = 0$ contradicts Assumption 19.6.

By the last Lemma, $uv' - vu'$ is never zero; by assumption (SL1) p is never zero; by Lagrange's Lemma, $(p \cdot (uv' - vu'))' = 0$ since $L_0 u = L_0 v = 0$. Hence $p \cdot (uv' - vu')$ must be a nonzero constant, and by multiplying u or v by a suitable constant we may and do assume that

$$p \cdot (uv' - vu') = -1 \quad \text{on } [a, b] \tag{3}$$

The **Green's function** G of the Sturm–Liouville system is now defined by

$$G(s, t) = \begin{cases} v(s)\, u(t) & (t \leqslant s) \\ u(s)\, v(t) & (t > s) \end{cases} \tag{4}$$

for $a \leqslant s \leqslant b$, $a \leqslant t \leqslant b$. Immediately from the definition follow two basic properties of G:

G is real-valued and $G(s, t) = G(t, s)$ (we say G is *symmetric*); (5)

G is continuous on the square $a \leqslant s \leqslant b$, $a \leqslant t \leqslant b$. (6)

To see (6) note that the only points where discontinuities could occur are on the line $s = t$, and both halves of the definition of G agree in giving the limiting value $u(t)\, v(t)$ as one approaches such a point.

19.10 Theorem For any $f \in \mathcal{C}[a, b]$ we have $y \in Y$ and $Ly = f$, where y is the function defined by

$$y(t) = \int_a^b G(s, t)\, f(s)\, ds \qquad a \leqslant t \leqslant b.$$

Proof By (4) one can rewrite this as

$$y(t) = \int_a^t u(s)\, v(t)\, f(s)\, ds + \int_t^b v(s)\, u(t)\, f(s)\, ds$$

that is

$$y = vU + uV \tag{7}$$

where

$$U(t) = \int_a^t u(s)\, f(s)\, ds, \quad V(t) = \int_t^b v(s)\, f(s)\, ds \tag{8}$$

By the definition of u, v one has

$$(pu')' = qu, \quad (pv')' = qv \tag{9}$$

and clearly by (8)

$$U' = uf, \quad V' = -vf \tag{10}$$

From (7), (10),

$$\begin{aligned} y' &= v'U + vU' + u'V + uV' \\ &= v'U + u'V + vuf - uvf, \end{aligned}$$

that is

$$y' = v'U + u'V. \tag{11}$$

Since v', u', U and V all have continuous derivatives, (11) shows in particular that $y \in \mathcal{C}^2[a, b]$. We now show that y belongs to the subspace Y of $\mathcal{C}^2[a, b]$. In fact since $U(a) = 0$ and $a_1 u(a) + a_2 u'(a) = 0$ we have

$$\begin{aligned} a_1 y(a) + a_2 y'(a) &= (a_1 v(a) + a_2 v'(a))\, U(a) + \\ &\quad + (a_1 u(a) + a_2 u'(a))\, V(a) \\ &= 0. \end{aligned}$$

Similarly $b_1\, y(b) + b_2 y'(b) = 0$.

Finally a routine but tedious computation shows that $Ly = f$:

$$\begin{aligned}(py')' &= (pv'U + pu'V)' \text{ by (11)} \\ &= (pv')'U + pv'U' + (pu')'V + pu'V' \\ &= (pv')'U + (pu')'V + p\cdot(uv' - vu')f, \text{ using (10)} \\ &= qvU + quV - f, \text{ using (3), (9)} \\ &= qy - f \text{ by (7).}\end{aligned}$$

Hence $(-py')' + qy = f$ which is what we wish to prove.

19.11 In summary: the differential operator L has the undesirable properties of being unbounded, and not having its range contained in its domain, while L^{-1} is the much pleasanter **integral operator** – let us denote it by $\hat{G}$ – defined by

$$(\hat{G}f)(s) = \int_a^b G(s, t)f(t)\,dt, \qquad s \in [a, b]. \tag{12}$$

At present we are thinking of $\hat{G}$ as an operator on $\mathcal{C}[a, b]$ with the sup norm $\|f\|_\infty$. However, for reasons which should become clear shortly, we now change our viewpoint and turn $\mathcal{C}[a, b]$ into an inner product space.

Let X denote $\mathcal{C}[a, b]$ with the inner product $\langle f, g\rangle = \int_a^b f(t)g(t)\,dt$ and the associated norm $\|f\|_2$. Note the inequality $\|f\|_2 \leqslant (b-a)^{\frac{1}{2}}\|f\|_\infty$ which follows at once from $\int_a^b |f(t)|^2\,dt \leqslant \int_a^b \|f\|_\infty^2\,dt = (b-a)\,\|f\|_\infty^2$, and implies that:

Any $\|\ \|_\infty$-bounded set in X is $\|\ \|_2$-bounded; (13)

Any $\|\ \|_\infty$-convergent sequence in X is $\|\ \|_2$-convergent (14)

19.12 Theorem

(i) The image $\hat{G}(A)$ of each bounded set A in X is $\|\ \|_\infty$-bounded and equicontinuous.

(ii) $\hat{G}$ is a compact operator on X.

(iii) $\hat{G}$ is self-adjoint – that is $\langle \hat{G}f, g\rangle = \langle f, \hat{G}g\rangle$ for all $f, g \in X$.

Proof (i) Since G is continuous on the square $a \leqslant s \leqslant b$, $a \leqslant t \leqslant b$ by 19.9 property (6), it is bounded and also uniformly continuous. Let A be a bounded subset of X – let us say $\|f\|_2 \leqslant m$ $(f \in A)$. For any $f \in A$ and $s \in [a, b]$ the Cauchy–Schwarz inequality gives

$$|\hat{G}f(s)| \leqslant \int_a^b |G(s,\, t)\, f(t)|\, dt \leqslant \left(\int_a^b |G(s,\, t)|^2\, dt\right)^{\frac{1}{2}} \left(\int_a^b |f(t)|^2\, dt\right)^{\frac{1}{2}}$$

$$\leqslant (b-a)^{\frac{1}{2}} \|G\|_\infty\, m, \tag{15}$$

where $\|G\|_\infty$ denotes the supremum of $|G(s,\, t)|$ on the square. Letting s vary yields $\|\hat{G}f\|_\infty \leqslant (b-a)^{\frac{1}{2}} \|G\|_\infty\, m$, $(f \in A)$, so $\hat{G}(A)$ is a $\| \;\|_\infty$ - bounded set of functions.

The uniform continuity of G implies that given $\epsilon > 0$ we may choose $\delta > 0$ so that $|G(s_1,\, t) - G(s_2,\, t)| < \epsilon$ whenever $s_1,\, s_2,\, t \in [a,\, b]$ and $|s_1 - s_2| < \delta$. Again by Cauchy–Schwarz, we have,

$$|\hat{G}f(s_1) - \hat{G}f(s_2)| = \left|\int_a^b (G(s_1,\, t) - G(s_2,\, t))\, f(t)\, dt\right|$$

$$\leqslant \left(\int_a^b |G(s_1,\, t) - G(s_2,\, t)|^2\, dt\right)^{\frac{1}{2}} \left(\int_a^b |f(t)|^2\, dt\right)^{\frac{1}{2}}$$

$$\leqslant ((b-a)\epsilon^2)^{\frac{1}{2}} \|f\|_2$$

$$\leqslant (b-a) m\, \epsilon$$

whenever $f \in A$ and $|s_1 - s_2| < \delta$. This shows $\hat{G}(A)$ is equicontinuous and completes the proof of (i).

(ii) By (i) and Ascoli's Theorem 6.12, if $\{f_n\}$ is a bounded sequence in X then $\{\hat{G}f_n : n = 1, 2, \ldots\}$ lies in a $\| \;\|_\infty$ -compact set. We can thus choose a uniformly convergent subsequence $\{\hat{G}f_{n_k}\}$ and this is also $\| \;\|_2$ -convergent, by (14). Hence $\hat{G}$ is a compact operator on X.

(iii) The facts that G is real and $G(s,\, t) = G(t,\, s)$ (19.9 property (5)) and that one can invert the order of integration for the iterated integral of a continuous function (cf. Problem 6N) give:

$$\langle \hat{G}f,\, g \rangle = \int_a^b \hat{G}f(s)\, \overline{g(s)}\, ds = \int_a^b \int_a^b G(s,\, t)\, f(t)\, \overline{g(s)}\, dt\, ds$$

$$= \int_a^b \int_a^b G(s,\, t)\, \overline{g(s)}\, f(t)\, ds\, dt$$

$$= \int_a^b f(t) \int_a^b \overline{G(t,\, s)\, g(s)}\, ds\, dt$$

$$= \int_a^b f(t)\, \overline{\hat{G}g(t)}\, dt = \langle f,\, \hat{G}g \rangle \tag{16}$$

and the proof is complete.

Finally in order to tie the operator $\hat{G}$ in with the theory of the next few pages one wants it to operate on a Hilbert space. Since, by 9.9, $X = \mathcal{C}[a, b]$ is dense in $L_2[a, b]$, the extension-by-continuity theorem (Problem 7L) gives the first part of:

19.13 Theorem $\hat{G}$ has a unique extension (also called $\hat{G}$) to a bounded operator in $L_2[a, b]$. Moreover the extended operator is also compact and self-adjoint, and its range is contained in $\mathcal{C}[a, b]$.

The self-adjointness is left as an exercise, while the rest comes from the result below which is worth stating on its own.

19.14 Proposition Let X be a dense subspace of a Banach space Y, let $T \in \mathcal{B}(X)$ be compact and let $\bar{T}$ be its extension by continuity to an element of $\mathcal{B}(Y)$. Then $\bar{T}$ is compact and $\text{im}\, \bar{T} \subset X$.

Proof Given a bounded sequence $\{y_n\}$ in Y one can choose $x_n \in X$ with $\|x_n - y_n\| < 1/n$ for each n. Clearly $\{x_n\}$ is bounded so there is a subsequence of indices $\{n_k\}$ such that $\{Tx_{n_k}\}$ converges to a point x of X. It is then easily verified that $\{\bar{T}y_{n_k}\}$ also converges to x, so $\bar{T}$ is compact. In particular for any $y \in Y$ let $\{y_n\}$ be the constant sequence with $y_n = y$ for all n. Then the limit point is clearly just $\bar{T}y$; this shows $\bar{T}y \in X$, so $\text{im}\, \bar{T} \subset X$.

There are no prizes for the (correct) guess that the extended operator $\hat{G}$ is given, like the original $\hat{G}$, by the integral formula (12). However to prove directly that $\hat{G}$, thus defined, exists and is compact and self-adjoint requires a bit more integration theory; the proof we have chosen is an instructive application of extension by continuity.

The spectral theorem for a compact self-adjoint operator

Throughout this section T will denote a nonzero compact self-adjoint operator on a Hilbert space H. The aim is to 'diagonalize' T in a way which directly generalizes the finite-dimensional situation discussed in 17.18–20. We recall from there that if $e_1, \ldots, e_n$ are a basis for a linear space, each e_j being an eigenvector of an operator T with eigenvalue λ_j, then the space essentially 'breaks up' into one-dimensional subspaces $N_j = \text{lin}\{e_j\}$, on each of which the effect of T is simply scalar multiplication by λ_j. Each vector x can be uniquely represented as $x = c_1 e_1 + \ldots + c_n e_n$, each $c_j e_j$ belonging to the corresponding N_j, and Tx is then the vector $\lambda_1 c_1 e_1 + \ldots + \lambda_n c_n e_n$: a pleasantly simple representation of T.

19.15 In order to obtain a similar representation (which in general will now involve *infinite* sums) we define, for each eigenvalue λ,

$$N_\lambda = \{x \in H : Tx = \lambda x\}.$$

Since N_λ can equivalently be described as $\ker(T - \lambda I)$ it is a nonzero closed linear subspace, called the **eigenspace** corresponding to λ. We also define

$$N_0 = \{x \in H : Tx = 0\} = \ker T,$$

whether or not 0 is an eigenvalue of T.

As yet there is no guarantee that T has *any* eigenvalues, a situation remedied by the next result. In fact the rather roundabout use to which we put this result has the effect of showing that there are plenty of eigenvalues, so many that the linear span of the N_λ's is dense in H. It will also transpire (in fact this is not hard to prove directly) that $sp(T) \sim \{0\}$ consists entirely of eigenvalues.

19.16 Proposition Either $\|T\|$ or $-\|T\|$ is an eigenvalue of T.

Proof Let $\alpha = \inf W(T)$, $\beta = \sup W(T)$. By Theorem 18.7(ii), $\|T\|$ is either β or $-\alpha$; suppose the latter. The argument of Theorem 18.8 shows that if $\{x_n\}$ is chosen in H with $\|x_n\| = 1$ and $\langle Tx_n, x_n\rangle \to \alpha$, then $\|Tx_n - \alpha x_n\|^2 \leqslant w(T - \alpha I)(\langle Tx_n, x_n\rangle - \alpha) \to 0$. By choosing a subsequence we may assume, since T is compact, that Tx_n converges to a vector y. But then $\alpha x_n = Tx_n - (Tx_n - \alpha x_n) \to y$ and by continuity it is clear that $\|y\| = \lim \|\alpha x_n\| = |\alpha| = \|T\| \neq 0$, and that $Ty = \lim T(\alpha x_n) = \alpha \lim Tx_n = \alpha y$, so that $\alpha = -\|T\|$ is an eigenvalue with eigenvector y. The proof in case $\|T\| = \beta$ is similar.

19.17 Proposition For each nonzero eigenvalue λ, N_λ is finite-dimensional and distinct N_λ's are mutually orthogonal.

Proof Let λ be a nonzero eigenvalue, and let B_λ denote the unit ball in N_λ. For any sequence $\{x_n\}$ in B_λ we have $x_n = \lambda^{-1} Tx_n$, and the compactness of T implies that $\{Tx_n\}$, and hence $\{x_n\}$, has a convergent subsequence. Hence B_λ is compact, which implies $\dim N_\lambda < \infty$ by Theorem 7.14.

If λ, μ are distinct eigenvalues and if $x \in N_\lambda$; $y \in N_\mu$ then $\lambda\langle x, y\rangle = \langle \lambda x, y\rangle = \langle Tx, y\rangle = \langle x, Ty\rangle = \langle x, \mu y\rangle = \mu\langle x, y\rangle$ using the fact that T is self-adjoint and has real spectrum (Theorem 18.8 and see Problem 18C). Hence $(\lambda - \mu)\langle x, y\rangle = 0$, so that $N_\lambda \perp N_\mu$.

19.18 Proposition The eigenvalues of T form a countable set whose only accumulation point, if any, is zero.

(Another way of saying this is that if we enumerate the eigenvalues in any way the resulting sequence either terminates or converges to zero.)

Proof Let Λ denote the set of all nonzero eigenvalues of T. Clearly it suffices to show that for each $\epsilon > 0$ the set $\{\lambda \in \Lambda : |\lambda| \geqslant \epsilon\}$ is finite. If this were not so then for some $\epsilon > 0$ there would exist a distinct sequence $\{\lambda_n\}$ in Λ with $|\lambda_n| \geqslant \epsilon$ for all n. Choose an eigenvector x_n of norm one for each λ_n; by the last Proposition the x_n are mutually orthogonal so by the Pythagorean identity,

$$\|Tx_m - Tx_n\|^2 = \|\lambda_m x_m - \lambda_n x_n\|^2 = |\lambda_m|^2 + |\lambda_n|^2 \geqslant 2\epsilon^2$$

whenever $m \neq n$, so $\{Tx_n\}$ can have no convergent subsequence, contradicting the compactness of T. This completes the proof.

The next result, which more-or-less completes the theory, shows that the subspaces N_λ behave roughly as a complete orthonormal set in H. Such a family of subspaces is often called an **orthogonal decomposition** of H.

19.19 Proposition If $x \perp N_\lambda$ for all λ (including $\lambda = 0$) then $x = 0$.

Proof If this is not so then $M = (\bigcup N_\lambda)^\perp$ is a nontrivial closed subspace, and since $M \perp N_0 = \ker T$, the restriction T_1 of T to M cannot be the zero operator. Now T_1 maps M into itself, for let $x \in M$ and let $y \in N_\lambda$ for some λ. Then $\langle x, y \rangle = 0$, so $\langle Tx, y \rangle = \langle x, Ty \rangle = \langle x, \lambda y \rangle = 0$. This shows $Tx \perp N_\lambda$ for all λ and hence $Tx \in M$.

It is thus clear that T_1 is a nonzero compact self-adjoint operator on the Hilbert space M, so by Proposition 19.16 there must be some nonzero $x \in M$ and some scalar λ such that $T_1 x = \lambda x$; that is, $Tx = \lambda x$. This contradicts the fact that M is orthogonal to all eigenvectors of T, and the proof is complete.

Before the main theorem we need some facts which are simple extensions of previous results on expansions relative to orthonormal sets. The proof makes use of those results, but it will be instructive for the student to construct a direct proof, by adapting the proof of Theorem 12.29.

19.20 Proposition Let $\{N_j\}_{j\in J}$ be a family of mutually orthogonal nonzero closed subspaces of H with the property described in the last Proposition.

(i) A series $\sum_J x_j$, where x_j is chosen from N_j for each j, converges iff $\sum \|x_j\|^2 < \infty$.

(ii) Let $x \in H$ and for each j let x_j be the projection of x on N_j (12.13). Then $\sum_J x_j$ converges to x.

Proof (i) For each j, since $N_j \neq \{0\}$, we can write $x_j = c_j e_j$ where $c_j = \|x_j\|$, $e_j \in N_j$, $\|e_j\| = 1$. The set $\{e_j\}$ is orthonormal, and the result is immediate from Theorem 12.29.

(ii) Recall that the projection x_j of x on N is characterized by $(x - x_j) \perp y$ $(y \in N)$, that is

$$\langle x, y \rangle = \langle x_j, y \rangle \qquad (y \in N) \tag{17}$$

Writing $x_j = c_j e_j$ as in (i) we have by (17)

$$\langle x, e_j \rangle = \langle x_j, e_j \rangle = \langle c_j e_j, e_j \rangle = c_j$$

so that $\sum_J x_j = \sum_J c_j e_j$ is the Fourier series of x relative to $\{e_j\}$. By Theorem 12.29 it converges to some $v \in H$. Using (17) we have for any k and any $y \in N_k$

$$\langle x - v, y \rangle = \langle x, y \rangle - \sum_{j\in J} \langle x_j, y \rangle = \langle x, y \rangle - \langle x_k, y \rangle = 0$$

since all terms of the sum vanish except that for $j = k$. By the hypothesis of the proposition this implies $x - v = 0$, $x = v = \sum_{j\in J} x_j$.

It is easy to see in the above that the 'decomposition' $x = \sum_{j\in J} x_j$ is unique for each x in H. Now combining the results above we have:

19.21 First form of the Spectral Theorem Let T be a compact self-adjoint operator on a Hilbert space H, and let Λ be the set of nonzero eigenvalues of T. For each $\lambda \in \Lambda \cup \{0\}$ let N_λ be the eigenspace $\{x \in H : Tx = \lambda x\}$. Then each $x \in H$ can be written (uniquely) as a finite or countably infinite sum

$$x = \sum_{\lambda\in\Lambda} x_\lambda + x_0 \tag{18}$$

where x_λ is the projection of x onto N_λ. One then has

$$Tx = \sum_{\lambda \in \Lambda} \lambda x_\lambda, \tag{19}$$

both series converging in the norm of H.

Proof The decomposition (18) comes from the last two propositions; (19) follows at once from the continuity of T and the fact that $Tx_\lambda = \lambda x_\lambda (\lambda \in \Lambda)$ and $Tx_0 = 0$. That the sum is at most countably infinite comes from Proposition 19.18.

This form of the theorem has the demerit that it is rather hard to visualize the process of summing over all the λ's, but on the other hand the merit that the decomposition $x = \Sigma\, x_\lambda + x_0$ is uniquely determined by the operator T and by nothing else. The summation is made easier to imagine at the cost of losing uniqueness in the following variant, whose connexion with the first form of the theorem should be self-explanatory, so the proof is left as an exercise.

19.22 Second form of the Spectral Theorem For each $\lambda \in \Lambda$ choose an orthonormal basis for N_λ (it must have only finitely many elements by Proposition 19.17) and enumerate all the vectors thus obtained as $e_1, e_2, \dots$ with corresponding eigenvalues $\lambda_1, \lambda_2, \dots$. Then $\{e_n\}$ is an orthonormal sequence; for each $x \in H$ we have

$$x = \sum_n \langle x, e_n \rangle e_n + x_0 \tag{20}$$

where $Tx_0 = 0$, and

$$Tx = \sum_n \lambda_n \langle x, e_n \rangle e_n. \tag{21}$$

The number of e_n's is finite or infinite according as T has a finite or infinite number of nonzero eigenvalues. Below, we shall assume there are infinitely many e_n's, since this is the case of most interest and the modification to the other case is obvious. It then follows from Proposition 19.18 that:

The λ_n form an infinite sequence converging to zero. (22)

Some deductions from these remarkable results will help the reader grasp their significance. First we show that for any nonzero scalar α that is not an eigenvalue one has an explicit formula for $(T - \alpha I)^{-1}$.

19.23 Proposition Let $\alpha \in \mathbf{C} \sim \Lambda$, $\alpha \neq 0$. Then $T - \alpha I$ is invertible, its inverse being the operator

$$R : x \mapsto \sum_{n=1}^{\infty} (\lambda_n - \alpha)^{-1} \langle x, e_n \rangle e_n - \alpha^{-1} x_0 \qquad (23)$$

where $x = \sum_{n=1}^{\infty} \langle x, e_n \rangle e_n + x_0$ according to (21).

Proof Since $\lambda_n \to 0$ the sequence $\{(\lambda_n - \alpha)^{-1}\}$ is bounded and it is then an easy consequence of Theorem 12.29 that the series (23) converges for each $x \in H$, so that R is properly defined. That R is linear is clear; that it is bounded is left as an exercise. We verify by direct substitution that $R(T - \alpha I)x = (T - \alpha I)Rx = x$, for each x, as follows:

Setting $y = (T - \alpha I)x$ one has $y = \sum_{n=1}^{\infty} \langle y, e_n \rangle e_n + y_0$ where, by (20, 21)

$$\langle y, e_n \rangle = (\lambda_n - \alpha) \langle x, e_n \rangle \quad \text{and} \quad y_0 = -\alpha x_0.$$

Thus

$$R(T - \alpha I)x = Ry = \sum_{n=1}^{\infty} (\lambda_n - \alpha)^{-1} \langle y, e_n \rangle e_n - \alpha^{-1} y_0$$

$$= \sum_{n=1}^{\infty} \langle x, e_n \rangle e_n + x_0 = x$$

and the computation for $(T - \alpha I)Rx$ is similar.

19.24 Corollary $sp(T) \sim \{0\}$ consists entirely of eigenvalues.

Proof We have just shown that $\alpha \neq 0$, α not an eigenvalue $\Rightarrow$ $\alpha \notin sp(T)$.

19.25 Many other 'functional calculus' properties of T follow in the same way: for instance $T^3 x = \sum \lambda_n^3 \langle x, e_n \rangle e_n$, and so on. Indeed there is a very strong connexion between the Spectral Theorem and the Functional Calculus Theorem 18.13 (see Problem 19H).

The final result of the section is particularly important from the point of view of applications, for it provides a rich source of nontrivial complete orthonormal sets in concrete Hilbert spaces, such as L_2 spaces on the real line. It shows how pleasant things are when 0 is not an eigenvalue of T, or in other words when the eigenspace N_0 reduces to $\{0\}$, or T is one-to-one.

19.26 Theorem For a compact self-adjoint operator T on H, the following are equivalent:

(i) 0 is not an eigenvalue of T;
(ii) im T is dense in H;
(iii) There is a complete countable orthonormal set $\{e_1, e_2, \ldots\}$ in H consisting entirely of eigenvalues of T with nonzero eigenvalues.

Proof An easy corollary of Theorem 19.22, and left as an exercise.

Back to Sturm–Liouville systems

19.27 The operator $\hat{G}$, which was constructed in 19.9–14 as the inverse to the Sturm–Liouville operator L, has some special properties not derivable from the general theory of the last section. To remind the reader of the situation: the **eigenfunctions** of the Sturm–Liouville system (in the present case the terms eigenfunction and eigenvector are synonymous), whose existence and properties we are investigating, are the elements e of the space of functions

$$Y = \{y \in \mathcal{C}^2[a, b] : a_1 y(a) + a_2 y'(a) = 0 = b_1 y(b) + b_2 y'(b)\} \quad (24)$$

such that $(L - \mu I)e = 0$ where μ is the corresponding eigenvalue and L is the operator on Y defined by

$$Ly = (-py')' + qy. \quad (25)$$

By Theorem 19.10 the inverse of L is the Green's function operator $\hat{G}$, restricted to $\mathcal{C}[a, b]$; from that theorem and Proposition 19.12 we obtain

$$\begin{array}{l} \hat{G} \text{ maps } L_2[a, b] \text{ into } \mathcal{C}[a, b] \\ \hat{G} \text{ maps } \mathcal{C}[a, b] \text{ onto } Y. \end{array} \quad (26)$$

The next result forms the bridge between the general theory and the main theorem below.

19.28 Proposition

(i) 0 is not an eigenvalue of $\hat{G}$.

(ii) Each eigenfunction of $\hat{G}$ lies in Y.

(iii) The eigenvalues of $\hat{G}$ form an infinite sequence converging to 0.

(iv) e is an eigenfunction of $\hat{G}$ with eigenvalue $\lambda \Leftrightarrow e$ is an eigenfunction of L with eigenvalue $\mu = \lambda^{-1}$.

Proof By (26), im $\hat{G}$ contains the subspace Y; it is left as an easy exercise to show that Y is a dense subset of $L_2[a, b]$ in the L_2 norm. Consequently by Theorem 19.26, 0 is not an eigenvalue of $\hat{G}$. Thus if e is an eigenfunction of $\hat{G}$ then $\hat{G}e = \lambda e$ for some $\lambda \neq 0$. By (27), $\hat{G}e \in \mathcal{C}[a, b]$, $\hat{G}^2 e \in Y$, so $e = \lambda^{-2} \hat{G}^2 e \in Y$, proving (ii). Part (iii) follows from Theorem 19.26(iii), Proposition 19.18 and the fact that each eigenspace of $\hat{G}$ is finite dimensional by Proposition 19.17 whereas $L_2[a, b]$ is infinite dimensional. Part (iv) is clear from (ii) and the fact that $\hat{G}|\mathcal{C}[a, b] = L^{-1}$.

19.29 Sturm–Liouville Theorem

(i) The Sturm–Liouville system defined in 19.3 has an infinite number of eigenvalues forming a countable set of real numbers λ_k with $|\lambda_k| \to \infty$.

(ii) Each eigenspace of the system is one-dimensional.

(iii) The choice of an eigenfunction e_k for each λ_k, normalized so that $\|e_k\|_2^2 = \int_a^b |e_k(t)|^2\, dt = 1$, yields a complete ortho-normal set in $L_2[a, b]$. One may choose e_k to be real-valued.

Proof (i) is clear from (iii) and (iv) of the last Proposition. (ii) By the Existence and Uniqueness Theorem 19.4, the subspace $Z_\mu = \{y \in \mathcal{C}^2[a, b] : (L_0 - \mu I)y = 0\}$ is two-dimensional for each scalar μ. (Recall that L_0 is the differential operator $y \to (-py')' + qy$ acting on all of $\mathcal{C}^2[a, b]$, and $L = L_0|_Y$.)

$$\text{Thus } \{y \in Y : (L - \mu I)y = 0\} = Y \cap Z_\mu$$
$$= \{y \in Z_\mu : a_1 y(a) + a_2 y'(a) = 0 = b_1 y(b) + b_2 y'(b)\}$$

is a proper subspace of Z_μ and so can *at most* have dimension one. In fact in general it reduces to $\{0\}$, the eigenvalues of L being precisely those μ for which it does not do so; (ii) follows from this.

Finally, (iii) follows from (ii), Theorem 19.26(iii), parts (i) and (iv) of the last proposition, and the last part of Theorem 19.4.

19.30 Example 1 To give a simple example of the theory above let us consider the Sturm–Liouville system on the interval $[0, \pi]$ formed by the differential equation

$$y'' + \lambda y = 0 \text{ in } 0 \leqslant x \leqslant \pi$$

with the boundary conditions

$$y(0) = y(\pi) = 0.$$

By elementary methods it is easily seen that the functions

$$s_n(t) = \sqrt{\frac{2}{\pi}} \sin nt \qquad (n = 1, 2, \ldots)$$

form a representative set of eigenfunctions according to part (iii) of the Sturm–Liouville Theorem, the corresponding eigenvalues being

$$\lambda_n = n^2 \qquad (n = 1, 2, \ldots).$$

We deduce that the s_n form a complete orthonormal set in the Hilbert space $L_2[0, \pi]$, a fact which has been derived already in Problem 13B from previous results on classical Fourier series. The Fourier series of a function $f \in L_2[0, \pi]$ relative to this orthonormal set can be written as

$$f \sim \sum_{1}^{\infty} c_n \sin nt$$

where

$$c_n = \frac{2}{\pi}\int_0^{\pi} f(t) \sin nt \, dt \qquad (n = 1, 2, \ldots)$$

and is called the *Fourier sine series* of f.

Example 2 Similarly the Sturm–Liouville system

$$y'' + \lambda y = 0 \text{ in } 0 \leqslant x \leqslant \pi$$

with $$y'(0) = y'(\pi) = 0$$

yields the complete orthonormal set of functions

$$c_n(t) = \sqrt{\frac{2}{\pi}} \cos nt \qquad (n = 0, 1, 2, \ldots)$$

which are associated with the *Fourier cosine series* of a function $f \in L_2[0, \pi]$ – see Problem 13B.

19.31 Important though such results are, they need to be extended before they are much use for the applied mathematician, and we briefly sketch a few further results.

First, one needs more exact information on the behaviour of the eigenfunctions e_k and eigenvalues λ_k. By the methods of differential equation theory one can prove that not simply does $|\lambda_k| \to \infty$ but

$$\lambda_k \to +\infty \text{ as } k \to \infty$$

and that if one enumerates the λ_k so that $\lambda_1 < \lambda_2 < \lambda_3 \ldots$ then

> the kth eigenfunction e_k has exactly $(k - 1)$ zeros in the open interval (a, b),

so that, roughly, higher eigenfunctions become rapidly oscillating. (Verify these facts in Examples 1, 2 above!) The positivity of the coefficient function p in the differential equation is the essential point here.

19.32 Second, many interesting problems require one to consider **singular** Sturm–Liouville systems, where the interval

$[a, b]$ is infinite, or p vanishes at one of the endpoints, or q is only continuous in the *open* interval (a, b), or the boundary conditions do not have such a simple form.

Example 3 The system

$$y'' + \left(\lambda - \frac{1}{4x^2}\right)y = 0 \text{ in } 0 < x \leqslant 1$$

with boundary conditions

$$y \text{ bounded as } x \to 0, \; y(1) = 0$$

has eigenfunctions $e_n(x) = x^{\frac{1}{2}} J_0(k_n x)$, where J_0 is the **Bessel function** of order zero:

$$J_0(x) = 1 - \frac{x^2}{2^2(1!)^2} + \frac{x^4}{2^4(2!)^2} - \frac{x^6}{2^6(3!)^2} + \dots$$

and $k_1, k_2, \dots$ are the positive roots of the equation $J_0(x) = 0$. It can be shown that these e_n's, multiplied by a suitable scalar constant, form a complete orthonormal set in $L_2[0, 1]$.

In more complicated cases the system has, not eigenvalues, but a 'continuous spectrum', and more advanced methods of analysing the structure of linear operators need to be used.

19.33 Third, the applied mathematician tends to prefer pointwise convergence, or uniform convergence, to convergence in the norm of $L_2[a, b]$. The theory of pointwise convergence of eigenfunction expansions is well developed and very similar to that for classical Fourier series.

The interested reader is referred to Titchmarsh (**14**), a famous work where these topics are dealt with in detail by classical methods.

Finally it should be mentioned that, though we have concentrated on the analysis of an individual operator, the compact operators as a class have many interesting algebraic and topological properties – such as are summed up for instance in the intriguing but deceptively simple statement that they form a *closed, two-sided ideal* in the Banach algebra $\mathcal{B}(X)$ – see Problem 19B. For a systematic discussion the reader is referred to Bonic (**3**) or Brown and Page (**4**).

Problems

Compact operators

19A Prove that every Fredholm or Volterra integral operator on $\mathcal{C}[a, b]$ (Problem 7O) is compact.

19B Let X be a Banach space, and let $\mathcal{K}(X)$ denote the set of compact operators on X. Prove:

(i) If $S, T \in \mathcal{K}(X)$ and $\alpha \in \boldsymbol{F}$ then $S + T, \alpha S \in \mathcal{K}(X)$.
(ii) If $S \in \mathcal{B}(X)$, $T \in \mathcal{K}(X)$ then $ST, TS \in \mathcal{K}(X)$.
(iii) If $T_1, T_2, \ldots \in \mathcal{K}(X)$ and $\|T_n - T\| \to 0$ then $T \in \mathcal{K}(X)$.

[A neat proof uses 4.14 – T is compact iff $T(B)$ has an ϵ-net for each $\epsilon > 0$, where B is the unit ball.]

In other words: $\mathcal{K}(X)$ is a closed two-sided ideal in $\mathcal{B}(X)$.

19C Prove that every finite rank operator is compact. Prove also that every Fredholm or Volterra integral operator on $\mathcal{C}[a, b]$, with continuous kernel, is the limit of a sequence of finite rank operators, thus giving another proof of Problem 19A. [Use Stone–Weierstrass.]

19D Let $\{e_n\}$ be an orthonormal sequence in a Hilbert space H and $\{\alpha_n\}$ a bounded sequence of scalars. Prove that the operator $T \in \mathcal{B}(H)$ defined by $Tx = \sum_{n=1}^{\infty} \alpha_n \langle x, e_n \rangle e_n$ is compact if, and only if, $\lim_n \alpha_n = 0$.

Sturm–Liouville systems

19E

(i) Suppose, in formula (1) of 19.5, that $p, q > 0$ on $[a, b]$ and that y satisfies $L_0 y = 0$, $y(a) = 0$, $y'(a) \neq 0$. Prove $y(b) \neq 0$. [We may suppose $y'(a) > 0$. Let $v = py'$, so that $v' = qy$. If $y(b) = 0$, define $c = \sup\{x \in [a, b]: y(t) > 0 \text{ for } a < t < x\}$; then $a < c \leqslant b$ and $y(c) = 0$ (why?). Juggling with Rolle's Theorem and the Mean Value Theorem gives some $x \in (a, c)$ with $y(x) = 0$, contradiction.]

(ii) Deduce, in the case when the boundary conditions have the simple form $y(a) = y(b) = 0$, that the set of eigenvalues of the Sturm–Liouville problem 19.3 is bounded below, and the problem can be replaced by an equivalent one for which Assumption 19.6 holds. (A variant of this argument will deal with the general case.)

19F Prove directly, using Fubini's theorem, that formula (12) of 19.11 defines a compact self-adjoint operator on $L_2[a, b]$.

19G The following 'soft analysis' proof that the operator L has a compact inverse makes an interesting contrast with the 'hard analysis' proof (explicit construction of $L^{-1} = \hat{G}$) given in the text. Notation as in 19.9.

(i) Show that for any $y \in \mathcal{C}^2[a, b]$ there are scalars α, β such that $y + \alpha u + \beta v \in Y$. Deduce, using the existence and uniqueness theorem, that L maps Y *onto* $\mathcal{C}[a, b]$.
(ii) Use the open mapping theorem to show that $L^{-1} \in \mathcal{B}(\mathcal{C}[a, b], Y)$ where Y carries the $\mathcal{C}^2[a, b]$ norm.
(iii) Deduce that L^{-1} maps the unit ball of $\mathcal{C}[a, b]$ to an equicontinuous set and hence that L^{-1} is compact as an operator on $\mathcal{C}[a, b]$ with the L_2 norm.

The spectral theorem

19H Let T be a compact self-adjoint operator on a Hilbert space H and let $\Sigma = sp(T)$. Deduce the spectral theorem 19.21 from the functional calculus of 18.13–15 as follows:

(i) Show directly that each nonzero point of Σ is an eigenvalue. [Use Problem 18L(i).]
(ii) By 19.18 each $\alpha \in \Sigma \sim \{0\}$ is an isolated point of Σ, so the function f_α defined to be 1 at α, 0 elsewhere, is in $\mathcal{C}(\Sigma)$. Let $P_\alpha = f_\alpha(T)$ in the sense of 18.13. Show that P_α is the orthogonal projection onto the eigenspace N_α of T.
(iii) Show that $T = \sum_{\lambda \in \Sigma \sim \{0\}} \lambda P_\lambda$, the series converging in norm, and deduce Theorem 19.21.
(iv) Show further that $f(T) = f(0) I + \sum_{\lambda \in \Sigma \sim \{0\}} (f(\lambda) - f(0)) P_\lambda$, for any $f \in \mathcal{C}(\Sigma)$, this series also converging in norm.

19I Determine as far as you can the nonzero eigenvalues and a basic set of eigenvectors corresponding to each eigenvalue, for each of the integral operators on $L_2[0, 1]$ determined by the following kernels $k(s, t)$:

(i) $s + t$
(ii) e^{s+t}
(iii) $\cos^2(s + t)$
(iv) $\sum_{r=-\infty}^{\infty} \alpha_r e^{2\pi i r(s-t)}$, where $\sum |\alpha_r| < \infty$.

[Hint: the first three have finite rank.]

19J Consider the Sturm–Liouville systems in Examples 1, 2, 3 of 19.30 and 19.32. Carry out the construction of the Green's function, as outlined in the text, for Examples 1, 3. Show that the construction breaks down for Example 2 since 0 is an eigenvalue, and carry it out for the modified system obtained by substituting $\lambda = \mu + \frac{1}{4}$. [Answers, given for $s \leqslant t$ only; $s > t$ by symmetry:

(1) $s - \dfrac{st}{\pi}$

(2) $-2 \cos \dfrac{s}{2} \sin \dfrac{t}{2}$

(3) $-(st)^{\frac{1}{2}} \log t$.]

19K Analyse, as in Problem 19I, the operator B of Problem 18A. [Turn the equation $Bf = \lambda f$ back into a differential equation.]

19L Let the operator A be as in 18A.

(i) Show $A^* A$ is the integral operator with kernel $k(s, t) = \min\{1 - s, 1 - t\}$.

(ii) Make up a Sturm–Liouville problem (by the way, is it unique?) of which k is the Green's function, and find its eigenvalues and eigenfunctions, and hence find those of $A^* A$.

(iii) Deduce that $\|A\| = 2/\pi$. For what f, if any, is it true that $\|Af\| = \|A\| \, \|f\|$?

19M It is natural to wonder whether all compact self-adjoint operators on $L_2[a, b]$ are integral operators. In fact, no. The following results, which need slight knowledge of two-dimensional Lebesgue measure, give a partial answer. An **L_2-kernel** on $[a, b]$ is a function $k(s, t)$ on the square $a \leqslant s \leqslant b$, $a \leqslant t \leqslant b$ such that $\int_a^b \int_a^b |k(s, t)|^2 < \infty$.

(i) Show that every L_2 kernel defines an integral operator T on $L_2[a, b]$ by the formula

$$Tf(s) = \int_a^b k(s, t) f(t) dt;$$

that T is always compact; and that T is self-adjoint iff $\overline{k(s, t)} = k(t, s)$.

(ii) Show that a necessary and sufficient condition for a compact self-adjoint operator on $L_2[a, b]$ to be derived from an L_2 kernel is that $\Sigma\, \{|\lambda|^2 : \lambda \in sp(T)\} < \infty$.

APPENDIX

Zorn's Lemma

Recall that a **partially ordered set** is a nonempty set P on which is defined a relation $\leqslant$ (a **partial ordering**) satisfying: (a) $x \leqslant x$ for each x; (b) if $x \leqslant y$ and $y \leqslant x$ then $x = y$; (c) if $x \leqslant y$ and $y \leqslant z$ then $x \leqslant z$. One writes $y \geqslant x$ synonymously with $x \leqslant y$; $x < y$ means $x \leqslant y$ and $x \neq y$; $y > x$ is defined similarly.

Examples

(1) $P = \{$all subsets of a given set $S\}$, and $\leqslant$ means $\subset$.
(2) P is $\boldsymbol{R}$, and $\leqslant$ has its usual meaning.
(3) P is $\boldsymbol{R}^2$, and $(x_1, x_2) \leqslant (y_1, y_2)$ means that $x_1 \leqslant y_1$ and $x_2 \leqslant y_2$.

Clearly any subset of a partially ordered set is a partially ordered set in its own right so

(4) Any subset of the above examples, with the inherited ordering.

Many important mathematical objects are defined by the property of being **maximal** elements of some partially ordered set P: such an element m, by definition, is one such that there is no $x \in P$ satisfying $x > m$. For instance the *hyperplanes* in a linear space X are the maximal elements of $P = \{$proper linear subspaces of $X\}$; the *connected components* of a topological space T are the maximal elements of $P = \{$all connected subsets of $T\}$ – with the ordering $\subset$ in either case. Note that maximal elements may not exist (Examples 2, 3 above) and 'maximal' is not the same as 'largest'. For let P be the half-plane $x_1 + x_2 \leqslant 0$ in $\boldsymbol{R}^2$ with the ordering of Example 3. Then P has no largest element, but lots of maximal ones – all points on the line $x_1 + x_2 = 0$ are maximal.

An **upper bound** for a subset A of a partially ordered set P is an $x \in P$ such that $x \geqslant a$ for all $a \in A$ – just as in the familiar case of subsets of the real line. A partially ordered set P is called a **chain (totally ordered set, linearly ordered set)** if for any $x, y \in P$ one has either $x \leqslant y$ or $y \leqslant x$; for instance Example 2 is a chain but Example 3 is not. A **chain in** P means a subset C of P which is a chain in its own right: for instance in Example 3 the graph of $y = x^3$ is a chain.

We can now state one of the most powerful tools for proving the existence of maximal elements in certain situations:

Zorn's Lemma A partially ordered set P, in which each chain has an upper bound, has at least one maximal element.

An intuitive 'proof' runs: choose any $x_0 \in P$; if possible choose $x_1 > x_0$, then $x_2 > x_1$, and so on. If this goes on indefinitely, then $\{x_0, x_1, x_2, \dots\}$ is a chain so (by hypothesis) has an upper bound x_0'. If x_0' is not maximal, pick $x_1' > x_0'$ and so on .... If you do this 'for long enough' you are 'bound to come to' a maximal element, and the process stops.

Put this way it looks highly suspect (it was meant to); but one can show Zorn's Lemma is equivalent to the **Axiom of Choice,** which says that given any family of nonempty sets $\{A_i\}$ one can form a set containing precisely one element from each A_i. Neither of these principles can be proved from the other axioms of set theory, but the Axiom of Choice looks intuitively clear to most people and is usually taken for granted. For a good introduction to these topics see Halmos (**7**, §§15, 16).

It is often instructive, and usually desirable, to give 'Zorn-free' proofs where possible (see Problem 8I).

REFERENCES AND SUGGESTED READING

Anyone who starts advanced work in functional analysis will probably have a research supervisor with his own ideas on the course he should follow, so it is pointless to make suggestions for more advanced reading. Among the list below the following are included especially because they fill gaps left by this book and are at about the same level. Roughly in order of difficulty:

Brown and Page (4): leisurely, with a detailed introduction to differentiation in Banach spaces; a simple vector-valued integral; many examples.

Simmons (13): excellently readable, with a good introduction to Banach algebras.

Halmos (6): angled towards the spectral analysis of normal operators by so-called spectral measures.

Bonic (3): a vast amount of material in a small space; plenty on representing operators by infinite series of rank-one operators.

Goffman and Pedrick (5): hard work but fascinating, a wealth of applications to classical analysis.

Schechter (12): discusses unbounded operators and their uses in differential equations; analytic functions of an operator; loosely organized but most stimulating.

(1) Bartle, R.G. *The elements of integration*. Wiley (1966).

(2) Berberian, S.K. *Introduction to Hilbert space*. Oxford University Press (1961).

(3) Bonic, R.A. *Linear functional analysis*. Gordon and Breach (1969).

(4) Brown, A.L., and Page, A. *Elements of functional analysis*. Van Nostrand (1970).

(5) Goffman, C., and Pedrick, G.B. *First course in functional analysis*. Englewood Cliffs (1965).

(6) Halmos, P.R. *Introduction to Hilbert space* (2nd ed.). Chelsea (1957).

(7) Halmos, P.R., *Naive set theory*. Van Nostrand (1960).

(8) Hardy, G.H., Littlewood, J.E., and Polya, G. *Inequalities* (2nd ed.). Cambridge University Press (1959).

(9) Hille, E. *Lectures on ordinary differential equations*. Addison–Wesley (1969).

(10) Molière, J.B.P. *Le bourgeois gentilhomme* (Ed. Gaston Hall). London University Press (1966).

(11) Rudin, W. *Real and complex analysis*. McGraw–Hill (1966).

(12) Schechter, M. *Principles of functional analysis*. Academic Press (1971).

(13) Simmons, G.F. *Introduction to topology and modern analysis*. McGraw–Hill (1963).

(14) Titchmarsh, E.C. *Eigenfunction expansions* (2nd ed.). Oxford University Press (1962).

INDEX OF SYMBOLS

roughly in order of appearance

$\boldsymbol{N}$, $\boldsymbol{Z}$, $\boldsymbol{Q}$, $\boldsymbol{R}$, $\boldsymbol{C}$ denote respectively the natural numbers, integers, rationals, reals and complex numbers. $\boldsymbol{R}^+$ denotes $\{x \in \boldsymbol{R}: x \geq 0\}$, and ph$(z)$ denotes the phase of a complex number z (also called its argument).

INDEX

page numbers in italic refer to problem